Problem Solving in Quantum Mechanics

Problem Solving in Quantum Mechanics

From Basics to Real-World Applications for Materials Scientists, Applied Physicists, and Devices Engineers

Marc Cahay

Spintronics and Vacuum Nanoelectronics Laboratory, University of Cincinnati, USA

Supriyo Bandyopadhyay

School of Engineering, Virginia Commonwealth University, USA

M. Cahay would like to thank the students who took his course on Quantum Systems taught at the University of Cincinnati over the last four years and who carefully checked many of the problems in this book, including (in alphabetic order): Joshua Alexander, Kelsey Baum, Graham Beach, Jordan Bishop, Ryan Blanford, Mike Bosken, Joseph Buschur, James Charles, Sumeet Chaudhary, Triet Dao, Aaron Diebold, Chelsey Duran, Adam Fornalczyk, Sthitodhi Ghosh, Adam Hauke, Erik Henderson, Phillip Horn, Alexander Jones, Henry Jentz, Jesse Kreuzmann, Alex Lambert, Craig Mackson, Ashley Mattson, James McNay, Thinh Nguyen, Francois Nyamsi, Jesse Owens, Nandhakishore Perulalan, Kaleb Posey, Logan Reid, Charles Skipper, Adam Steller, Matthew Strzok, William Teleger, Nicole Wallenhorst, Brian Waring, Aaron Welton, Samuel Wenke, and Robert Wolf.

M. Cahay would like to thank his wife, Janie, for her support and patience during this time-consuming project.

Supriyo Bandyopadhyay dedicates this book to his family and students.

Contents

About the Authors

Marc Cahay is a professor in the Department of Electrical Engineering and Computing Systems at the University of Cincinnati. His research interests include modeling of carrier transport in semiconductors, quantum mechanical effects in heterostructures, heterojunction bipolar transistors, spintronics, and cold cathodes. He has also been involved in experimental investigations of cold cathodes, and more recently organic light-emitting diodes based on rare-earth monosulfide materials. He has published over 200 papers in journals and conference proceedings in these areas. He also has organized many national and international symposia and conferences on his areas of expertise. Together with Supriyo Bandyopadhyay, he has authored a book titled *Introduction to Spintronics*. A second edition of this book was released by CRC Press in 2015.

Supriyo Bandyopadhyay is a professor in the Department of Electrical and Computer Engineering at Virginia Commonwealth University. His research interests include spintronics, nanomagnetism, straintronics, self-assembly of nanostructures, quantum dot devices, carrier transport in nanostructures, quantum networks, and quantum computing. He has published over 300 journal articles and conference papers in these fields, serves on the editorial boards of nine journals devoted to these fields, and has served on the organizing and program committees of many international conferences in these areas of research. He has authored a book titled *Physics of Nanostructured Solid State Devices* published by Springer in 2012.

The authors encourage comments about the book contents via email to marc.cahay@uc.edu and sbandy@vcu.edu.

Preface

Over the last two decades, there has been a dramatic increase in the study of physical and biological systems at the nanoscale. In fact, this millenium has been referred to as the "nanomillenium." The fields of nanoscience and nanoengineering have been fuelled by recent spectacular discoveries in mesoscopic physics, a new understanding of DNA sequencing, the advent of the field of quantum computing, tremendous progress in molecular biology, and other related fields. A fundamental understanding of physical phenomena at the nanoscale level will require future generations of engineers and scientists to grasp the intricacies of the quantum world and master the fundamentals of quantum mechanics developed by many pioneers since the 1920s. For electrical engineers, condensed matter physicists, and materials scientists who are involved with electronic and optical device research, quantum mechanics will assume a special significance. For instance, progress in the semiconductor industry has tracked Gordon Moore's prediction in 1965 regarding continued downscaling of electronic devices on a chip [1]. The density of transistors in a semiconductor chip has increased ever since in a geometric progression, roughly doubling every 18 months. In state-of-the-art semiconductor chips, the separation between the source and drain in currently used fin field effect transistors (FinFETs) is below 10 nm. All future devices for semiconductor chip applications are likely to be strongly affected by the laws of quantum mechanics, and an understanding of these laws and tenets must be added to the repertoire of a device engineer and scientist [2].

Another challenge is to understand the quantum mechanical laws that will govern device operation when the projected density of 10^{13} transistors per cm^2, anticipated by 2017, is finally reached. Density increase, however, comes with a cost: if energy dissipation does not scale down concomitantly with device dimensions there will be thermal runaway, resulting in chip meltdown. This doomsday scenario has been dubbed the "red brick wall" by the International Technology Roadmap for Semiconductors [3]. The foremost challenge is to find alternatives to the current semiconductor technology that would lead to a drastic reduction in energy dissipation during device operation. Such a technology, if and when it emerges, will very likely draw heavily on quantum mechanics as opposed to classical physics. Alternatives based on semiconductor heterostructures employing AlGaAs/GaAs or other III–V or II–VI materials have been investigated for several decades and have led to myriad quantum mechanical devices and architectures exploiting the special properties of quantum wells, wires, and dots [4–7]. Future device engineers, applied physicists, and material scientists will therefore need to be extremely adept at quantum mechanics.

The need for reform in the teaching of quantum mechanics at both the undergraduate and graduate levels is now evident [8], and has been discussed in many articles over the last few years [9–16] and at dedicated conferences on the subject, including many recent Gordon Research conferences. There are already some efforts under way at academic institutions to better train undergraduate students in this area. Many curricula have been modified to include more advanced classes in quantum mechanics for students in the engineering disciplines [10, 11]. This initiative has been catalyzed by the recent enthusiasm generated by the prospects of quantum computing and quantum communication [17]. This is a discipline that embraces knowledge in four different fields: electrical engineering, physics, materials science, and computer science.

Many textbooks have been written on quantum mechanics [18–30]. Only a few have dealt with practical aspects in the field suitable for a wide audience comprised of device engineers, applied physicists, and materials scientists [31–44]. In fact, quantum mechanics is taught very differently by high energy physicists and electrical engineers. In order for the subject to be entertaining and understandable to either discipline, they must be taught by their own kind to avoid a culture shock for the uninitiated students. Carr and McKagan have recently discussed the significant problems with graduate quantum mechanics education [13]. Typically, most textbooks are inadequate or devote too little time to exploring topics of current exciting new research and development that would prepare graduate students for the rapidly growing fields of nanoscience, nanoengineering and nanotechnology. As pointed out by Carr and McKagan, from a purely theoretical point of view, the history of quantum mechanics can be divided into four periods. In the first ten years following the 1926 formulation of the famous equation by Schrödinger, the early pioneers in the field developed the formalism taught in many undergraduate classes, including wave mechanics, its matrix formulations, and an early version of its interpretation with the work of Bohm and Bohr, among others. Then, until the mid 1960s, new concepts were developed, mostly addressing many-body aspects, with landmark achievements such as a formulation of density functional theory. This was accompanied by quantum electrodynamics and a successful explanation of low temperature superconductivity by Bardeen, Cooper, and Schrieffer. The third period began in 1964 with the pioneering work of Bell. The question of interpretation of quantum mechanics reached a deeper level with many theoretical advances, which eventually led to the fourth period in the field starting with the pioneering work of Aspect et al. in 1982 and the first successful experimental proof of Bell's inequality. Fundamental research in quantum mechanics now includes the fields of quantum computing and quantum communication, which have progressed in large strides helped by the rapid technological advances in non-linear optics, spintronic devices, and other systems fabricated with sophisticated techniques such as molecular beam epitaxy, metal organic chemical vapor deposition, atomic layer epitaxy, and various self-assembly techniques. The tremendous progress in the field has also been accelerated with the development of new characterization techniques including scanning and tunneling electron microscopy, atomic force microscopy, near field scanning optical microscopy, single photon detection, single electron detection, and others.

Many books dedicated to problems in quantum mechanics have appeared over the years. Most of them concentrate on exercises to help readers master the principles and fundamentals of the theory. In contrast, this work is a collection of problems for students, researchers, and practitioners interested in state-of-the-art material and device applications. It is not a textbook filled with precepts. Since examples are always better than precepts, this book is a collection of practical problems in quantum mechanics with solutions. Every problem is relevant either to a new device or a device concept, or to topics of current material relevant to the most recent research and development in practical quantum mechanics that could lead to new technological developments. The collection of problems covered in this book addresses topics that are covered in quantum mechanics textbooks but whose practical applications are often limited to a few end-of-the-book problems, if even that.

The present book should therefore be an ideal companion to a graduate-level textbook (or the instructor's personal lecture notes) in an engineering, condensed matter physics, or materials science curriculum. This book can not only be used by graduate students preparing for qualifying exams but is an ideal resource for the training of professional engineers in the fast-growing field of nanoscience. As such, it is appealing to a wide audience comprised not only of future generations of engineers, physicists, and material scientists but also of professionals in need of refocusing their areas of expertise toward the rapidly burgeoning areas of nanotechnology in our everyday life. The student is expected to have some elementary knowledge of quantum mechanics gleaned from modern physics classes. This includes a basic exposure to Planck's pioneering work, Bohr's concept of the atom, the meaning of the de Broglie wavelength, a first exposure to Heisenberg's uncertainty principle, and an introduction to the Schrödinger equation, including its solution for simple problems such as the particle in a box and the analysis of tunneling through a simple rectangular barrier. The authors have either organized or served on panels of many international conferences dedicated to the field of nanoscience and nanotechnology over the last 25 years. They have given or organized many short courses in these areas and given many invited talks in their field of expertise spanning nanoelectronics, nano-optoelectronics, nanoscale device simulations, spintronics, and vacuum nanoelectronics, among others. They also routinely teach graduate classes centered on quantum mechanical precepts, and therefore have first-hand experience of student needs and where their understanding can fall short.

The problems in this book are grouped by theme in 13 different chapters. At the beginning of each chapter, we briefly describe the theme behind the set of problems and refer the reader to specific sections of existing books that offer some of the clearest exposures to the material needed to tackle the problems. The level of difficulty of each problem is indicated by an increasing number of asterisks. Most solutions are typically sketched with an outline of the major steps. Intermediate and lengthy algebra steps are kept to a minimum to keep the size of the book reasonable. Additional problems are suggested at the end of each chapter and are extensions of or similar to those solved explicitly.

Each chapter contains a section on further reading containing references to articles where some of the problems treated in this book were used to investigate specific practical applications. There are several appendices to complement the set of problems. Appendix A reviews the postulates of quantum mechanics. Appendix B reviews some basic properties of the one-dimensional harmonic oscillator. Appendix C reviews some basic definitions and properties of quantum mechanical operators. Appendix D reviews the concept of Pauli matrices and their basic properties. Appendix E is a derivation of an analytical expression for the threshold voltage of a high electron mobility transistor. Appendix F is a derivation of Peierl's transformation, which is crucial to the study of the properties of a particle in an external electromagnetic field. Finally, Appendix G contains some of the Matlab code necessary to solve some of the problems and generate figures throughout the book.

The problems in this book have been collected by the authors over a period of 25 years while teaching different classes dealing with the physics and engineering of devices at the submicron and nanoscale levels. These problems were solved by the authors as part of several classes taught at the undergraduate and graduate levels at their respective institutions. For instance, some of the exercises have been assigned as homework or exam questions as part of first-year graduate courses on High-Speed Electronic Devices and Quantum Systems taught by M. Cahay at the University of Cincinnati. Since 2003, M. Cahay has also taught a class on Introduction to Quantum Computing with his colleagues in the Physics Department at the University of Cincinnati. S. Bandyopadhyay has taught a multi-semester graduate level course in Quantum Theory of Solid State Devices in three different institutions: University of Notre Dame, University of Nebraska, and Virginia Commonwealth University.

Should this edition be a success, we intend to upgrade future editions of this book with solutions to all the suggested problems. This book could not obviously cover all aspects of current research. For instance, topics left out are quantization of phonon modes, Coulomb and spin blockaded transport in nanoscale devices, and carrier transport in carbon nanotubes and graphene, among others. Future editions will include new sets of problems on these topics as well as others based on suggestions by readers, keeping pace with the most recent topics which will, without a doubt, bloom in the exciting fields of nanoscience, nanoengineering, and nanotechnology.

The contents of this book are as follows:

Chapter 1: General Properties of the Schrödinger Equation This chapter describes some general properties of the time-independent *effective mass Schrödinger equation* (EMSE), which governs the steady-state behavior of an electron in a solid with spatially varying potential profile. The solid may consist of one or more materials (e.g., a heterostructure or superlattice); hence the effective mass of the electron may vary in space. The EMSE is widely used in studying the electronic and optical properties of solids. This chapter also discusses some general properties of the EMSE, including the concepts of linearly independent solutions and their Wronskian. It is shown that in the presence of a spatially varying effective mass, the Ben Daniel–Duke boundary conditions must be satisfied. The concept of quantum mechanical wave impedance is introduced to point out the similarity between solutions to the time-independent Schrödinger equation and transmission line theory in classical electrodynamics and microwave theory.

Chapter 2: Operators All quantum mechanical operators describing physical variables are Hermitian. This chapter derives several useful identities involving operators. This includes derivations of the shift operator, the Glauber identity, the Baker–Hausdorff formula, the hypervirial theorem, Ehrenfest's theorem, and various quantum mechanical sum rules. The concept of unitary transformation is also introduced and illustrated through a calculation of the polarizability of the one-dimensional harmonic oscillator. Usage of the operator identities and theorems derived in this chapter is illustrated in other chapters. Some general definitions and properties of operators are reviewed in Appendix C, which the reader should consult before trying out the problems in this chapter.

Chapter 3: Bound States The problems in this chapter deal with one-dimensional bound state calculations, which can be performed analytically or via the numerical solution of a transcendental equation. These problems give some insight into more complicated three-dimensional bound state problems whose solutions typically require numerically intensive approaches.

Chapter 4: Heisenberg Principle This chapter starts with three different proofs of the generalized Heisenberg uncertainty relations followed by illustrations of their application to the study of some bound state and scattering problems, including diffraction from a slit in a screen and quantum mechanical tunneling through a potential barrier.

Chapter 5: Current and Energy Flux Densities This set of problems introduces the current density operator, which is applied to the study of various tunneling problems, including the case of a general one-dimensional heterostructure under bias (i.e., subjected to an electric field), the tunneling of an electron through an absorbing one-dimensional delta scatterer and potential well, and the calculation of the dwell time above a quantum well (QW). The dwell time is the time that an electron traversing a QW potential, with energy above the well's barrier, lingers within the well region. This chapter also includes an introduction to a quantum mechanical version of the energy conservation law based on the concept of quantum mechanical energy flux derived from the Schrödinger equation. Some basic tunneling problems are revisited using the conservation of energy flux principle.

Chapter 6: Density of States This chapter introduces the important concept of density of states (DOS) going from bulk to quantum confined structures.

Applications of DOS expressions are illustrated by studying the onset of degeneracy in quantum confined structures, by calculating the intrinsic carrier concentration in a two-dimensional electron gas, and by establishing the relation between the three- and two-dimensional DOS. We also illustrate the use of the DOS concept to calculate the electron density due to reflection from an infinite potential wall, the electron charge concentration in a QW in the presence of carrier freeze-out, and the threshold voltage, gate capacitance, and current–voltage characteristics of a high electron mobility transistor. Finally, the DOS concept is applied to study the properties of the blackbody radiation in three- and one-dimensional cavities.

Chapter 7: Transfer Matrix In this chapter, the use of the transfer matrix approach to solving the time-independent Schrödinger equation is illustrated for simple examples such as tunneling through a one-dimensional delta scatterer and through a square potential barrier. Using the cascading rule for transfer matrices, a general expression for the reflection and tunneling coefficient through an arbitrary potential energy profile is then derived in the presence of an applied bias across the structures. A derivation of the Floquet and Bloch theorems pertaining to an infinite repeated structure is then given based on the transfer matrix technique. The approach is also used to develop the Kroenig–Penney model for an infinite lattice with an arbitrary unit cell potential. Properties of the tunneling coefficient through finite repeated structures are then discussed, as well as their connection to the energy band structure of the infinite periodic lattice. The chapter concludes with the connection between the bound state and the tunneling problem for an arbitrary one-dimensional potential energy profile and a calculation of the dwell time above an arbitrary potential well.

Chapter 8: Scattering Matrix The concept of a scattering matrix to solve tunneling problems is first described, including their cascading rule. Explicit analytical expression of the scattering matrix through a one-dimensional delta scatterer, two delta scatterers in series separated by a distance L, a simple potential step, a square barrier, and a double barrier resonant tunneling diode are then derived. The connection between transfer and scattering matrices is then discussed, as well as applications of these formalisms to the study of electron wave propagation through an arbitrary one-dimensional energy profile.

Chapter 9: Perturbation Theory This chapter starts with a brief introduction to first-order time-independent perturbation theory and applies it to the study of an electro-optic modulator and calculation of band structure in a crystal. It then introduces Fermi's Golden Rule, which is a well-known result of time-dependent perturbation theory, and applies it to calculate the scattering rate of electrons interacting with impurities in a solid. Such rates determine the carrier mobility in a solid at low temperatures when impurity scattering dominates over phonon scattering. Fermi's Golden Rule is also applied to calculate the electron–photon interaction rate in a solid, and the absorption coefficient quantifying absorption of light as a function of light frequency.

Chapter 10: Variational Approach   Another important approach to finding approximate solutions to the Schrödinger equation is based on the Rayleigh–Ritz variational principle. For a specific problem, if the wave function associated with the ground or first excited states of a Hamiltonian cannot be calculated exactly, a suitable guess for the general shape of the wave functions associated with these states can be inferred using some symmetry properties of the system and the general properties of the Schrödinger equation as studied in Chapter 1. In this chapter, we first briefly describe the Rayleigh–Ritz variational procedure and apply it to the calculation of the energy of the ground and first excited states of problems for which an exact solution is known. Next, some general criteria for the existence of a bound state in a one-dimensional potential with finite range are derived.

Chapter 11: Electron in a Magnetic Field   Many important phenomena in condensed matter physics, such as the quantum Hall effect, require an understanding of the quantum mechanical behavior of electrons in a magnetic field. In this chapter, we introduce the concept of a vector potential and gauge to incorporate magnetic fields in the Hamiltonian of an electron. We then study quantum confined systems and derive the eigenstates of an electron in such systems subjected to a magnetic field, an example being the formation of Landau levels in a two-dimensional electron gas with a magnetic field directed perpendicular to the plane of the electron gas. The effect of a magnetic field (other than lifting spin degeneracy via the Zeeman effect) is to modify the momentum operator through the introduction of a magnetic vector potential. We study properties of the transformed momentum operator and conclude by deriving the polarizability of a harmonic oscillator in a magnetic field.

Chapter 12: Electron in an Electromagnetic Field and Optical Properties of Nanostructures   This chapter deals with the interaction between an electron and an electromagnetic field. We derive the electron–photon interaction Hamiltonian and apply it to calculate absorption coefficients. Some problems dealing with emission of light are also examined, concluding with the derivation of the Schrödinger equation for an electron in an intense laser field.

Chapter 13: Time-Dependent Schrödinger Equation   This chapter examines several properties of one-dimensional Gaussian wave packets, including a calculation of the spatio-temporal dependence of their probability current and energy flux densities and a proof that their average kinetic energy is a constant of motion. An algorithm to study the time evolution of wave packets based on the Crank–Nicholson scheme is discussed for the cases of totally reflecting and absorbing boundary conditions at the ends of the simulation domain.

This book should be of interest to any reader with a preliminary knowledge of quantum mechanics as taught in a typical modern physics class in undergraduate curricula. It should be a strong asset to professionals refocusing their expertise on many different areas of nanotechnology that affect our daily life. If successful, future editions of the book will be geared toward practical aspects of quantum mechanics useful to chemists, chemical engineers, and researchers and practitioners in the field of nanobiotechnology.

This book will be an ideal companion to a graduate-level textbook (or the instructor's personal lecture notes) in an engineering, physics, or materials science curriculum. It can not only be used by graduate students eager to better grasp the field of quantum mechanics and its applications, but should also help faculty develop teaching materials. Moreover, it will be an ideal resource for the training of professional engineers in the fast-growing fields of nanoscience, nanoengineering, and nanotechnology. As such it should be appealing to a wide audience of future generations of engineers, physicists, and material scientists.

References

[1] Moore, G. E. (1965) *Electronics Magazine*, McGraw Hill, New York.

[2] See, for instance, Iwai, H. and Ohmi, S. (2002) A review of silicon integrated circuit technology from past to future. *Microelectronics Reliability* 42, pp. 465–491.

[3] The current International Technology Roadmap for Semiconductors is available at http://www.itrs2.net.

[4] Goldhaber-Gordon, D. et al., (1997) Overview of nanoelectronic devices. *Proceedings of the IEEE* 85, pp. 521–540.

[5] Jacak, L. et al. (1998) *Quantum Dots*, Springer, New York.

[6] Gaylord, T. K. et al. (1991) Quantum interference effects in semiconductors: A bibliography. *Proceedings of the IEEE* 79, pp. 1159–1180.

[7] Cela, D., Dresselhaus, M., Zeng, T. H., Souza Filho, A. G., and Ferreira, O. P. (2014) Resource letter: N-1: Nanotechnology. *American Journal of Physics* 82, pp. 8–22.

[8] Singh, C. (2001) Student understanding of quantum mechanics. *American Journal of Physics* 69, pp. 885–895.

[9] Thacker, B.-A., Leff, H. S. and Jackson, D. P. (2002) An introduction to the theme issue. *American Journal of Physics* 70, p. 199 [special issue on teaching quantum mechanics].

[10] Müller, R. and Weisner, H. (2002) Teaching quantum mechanics on an introductory level. *American Journal of Physics* 70, pp. 200–209.

[11] Singh, C. (2006) Assessing and improving student understanding of quantum mechanics. *AIP Conf.* 818, pp. 69–72.

[12] Singh, C. (2008) Student understanding of quantum mechanics at the beginning of graduate instruction. *American Journal of Physics* 76, pp. 277–287.

[13] Carr, L. D. and Mckagan, S. B. (2009) Graduate quantum mechanics reform. *American Journal of Physics* 77, pp. 308–319.

[14] Lin, S. Y. and Singh, C. (2010) Categorization of quantum mechanics problems by professors and students. *Eur. J. Phys.* 31, pp. 57–68.

[15] Alonso, M. (2002) Emphasize applications in introductory quantum mechanics courses. *American Journal of Physics* 70, p. 887.

[16] Zhu, G. and Singh, C. (2012) Surveying students' understanding of quantum mechanics in one spatial dimension. *American Journal of Physics* 80, pp. 252–259.

[17] Nielsen, M. A. and Chuang, I. L. (2003) *Quantum Computation and Quantum Information*, Cambridge University Press, Cambridge.

[18] Lindsay, P. A. (1967) *Introduction to Quantum Mechanics for Electrical Engineers*, McGraw-Hill Book Company, New York.

[19] Pohl, H. A. (1967) *Quantum Mechanics for Science and Engineering*, Prentice-Hall, Englewood Cliffs, NJ.

[20] Schiff, L. I. (1968) *Quantum Mechanics*, McGraw-Hill, New York.

[21] Gillepsie, D. (1970) *A Quantum Mechanics Primer*, International Textbook Company, Scranton, PA.

[22] Alonso, M. and Valk, H. (1973) *Quantum Mechanics: Principles and Applications*, Addison-Wesley, Reading, MA.

[23] Flügge, S. (1974) *Practical Quantum Mechanics*, Springer-Verlag, New York.

[24] Cohen-Tannoudji, C., Diu, B., and Laloë, F. (1977) *Quantum Mechanics*, Wiley, New York.

[25] Ridley, B. K. (1988) *Quantum Processes in Semiconductors*, Oxford Science Publications, Oxford.

[26] Ohanian, H. C. (1990) *Principles of Quantum Mechanics*, Prentice Hall, Englewood Cliffs, NJ.

[27] Griffiths, D. J. (1995) *Introduction to Quantum Mechanics*, Prentice Hall, Upple Saddle River, NJ.

[28] Peres, A. (1995) *Quantum Theory: Concepts and Methods*, Kluwer Academic, Norwell, MA.

[29] Brandt, S. and Dahmen, H. D. (1995) *The Picture Book of Quantum Mechanics*, Springer-Verlag, New York.

[30] Gasiorowicz, S. (1996), *Quantum Physics*, Wiley, New York.

[31] Fromhold, A. T. (1981) *Quantum Mechanics for Applied Physics and Engineering*, Academic Press, New York.

[32] Datta, S. (1989) *Quantum Phenomena*, Volume III, Modular Series, Addison-Wesley, Reading, MA.

[33] Bohm, A., (1993) *Quantum Mechanics: Foundations and Applications*, Springer-Verlag, New York.

[34] Kroemer, H. (1994) *Quantum Mechanics for Engineering, Materials Science, and Applied Physics*, Prentice-Hall, Englewood Cliffs, NJ.

[35] Datta, S. (1995) *Electronic Transport in Mesoscopic Systems*, Cambridge University Press, Cambridge.

[36] Singh, J. (1997) *Quantum Mechanics: Fundamentals and Applications to Technology*, Wiley, New York.

[37] Basdevant, J. L. (2000) *The Quantum Mechanics Solver: How to Apply Quantum Theory to Modern Physics*, Springer, Berlin.

[38] Townsend, J. S. (2000) *A Modern Approach to Quantum Mechanics*, University Science Books, CA.

[39] Ferry, D. K. (2001) *Quantum Mechanics: An Introduction for Device Physicists and Electrical Engineers*, Institute of Physics Publishing, London.

[40] Rae, A. I. M. (2002) *Quantum Mechanics*, Taylor and Francis, New York.

[41] Gottfried, K. and Yan, T.-M. (2004) *Quantum Mechanics*, Springer, New York.

[42] Basdevant, J. L. and Dalibard, J. (2005) *Quantum Mechanics*, Springer, New York.

[43] Levi, A. F. J. (2006) *Applied Quantum Mechanics*, Cambridge University Press, New York.

[44] Le Bellac, M. (2006) *Quantum Physics*, Cambridge University Press, New York.

Suggested Reading

- Kogan, V. I. and Galitsiv, V. M. (1963) *Problems in Quantum Mechanics*, Prentice Hall, Englewood Cliffs, New Jersey.

- ter Haar, D. (1964) *Selected Problems in Quantum Mechanics*, Academic Press, New York.

- Constantinesu, F. and Magyari, E. (1976) *Problems in Quantum Mechanics*, Pergamon Press, New York.

- Peleg, Y., Pnini, R., and Zaarir, Z. (1998) *Schaum's Outline of Theory and Problems of Quantum Mechanics*, McGraw Hill, New York.

- Mavromatis, H. A. (1992) *Exercises in Quantum Mechanics: A Collection of Illustrative Problems and Their Solutions*, Kluwer Academic, New York.

- Squires, G. L. (1995) *Problems in Quantum Mechanics with Solutions*, Cambridge University Press, New York.

- Lim, Y.-K. (ed.) (1998) *Problems and Solutions on Quantum Mechanics Compiled by the Physics Coaching Class, University of Science and Technology of China*, World Scientific, Singapore.

- Basdevant, J.-L. and Dalibard, J. (2000) *The Quantum Mechanics Solver: How to Apply Quantum Theory to Modern Physics*, Springer, New York.

- Capri, A. Z. (2002) *Problems and Solutions in Non-Relativistic Quantum Mechanics*, World Scientific, River Edge, NJ.

- Tamvakis, K. (2005) *Problems and Solutions in Quantum Mechanics*, Cambridge University Press, Cambridge.

- Gold'man, I. I. and Krivchenkov, V. D. (2006) *Problems in Quantum Mechanics*, Dover Publications, New York.

- d'Emilio, E. and Picasso, L. E. (2011) *Problems in Quantum Mechanics with Solutions*, Springer, New York.

- Cini, M., Fucito, R., and Sbragaglia, M. (2012) *Solved Problems in Quantum and Statistical Mechanics*, Springer, New York.

- Flügge, S. (1999) *Practical Quantum Mechanics*, Springer, New York.

- Harrison, W. A. (2000) *Applied Quantum Mechanics*, World Scientific, River Edge, NJ.

- Marchildon, L. (2002) *Quantum Mechanics: From Basic Principles to Numerical Methods and Applications*, Springer, New York.

- Blinder, S. M. (2004) *Introduction to Quantum Mechanics: In Chemistry, Materials Science, and Biology*, Elsevier, Boston.

- Tang, C. L. (2005), *Fundamentals of Quantum Mechanics: For Solid State Electronics and Optics*, Cambridge University Press, New York.

- Harrison, P. (2005) *Quantum Wells, Wires, and Dots*, Wiley, Chichester.

- Robinett, R. W. (2006) *Quantum Mechanics, Classical Results, Modern Systems, and Visualized Examples*, Oxford University Press, New York.

- Miller, D. A. B. (2008) *Quantum Mechanics for Scientists and Engineers*, Cambridge University Press, New York.

- Zettili, N. (2009) *Quantum Mechanics: Concepts and Application*, Wiley, Chichester.

- Kim, D. M. (2010) *Introductory Quantum Mechanics for Semiconductor Nanotechnology*, Wiley-VCH, Weinheim.

- Ahn, D. and Park, S.-H. (2011) *Engineering Quantum Mechanics*, Wiley-IEEE, Hoboken, NJ.

- Sullivan, D. M. (2012) *Quantum Mechanics for Electrical Engineers*, Wiley-Blackwell, Oxford.

- Band, Y. B. and Avishai, Y. (2013) *Quantum Mechanics with Applications to Nanotechnology and Information Science*, Academic Press, New York.

Chapter 1: General Properties of the Schrödinger Equation

The following set of problems deals with the time-independent *effective-mass Schrödinger equation* (EMSE), which governs the steady-state behavior of an electron in a solid with spatially varying potential profile. The solid may consist of one or more materials (e.g., a heterostructure or superlattice); hence, the effective mass of the electron may vary in space. The EMSE is widely used in studying the electronic and optical properties of solids. The reader should first consult textbooks on EMSE if unfamiliar with the concept [1, 2]. The concept of quantum mechanical wave impedance is introduced to point out the similarity between solutions to the time-independent Schrödinger equation and transmission line theory in classical electrodynamics [3] and microwave theory [4–6].

---

* Problem 1.1: The effective mass Schrödinger equation for arbitrary spatially varying effective mass $m^*(z)$ and potential $E_c(z)$ profiles

(a) Consider an electron in a solid, which could be a semiconductor or an insulator. We will exclude a metal since the potential inside a metal is usually spatially invariant as a metal cannot sustain an electric field. The potential that an electron sees inside a semiconductor or insulator is the conduction band profile. We will assume that it is time independent and varies only along one direction, which we call the z-direction (see Figure 1.1). The effective mass also varies along that direction.

Show that the stationary solutions of the Schrödinger equation obey the equation

$$-\frac{\hbar^2}{2m^*(z)}\left(\frac{\mathrm{d}^2}{\mathrm{d}x^2}+\frac{\mathrm{d}^2}{\mathrm{d}y^2}\right)\psi(x,y,z)-\frac{\hbar^2}{2}\frac{\mathrm{d}}{\mathrm{d}z}\left[\frac{1}{m^*(z)}\frac{\mathrm{d}}{\mathrm{d}z}\right]\psi(x,y,z)$$

$$+E_c(z)\psi(x,y,z)=E\psi(x,y,z), \tag{1.1}$$

where $E_c(z)$ is the conduction band edge and E is the total energy of the electron, which is independent of z because the total energy is a good quantum number in the absence of dissipation. In Equation (1.1),

$$E=E_{\mathrm{kin}}(z)+E_c(z)\quad and$$

$$E_{\mathrm{kin}}(z)=\frac{\hbar^2}{2m_c^*}\left(k_z^2(z)+k_t^2\right), \tag{1.2}$$

where k_z and $k_t=\sqrt{k_x^2+k_y^2}$ are the longitudinal and transverse components of the electron's wave vector, respectively, while $m_c^*=m^*(0)$ is the effective mass in the

Problem Solving in Quantum Mechanics: From Basics to Real-World Applications for Materials Scientists, Applied Physicists, and Devices Engineers, First Edition.
Marc Cahay and Supriyo Bandyopadhyay.
© 2017 John Wiley & Sons Ltd. Published 2017 by John Wiley & Sons Ltd.

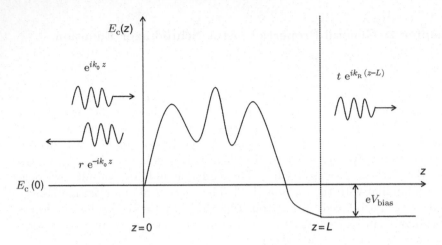

Figure 1.1: Illustration of electron impinging on the left on an arbitrary conduction band energy profile under bias. V_{bias} is the potential difference between the two contacts.

contacts. Since the potential varies only in the z-direction, k_t is spatially invariant, but k_z varies with the z-coordinate.

(b) Because the potential does not vary in the x–y plane, the transverse component of the wave function is a plane wave and we can write the wave function in Equation (1.1) as

$$\psi(x, y, z) = \phi(z) e^{i\vec{k}_t \cdot \vec{\rho}}. \tag{1.3}$$

Show that the z-component of the wave function $\phi(z)$ satisfies the following EMSE:

$$\frac{d}{dz}\left[\frac{1}{\gamma(z)}\frac{d}{dz}\right]\phi(z) + \frac{2m_c^*}{\hbar^2}\left[(E_p + E_t[1 - \gamma^{-1}(z)] - E_c(z))\right]\phi(z) = 0, \tag{1.4}$$

where $\gamma(z) = \frac{m^*(z)}{m_c^*}$, $E_t = \frac{\hbar^2 k_t^2}{2m_c^*}$, and $E_p = \frac{\hbar^2 k_z^2}{2m_c^*}$.

Solution: Taking into account a spatially varying effective mass along the z-direction, the time-dependent Schrödinger equation describing an electron moving in an arbitrary potential $E_c(z)$ is given by

$$\left[-\frac{\hbar^2}{2m^*(z)}\left(\frac{d^2}{dx^2} + \frac{d^2}{dy^2}\right) - \frac{\hbar^2}{2}\frac{d}{dz}\left(\frac{1}{m^*(z)}\frac{d}{dz}\right) + E_c(z)\right]\Psi(x, y, z, t)$$

$$= i\hbar\frac{d\Psi(x, y, z, t)}{dt}. \tag{1.5}$$

Since all quantities on the left-hand side are time independent by virtue of the fact that the potential is time invariant, we can write the wave function in a product form:

$$\Psi(x, y, z, t) = \psi(x, y, z)\xi(t). \tag{1.6}$$

Substitution of this form in Equation (1.5) immediately yields that $\xi(t) = \mathrm{e}^{-iEt/\hbar}$, and $\psi(x, y, z)$ obeys Equation (1.1).

Next, we note that in Equation (1.5) no quantity in the Hamiltonian on the left-hand side (i.e., the terms within the square brackets) depends on more than one coordinate (x, y, or z). Hence, the wave function $\psi(x, y, z)$ can be written as the product of an x-dependent term, a y-dependent term, and a z-dependent term. Furthermore, no quantity depends on the x or y coordinate. Therefore, the x-dependent and y-dependent terms must be plane waves. The z-dependent term will not be a plane wave since both $E_c(z)$ and $m^*(z)$ depend on the z-coordinate. Consequently, we write $\psi(x, y, z)$ as

$$\psi(x, y, z) = \mathrm{e}^{ik_x x}\mathrm{e}^{ik_y y}\phi(z). \tag{1.7}$$

Plugging this last expression into Equation (1.1) leads to the EMSE for the envelope function $\phi(z)$:

$$\left[-\frac{\hbar^2}{2m^*(z)}\left(k_x^2 + k_y^2\right) + \frac{\hbar^2}{2}\frac{\mathrm{d}}{\mathrm{d}z}\left(\frac{1}{m^*(z)}\frac{\mathrm{d}}{\mathrm{d}z}\right)\right]\phi(z) = [E - E_c(z)]\,\phi(z). \tag{1.8}$$

Multiplying both sides of the equation by $\frac{2m_c^*}{\hbar^2}$, we obtain

$$\left[-\frac{2m_c^*}{\hbar^2}\frac{E_t}{\gamma(z)} + \frac{\mathrm{d}}{\mathrm{d}z}\left(\frac{1}{\gamma(z)}\frac{\mathrm{d}}{\mathrm{d}z}\right)\right]\phi(z) = \frac{2m_c^*}{\hbar^2}\left(E - E_c(z)\right)\phi(z). \tag{1.9}$$

Hence,

$$\frac{\mathrm{d}}{\mathrm{d}z}\left[\frac{1}{\gamma(z)}\frac{\mathrm{d}}{\mathrm{d}z}\phi(z)\right] + \frac{2m_c^*}{\hbar^2}\left[E - \frac{E_t}{\gamma(z)} - E_c(z)\right]\phi(z) = 0, \tag{1.10}$$

where $E = E_t + E_p$, $E_p = \frac{\hbar^2 k_z^2}{2m_c^*}$ is the longitudinal kinetic energy (i.e., the kinetic energy associated with the z-component of the motion), and $E_t = \frac{\hbar^2(k_x^2 + k_y^2)}{2m_c^*}$ is the transverse kinetic energy.

Problem 1.2: The Ben Daniel–Duke boundary condition

Starting with the one-dimensional EMSE derived in the previous problem, show that in addition to the continuity of the wave function required by the postulates of quantum mechanics (see Appendix A), the following quantity must be continuous when taking into account the spatial variation of the effective mass:

$$\frac{1}{m^*(z)}\frac{\mathrm{d}\phi(z)}{\mathrm{d}z}. \tag{1.11}$$

The continuity of this quantity generalizes the continuity of the first derivative of the wave function typically used in solving quantum mechanical problems. Together with the continuity of the wave function, imposing continuity of this quantity is referred to as the Ben Daniel–Duke boundary conditions [7]. The application of these boundary conditions will be illustrated in several problems throughout the book.

Solution: Our starting point is the time-independent effective one-dimensional Schrödinger equation derived in the previous problem:

$$\frac{\mathrm{d}}{\mathrm{d}z}\left[\frac{1}{\gamma(z)}\frac{\mathrm{d}}{\mathrm{d}z}\phi(z)\right] + \frac{2m_c^*}{\hbar^2}\left[E - \frac{E_t}{\gamma(z)} - E_c(z)\right]\phi(z) = 0, \qquad (1.12)$$

where $E = E_t + E_p$, $E_p = \frac{\hbar^2 k_z^2}{2m_c^*}$ is the longitudinal kinetic energy, $\gamma(z) = m^*(z)/m_c^*$, where m_c^* is the effective mass in the contacts, and $E_t = \frac{\hbar^2(k_x^2 + k_y^2)}{2m_c^*}$ is the transverse kinetic energy.

Assuming that both $\gamma(z)$ and $E_c(z)$ are either continuous or have finite jumps for all z, we integrate the last Schrödinger equation on both sides from $z_0^- = z_0 - \epsilon$ to $z_0^+ = z_0 + \epsilon$, where ϵ is a small positive quantity. This leads to

$$\frac{1}{\gamma(z)}\frac{\mathrm{d}\phi(z)}{\mathrm{d}z}\bigg|_{z=z_0^-} = \frac{1}{\gamma(z)}\frac{\mathrm{d}\phi(z)}{\mathrm{d}z}\bigg|_{z=z_0^+}. \qquad (1.13)$$

As $\epsilon \to 0$, this last equation shows that the quantity $\frac{1}{m^*(z)}\frac{\mathrm{d}\phi(z)}{\mathrm{d}z}$ is continuous everywhere [7].

When the effective mass varies in space, the correct boundary condition (involving the spatial derivative of the wave function) to use has been a hotly debated topic. There is some controversy regarding the appropriate form of the Hamiltonian to use in the case of spatially varying effective mass, and the reader is referred to the literature on this topic [8–10].

Preliminary: Linearly independent solutions of the Schrödinger equation

We consider the three-dimensional time-independent Schrödinger equation associated with an electron moving in an arbitrary potential energy profile (such as the conduction band of a semiconductor structure) $E_c(z)$ varying along the z-direction only, as shown in Figure 1.1. If the effective mass of the electron is assumed to be independent of z, the Schrödinger equation for the envelope function component along the z-direction is given by (see Problem 1.1)

$$-\frac{\hbar^2}{2m^*}\ddot{\phi}(z) + E_c(z)\phi(z) = E_p\phi(z), \qquad (1.14)$$

where $\ddot{\phi}(z)$ stands for $\frac{\mathrm{d}^2}{\mathrm{d}z^2}\phi(z)$, the second derivative with respect to z, and E_p is the longitudinal kinetic energy, i.e., the kinetic energy component associated with motion in the z-direction.

The general solution of this second-order differential equation for $\phi(z)$ can be written as a linear combination of two linearly independent solutions [11]. Two solutions $\phi_1(z)$ and $\phi_2(z)$ of a differential equation are linearly independent if the equality $c_1\phi_1(z) + c_2\phi_2(z) = 0$ cannot be satisfied for all z for any choice of (c_1, c_2) except for $c_1 = c_2 = 0$. If non-zero solutions (c_1, c_2) exist, then $\phi_1(z)$ and $\phi_2(z)$ are said to be linearly dependent.

Concept of Wronskian: If ϕ_1 and ϕ_2 are linearly dependent, then $c_1\phi_1(z) + c_2\phi_2(z) = 0$ and $c_1\dot{\phi}_1(z) + c_2\dot{\phi}_2(z) = 0$, where the dot stands for the first derivative with respect to z. Hence, in matrix form, we have

$$\left[\begin{array}{cc} \dot{\phi}_1(z) & \dot{\phi}_2(z) \\ \phi_1(z) & \phi_2(z) \end{array} \right] \left[\begin{array}{c} c_1 \\ c_2 \end{array} \right] = \left[\begin{array}{c} 0 \\ 0 \end{array} \right]. \tag{1.15}$$

This means that the Wronskian $W(z) = \dot{\phi}_1(z)\phi_2(z) - \phi_1(z)\dot{\phi}_2(z)$, which is the determinant of the 2×2 matrix in the last equation, must be identically zero since otherwise only the trivial solution $c_1 = c_2 = 0$ in the last equation would be admissible. Stated differently, for two solutions to be linearly independent, their Wronskian must be non-zero for all z.

Even if $E_c(z)$ has finite discontinuities, it is obvious from Equation (1.14) that $\ddot{\phi}(z)$ exists throughout and, hence, $\phi(z)$ and $\dot{\phi}(z)$ must be continuous.

Since $\phi_1(z), \phi_2(z)$ satisfy the Schrödinger equation in Equation (1.14), it is easy to see that

$$\ddot{\phi}_1(z)\phi_2(z) - \ddot{\phi}_2(z)\phi_1(z) = 0, \tag{1.16}$$

or

$$\frac{\mathrm{d}}{\mathrm{d}z}\left[\dot{\phi}_1(z)\phi_2(z) - \dot{\phi}_2(z)\phi_1(z) \right] = \frac{\mathrm{d}W}{\mathrm{d}z} = 0. \tag{1.17}$$

Thus, $W(z)$ is a constant independent of z. It is obviously independent of z when it is zero, but it is also independent of z when it is non-zero.

In summary, if the Wronskian $W(z) = 0$, then (ϕ_1, ϕ_2) are linearly dependent, and, if $W(z) \neq 0$, they are are linearly independent. If we can find these two linearly independent solutions, their linear combination is the most general solution to the Schrödinger equation. This general solution is

$$\phi(z) = c_1\phi_1(z) + c_2\phi_2(z). \tag{1.18}$$

Typically, two solutions $\phi_1(z)$ and $\phi_2(z)$ of the Schrödinger equation are found such that they obey the boundary conditions

$$\phi_1(0) = 0, \quad \dot{\phi}_1(0) = 1, \tag{1.19}$$

and

$$\phi_2(0) = 1, \quad \dot{\phi}_2(0) = 0. \tag{1.20}$$

These solutions are indeed linearly independent since their Wronskian, which is independent of z, is equal to

$$W(z) = W(0) = \dot{\phi}_1(0)\phi_2(0) - \phi_1(0)\dot{\phi}_2(0) = 1. \tag{1.21}$$

In Chapter 7, we will show that the concept of Wronskian and linearly independent solutions can be used to introduce the concept of a transfer matrix, which is a very powerful approach to solving both bound state and tunneling problems in spatially varying potentials.

** Problem 1.3: General properties of the one-dimensional time-independent Schrödinger equation

Suppose two wave functions $\phi_1(z)$ and $\phi_2(z)$ satisfy the following two EMSEs:

$$-\frac{\hbar^2}{2m^*}\ddot{\phi}_1(z) + E_{c1}(z)\phi_1(z) = E_1\phi_1(z), \qquad (1.22)$$

and

$$-\frac{\hbar^2}{2m^*}\ddot{\phi}_2(z) + E_{c2}(z)\phi_2(z) = E_2\phi_2(z), \qquad (1.23)$$

respectively.

First, multiply Equation (1.22) by $\phi_2(z)$ and Equation (1.23) by $\phi_1(z)$, then subtract one from the other, and finally integrate the difference from z_1 to z_2. This will yield

$$[\phi_2(z)\phi'_1(z) - \phi_1(z)\phi'_2(z)]_{z_1}^{z_2}$$
$$= \frac{2m^*}{\hbar^2}\int_{z_1}^{z_2}[(E_2 - E_{c2}(z)) - (E_1 - E_{c1}(z))]\,\phi_1(z)\phi_2(z)\mathrm{d}z. \qquad (1.24)$$

Starting with the last equation, prove the following statements [11]:

(a) If $E_2 - E_{c2}(z) \geq E_1 - E_{c1}(z)$, there is a node of $\phi_2(z)$ between any two nodes of $\phi_1(z)$.

(b) If $E_{c2}(z) = E_{c1}(z) = E_c(z)$, $E_2 > E_1$, and $\phi_{1,2}(\pm\infty) = \phi'_{1,2}(\pm\infty) = 0$, then ϕ_2 has more nodes than ϕ_1.

(c) If ϕ_1 and ϕ_2 are solutions of Equations (1.22)–(1.23) with eigenvalues E_1 and E_2, respectively, and are such that ϕ_1, ϕ'_1, ϕ_2, and ϕ'_2 vanish at either $+\infty$ or $-\infty$, and ϕ_1 and ϕ_2 are linearly independent, then $E_1 \neq E_2$.

Solution:
(a) We prove these assertions indirectly. Let us assume that z_1 and z_2 are the locations of two consecutive nodes of $\phi_1(z)$ and that $\phi_2(z)$ has no node in the interval $[z_1, z_2]$. Since the wave functions can have nodes, both $\phi_1(z)$ and $\phi_2(z)$ must be real (since no solution of the Schrödinger equation that has an imaginary component can vanish anywhere). Furthermore, let us assume (without loss of generality) that $\phi_1(z)$ and $\phi_2(z)$ are both positive within the interval $[z_1, z_2]$ (the proofs for the cases when one is positive and the other negative, or both are negative, are no different from what follows and can be worked out by the reader following the derivation that ensues). Since $\phi_1(z_1) = \phi_1(z_2) = 0$ by assumption, Equation (1.24) leads to

$$\phi_2(z_2)\phi'_1(z_2) - \phi_2(z_1)\phi'_1(z_1)$$
$$= \frac{2m^*}{\hbar^2}\int_{z_1}^{z_2}[(E_2 - E_{c2}(z)) - (E_1 - E_{c1}(z))]\,\phi_1(z)\phi_2(z)\mathrm{d}z. \qquad (1.25)$$

Now, since $\phi_1(z)$ vanishes at z_1 and z_2, and in between these two locations is positive, clearly $\phi_1(z_1 + \epsilon) - \phi_1(z_1) > 0$ and $\phi_1(z_2) - \phi_1(z_2 - \epsilon) < 0$, where ϵ is an infinitesimally small positive quantity. Therefore the spatial derivatives $\phi'_1(z_1)$ and $\phi'_1(z_2)$ are, respectively, positive and negative. Consequently, the left-hand side of the last equation will be negative while the right-hand side is positive, which leads to a mathematical contradiction. Since our original assumption leads to this contradiction, the statement (a) above is proved by *reductio ad absurdum*.

(b) The conditions $\phi_1(\pm\infty) = 0$ and $\phi'_1(\pm\infty)$ imply that the particle is confined to a finite region of the z-axis. The corresponding states are called *bound states*. Letting $E_{c2}(z) = E_{c1}(z)$ in Equation (1.25), we get

$$\phi_2(z_2)\phi'_1(z_2) = \frac{2m^*}{\hbar^2}(E_2 - E_1)\int_{z_1}^{z_2}\phi_1(z)\phi_2(z)\mathrm{d}z. \tag{1.26}$$

If $z_1 = -\infty$ and z_2 is the first node of $\phi_1(z)$ from the left, we can, without loss of generality, take $\phi_1(z) > 0$ in the interval $]-\infty, z_2]$. Since $\phi_1(z_2) = 0$ (it is a node) and $\phi_1(z_2 - \epsilon) > 0$ (ϵ is an infinitesimally small positive quantity), obviously the spatial derivate of ϕ_1 at z_2 is negative, i.e., $\phi'_1(z_2) < 0$.

If $\phi_2(z)$ did not have a node in the interval $]-\infty, z_2]$, then its sign will not change within that interval, i.e. the sign will be constant. Since $\phi_{1,2}(\pm\infty) = \phi'_{1,2}(\pm\infty) = 0$ and $\phi_1(z_2) = 0$, we obtain

$$\phi_2(z_2)\phi'_1(z_2) = \frac{2m^*}{\hbar^2}(E_2 - E_1)\int_{-\infty}^{z_2}\phi_1(z)\phi_2(z)\mathrm{d}z. \tag{1.27}$$

Let us assume that the constant sign of $\phi_2(z)$ in the interval is positive. Then, since $\phi'_1(z_2) < 0$, the left-hand side of the last equation becomes negative while the right-hand side remains positive, leading to an absurdity. The reader can verify that the same absurdity would have arisen if we had assumed that the constant sign of $\phi_2(z)$ in the interval was negative, instead of positive. Therefore, the assumption that $\phi_2(z)$ has a constant sign in the interval $]-\infty, z_2]$ is invalid. As a consequence, the wave function $\phi_2(z)$ must have at least one node to the left of the first node of $\phi_1(z)$. According to part (a), there must be at least one node of $\phi_2(z)$ between any two nodes of $\phi_1(z)$. In a similar fashion, it can be shown that there is at least one node of $\phi_2(z)$ to the right of the last node of $\phi_1(z)$. Thus, $\phi_2(z)$ has at least one more node than $\phi_1(z)$, and part (b) is proved.

(c) Suppose $E_1 = E_2$. Then, since $\phi_1(z)$ and $\phi_2(z)$ are solutions of the same EMSE with the same energy $E_1 = E_2$, we can define a Wronskian W. Furthermore, since $\phi_1(z)$ and ϕ_2 are linearly independent, their Wronskian, $W(z) = \phi'_1(z)\phi_2(z) - \phi_1(z)\phi'_2(z)$, is a constant *different* from zero. But W is independent of the z-coordinate and, because either $W(+\infty)$ or $W(-\infty)$ vanishes, W must be exactly zero. Once again, we have a contradiction, which tells us that our original assumption must have been incorrect and therefore $E_1 \neq E_2$. Two (or more) linearly independent states having the same energy eigenvalues are said to be *degenerate*. This last property shows that *bound states of a particle in a one-dimensional potential are always non-degenerate*.

* **Problem 1.4: Bound states of a particle moving in a one-dimensional potential of even parity**

Show that the bound state solutions of the one-dimensional Schrödinger equation in a one-dimensional potential of even parity have either odd or even parity.

Solution: If the one-dimensional potential energy profile has even parity, then it is symmetric about $z = 0$, i.e., $E_c(z) = E_c(-z)$. In that case, if $\phi(z)$ is solution of the one-dimensional EMSE in Equation (1.14), then clearly so is $\phi(-z)$.

Any wave function $\phi(z)$ can be written as a sum of a symmetric part $\phi_+(z)$ and an antisymmetric part $\phi_-(-z)$:

$$\phi(z) = \frac{1}{2}[\phi(z) + \phi(-z)] + \frac{1}{2}[\phi(z) - \phi(-z)] = \phi_+(z) + \phi_-(-z). \qquad (1.28)$$

For an even $E_c(z)$, since both $\phi(z)$ and $\phi(-z)$ are solutions of Equation (1.14), so must be $\phi_+(z)$ and $\phi_-(z)$ since they are linear combinations of $\phi(z)$ and $\phi(-z)$.

From the results of Problem 1.3, the bound states of a general one-dimensional $E_c(z)$ are non-degenerate. Hence, the bound states of a particle moving in an even $E_c(z)$ must have a definite parity, i.e., must be either odd (antisymmetric) or even (symmetric).

---

* **Problem 1.5: Quantum measurement**

The "particle in a one-dimensional box" problem is one of the simplest problems in quantum mechanics. It refers to an electron confined within a one-dimensional box with infinite barriers. The eigenstates (or allowed wave functions) of the electron are given by (see Problem 3.5):

$$\phi_n(z) = \sqrt{\frac{2}{W}} \sin\left(\frac{n\pi z}{W}\right), \qquad (1.29)$$

where n is an integer ($n = 1, 2, 3, \ldots$) and W is the width of the box. Each value of n defines an eigenstate.

The corresponding energy eigenvalues are given by

$$E_n = n^2 \frac{\hbar^2 \pi^2}{2m^* W^2}, \qquad (1.30)$$

and they are all distinct or non-degenerate, in keeping with what was proved in Problem 1.2.

For an electron injected into a quantum box of width W at time $t = 0$ in the state

$$\phi(0) = A\left[i\sqrt{2}\sin\left(\frac{3\pi z}{W}\right) - \sin\left(\frac{7\pi z}{W}\right) + 2\sin\left(\frac{9\pi z}{W}\right)\right] e^{i(k_x x + k_y y)}, \qquad (1.31)$$

where the plane of the quantum well is the (x, y) plane:

(a) *Find the value of A.*

(b) *Find the wave function at time t.*

(c) *What is the expectation value of the electron's energy?*

Solution: (a) Normalization of the wave function requires $A^2 \frac{W}{2}(2 + 1 + 4) = 1$, hence $A = \sqrt{\frac{2}{7W}}$.

(b) The wave function at time t is

$$
\phi(t) = \sqrt{\frac{2}{7W}} \left[i\sqrt{2} \sin\left(\frac{3\pi z}{W}\right) e^{\frac{iE_3 t}{\hbar}} - \sin\left(\frac{7\pi z}{W}\right) e^{\frac{iE_7 t}{\hbar}} \right.
$$
$$
\left. + 2 \sin\left(\frac{9\pi z}{W}\right) e^{\frac{iE_9 t}{\hbar}} \right] e^{i(k_x x + k_y y)}, \tag{1.32}
$$

where E_n is given by Equation (1.30).

(c) The expectation value of the energy is independent of time and given by

$$
\langle E \rangle = \frac{2E_3 + E_7 + 4E_9}{7}
$$
$$
= \frac{\hbar^2}{14m^*} \left[2 \times 3^2 + 7^2 + 4 \times 9^2\right] \left(\frac{\pi}{W}\right)^2
$$
$$
= \frac{391\hbar^2}{14m^*} \left(\frac{\pi}{W}\right)^2. \tag{1.33}
$$

** Problem 1.6: Concept of quantum mechanical wave impedance

Starting with Problem 1.1, show that the Schrödinger equation for a particle moving in a general potential energy profile $E_c(z)$ (assuming constant effective mass)

$$
\frac{d^2}{dz^2}\phi(z) + \beta^2 \phi(z) = 0, \tag{1.34}
$$

where

$$
\beta^2 = \frac{2m^*}{\hbar^2}\left[E - E_c(z) - \frac{\hbar}{2m^*}\left(k_x^2 + k_y^2\right)\right] \tag{1.35}
$$

and (k_x, k_y) are the components of the transverse momentum, can be rewritten as two first-order differential equations

$$
\frac{dv(z)}{dz} = -Zu(z) \tag{1.36}
$$

and

$$\frac{du(z)}{dz} = -Yv(z) \tag{1.37}$$

if we introduce the two variables

$$u = \phi(z) \tag{1.38}$$

and

$$v = -\frac{2\hbar}{m^*i}\frac{d\phi}{dz}. \tag{1.39}$$

What are the expressions for Z and Y?

   This is equivalent to defining a quantum mechanical wave impedance $Z(z)$ as follows [12]:

$$Z(z) = \frac{v(z)}{u(z)} = \left(\frac{2\hbar}{im^*}\right)\frac{\frac{d\phi(z)}{dz}}{\phi(z)}. \tag{1.40}$$

Solution: Based on the definitions in Equations (1.38) and (1.39), we get

$$\frac{du}{dz} = \frac{d\phi}{dz} = -\frac{m^*}{2i\hbar}v \tag{1.41}$$

and

$$\frac{dv}{dz} = -\frac{2i\hbar}{m^*}\frac{d^2\phi}{dz^2} = \left(\frac{2i\hbar\beta^2}{m^*}\right)u, \tag{1.42}$$

which can be recast as

$$\frac{dv}{dz} = -Zu \tag{1.43}$$

and

$$\frac{du}{dz} = -Yv \tag{1.44}$$

by introducing the quantities

$$Y = \frac{m^*}{2i\hbar} \tag{1.45}$$

and

$$Z = -\frac{2i\hbar\beta^2}{m^*}. \tag{1.46}$$

Equations (1.43) and (1.44) look similar to the transmission line equations in electromagnetic field theory [3–5] if we use the transformations

$$V(z) \to -\frac{2i\hbar}{m^*}\frac{d\phi}{dz} \tag{1.47}$$

and

$$I(z) \to \phi(z). \tag{1.48}$$

The Schrödinger Equation (1.34) is therefore equivalent to two first-order coupled differential equations:

$$\frac{dv(z)}{dz} = -ZI(z), \tag{1.49}$$

$$\frac{dI(z)}{dz} = -YV(z). \tag{1.50}$$

This is equivalent to defining a quantum mechanical wave impedance $Z_{QM}(z)$ as follows [12]:

$$Z_{QM}(z) = \frac{2\hbar}{im^*}\frac{(d\phi(z)/dz)}{\phi(z)}. \tag{1.51}$$

The quantity $Z_{QM}(z)$ does not have the unit of ohms (it has the unit of velocity), but is a useful concept to solve some bound state and tunneling problems, as illustrated in Chapter 7.

Suggested problems

- Consider a particle with the one-dimensional wave function

$$\phi(z) = N\left(a^2 + z^2\right)^{-1/2} e^{\frac{ip_0 z}{\hbar}}, \tag{1.52}$$

where a, p_0, and N are real constants.

(1) Find the normalization constant N.

(2) Determine the probability of finding the particle in the interval $-\left(a/\sqrt{5}\right) \leq z \leq \left(a/\sqrt{5}\right)$.

(3) What is the expectation value of the momentum?

- Consider a particle in a box defined by the potential profile:

$$V(z) = 0 \text{ if } |z| \leq W \text{ and } V(z) = \infty \text{ if } |z| \geq W.$$

At time $t = 0$ the wave function $\phi_0(z)$ is an even function of z.

(1) What are the possible values resulting from a measurement of the kinetic energy?

(2) How soon after $t = 0$ will the particle return to its initial state if left undisturbed? (If the particle is left undisturbed, it periodically visits the initial state. This is known as Poincaré recurrence).

- Consider a particle described by the one-dimensional harmonic oscillator Hamiltonian (see Appendix B)

$$H = \frac{p^2}{2m^*} + \frac{1}{2}m^*\omega^2 z^2.$$ (1.53)

At time $t = 0$, the particle is described by the following wave function:

$$\phi(z) = N(1 + 3\alpha^2 z^2)e^{-\frac{1}{2}\alpha^2 z^2},$$ (1.54)

where $\alpha = m^*\omega/\hbar^{1/2}$ and N is the normalization constant.

(1) Find the value of N.

(2) What are the possible outcomes of the measurement of the total energy of the particle and with what probabilities?

(3) Write down the analytical expression of the wave function of the particle at time t.

- When the mass of a particle varies with position, the mass and momentum operators do not commute. The kinetic operator must be modified. Von Ross proposed the following kinetic operator [10]:

$$T(z) = \frac{1}{4}(m^\alpha p m^\beta p m^\gamma + m^\gamma p m^\beta p m^\alpha),$$ (1.55)

with $\alpha + \beta + \gamma = -1$.

In this case, show that the Schrödinger equation becomes

$$\frac{d^2}{dz^2}\phi(z) - \frac{m'(z)}{m(z)}\frac{d}{dz}\phi(z)$$
$$+ \left[\frac{1}{2}\left(rm'' - s\frac{m'^2}{m^2}\right) + \frac{2m(z)}{\hbar^2}(E - E_c(z))\right]\phi(z) = 0,$$ (1.56)

where $r = \alpha + \gamma$, $s = \alpha(\gamma+2) - \gamma(\alpha+2)$. Furthermore, $m' = \frac{dm}{dz}$ and $m'' = \frac{d^2m}{dz^2}$.

- Starting with the results of the previous problem, show that the first derivative of $\phi(z)$ can be eliminated by making the transformation $\phi(z) = \sqrt{m(z)}\psi(z)$. Show that the resulting effective Schrödinger equation for $\psi(z)$ is given by

$$\frac{d^2}{dz^2}\phi(z) + \left[(1+r)\left\{\frac{m'(z)}{2m(z)}\right.\right.$$
$$\left.\left. - \left(\frac{3}{4} + \frac{s}{2}\right)\frac{m'^2}{m^2} + m(E - E_c(z))\right\}\right]\psi(z) = 0.$$ (1.57)

References

[1] Datta, S. (1989) *Quantum Phenomena*, Vol. III, Modular Series on Solid State Devices, eds. R. F. Pierret and G. W. Neudeck, Addison-Wesley, Reading, MA.

[2] Levi, A. F. J. (2006) *Applied Quantum Mechanics*, 2nd edition, Section 4.9.4, Cambridge University Press, Cambridge.

[3] Jackson, J. D. (1975) *Classical Electrodynamics*, 2nd edition, Wiley, New York.

[4] Ramo, S., Whinnery, T. R. and van Duzer, T. (1965) *Fields and Waves in Communication Electronics*, John Wiley & Sons, New York.

[5] Pozar, D. M. (1990) *Microwave Engineering*, Addison-Wesley, New York.

[6] Gonzalez, G. (1997) *Microwave Transistor Amplifiers, Analysis and Design*, 2nd edition, Prentice Hall, Upper Saddle River, NJ.

[7] Bastard, G. (1981) Superlattice band structure in the envelope-function approximation. *Physical Review B* 24, pp. 5693–5697.

[8] Morrow, R. A. and Brownstein, K. R. (1984) Model effective-mass Hamiltonians for abrupt heterojunctions and the associated wavefunction matching conditions. *Physical Review B* 30, pp. 678–680.

[9] Morrow, R. A. (1987) Establishment of an effective-mass Hamiltonian for abrupt heterojunctions. *Physical Review B* 35, pp. 8074–8079.

[10] von Ross, O. (1983) Position-dependent effective masses in semiconductor theory. *Physical Review B* 27, pp. 7547–7552.

[11] Rodriguez, S. (1983) *Notes on Quantum Mechanics*, Purdue University (unpublished).

[12] Khondker, A. N., Rezwan Khan, M., and Anwar A. F. M. (1988) Transmission line analogy of resonance tunneling phenomena: the generalized impedance concept. *Journal of Applied Physics* 63(10), pp. 5191–5193.

Suggested Reading

- Moriconi, M. (2007) Nodes of wave functions. *American Journal of Physics*, 75(3), pp. 284–285.

- Kwong, W., Rosner, J. L., Schonfeld, J. F., Quig, C., and Thacker, H. B. (1980) Degeneracy in one-dimensional quantum mechanics. *American Journal of Physics* 48(11), pp. 926–930.

- Haley, S. B. (1997) An underrated entanglement: Ricatti and Schrödinger equations. *American Journal of Physics*, 65(3), pp. 237–243.

- Rosner, J. L. (1993) The Smith chart and quantum mechanics. *American Journal of Physics*, 61(4), pp. 310–316.

- Fazlul Kalvi, S. M., Khan, M. R., and Alam, M. A. (1991) Application of quantum mechanical wave impedance in the solution of Schrödinger's equation in quantum wells. *Solid-State Electronics*, 34(12), pp. 1466–1468.

- Di Ventra, M. and Fall, C. J. (1998) General solution scheme for second-order differential equations. Application to quantum transport. *Computer Simulations*, 12(3), pp. 248–253.

Chapter 2: Operators

Some general definitions and properties of operators are reviewed in Appendix C, which the reader should consult before trying out the following set of problems.

** Problem 2.1: Operator identities

If A and B are general operators (which do not necessarily commute), show that the following identities hold:

(a) $e^{-A}B^n e^{+A} = (e^{-A}Be^A)^n$, *where n is an integer.*

(b) $B^{-1}e^A B = e^{B^{-1}AB}$.

(c) $e^{\xi A}F(B)e^{-\xi A} = F(e^{\xi A}Be^{-\xi A})$, *where F is any arbitrary function and ξ is an arbitrary complex number.*

Solution: (a)

$$e^{-A}B^n e^A = (e^{-A}Be^A)(e^{-A}Be^A)\cdots(e^{-A}Be^A)(e^{-A}Be^A), \tag{2.1}$$

because $e^{-A}e^A = I$, where I is the identity operator.

Therefore

$$e^{-A}B^n e^A = (e^{-A}Be^A)^n. \tag{2.2}$$

(b)

$$e^{B^{-1}AB} = \sum_{n=0}^{\infty}\frac{1}{n!}(B^{-1}AB)^n = \sum_{n=0}^{\infty}\frac{1}{n!}(B^{-1}AB)(B^{-1}AB)\cdots(B^{-1}AB). \tag{2.3}$$

Making use of the fact that $BB^{-1} = I$, we obtain

$$e^{B^{-1}AB} = B^{-1}\left(\sum_{n=0}^{\infty}\frac{1}{n!}A^n\right)B = B^{-1}e^A B. \tag{2.4}$$

(c) If $F(\xi)$ has an expression of the form $\sum_n f_n \xi^n$, then

$$F(e^{\xi A}Be^{-\xi A}) = \sum_n f_n (e^{\xi A}Be^{-\xi A})(e^{\xi A}Be^{-\xi A})\cdots(e^{\xi A}Be^{-\xi A}). \tag{2.5}$$

Using the results of parts (a) and (b), we get

$$F(e^{\xi A}Be^{-\xi A}) = e^{\xi A}\left(\sum_n f_n B^n\right)e^{-\xi A} = e^{\xi A}F(B)e^{-\xi A}. \tag{2.6}$$

Problem Solving in Quantum Mechanics: From Basics to Real-World Applications for Materials Scientists, Applied Physicists, and Devices Engineers, First Edition.
Marc Cahay and Supriyo Bandyopadhyay.
© 2017 John Wiley & Sons Ltd. Published 2017 by John Wiley & Sons Ltd.

** Problem 2.2: More identities

If A and B both commute with the commutator [A, B], prove the following equalities [1, 2]:

$$[A, F(B)] = [A, B]F'(B) \tag{2.7}$$

and

$$[G(A), B] = [A, B]G'(A). \tag{2.8}$$

Solution: We first prove the following equality by induction:

$$[A, B^n] = n[A, B]B^{n-1}. \tag{2.9}$$

Using the operator identity in Appendix B, we get

$$[A, B^n] = [A, BB^{n-1}] = [A, B]B^{n-1} + B[A, B^{n-1}]$$
$$= [A, B]B^{n-1} + (n-1)B[A, B]B^{n-2}. \tag{2.10}$$

Hence,

$$[A, B^n] = [A, B]B^{n-1} + (n-1)[A, B]B^{n-1} = n[A, B]B^{n-1}. \tag{2.11}$$

Since, by Taylor series expansion, $F(B) = F(0) + \frac{1}{1!}F'(0)B + \frac{1}{2!}F''(0)B^2 + \cdots$, we get

$$[A, F(B)] = [A, B]F'(0) + [A, B]F''(0)B + \frac{1}{2!}[A, B]F'''(0)B^2 + \cdots \tag{2.12}$$

Furthermore, since

$$F'(B) = F'(0) + \frac{1}{1!}F''(0)B + \frac{1}{2!}F'''(0)B^2 + \cdots, \tag{2.13}$$

we have

$$[A, F(B)] = [A, B]F'(B). \tag{2.14}$$

Now, using this relation, we get

$$[G(A), B] = -[B, G(A)] = -[B, A]G'(A) = [A, B]G'(A). \tag{2.15}$$

** Problem 2.3: Glauber identity

If A and B both commute with [A, B], prove that the following relation is true [1]:

$$e^A e^B = e^{A+B}e^{\frac{1}{2}[A,B]}. \tag{2.16}$$

Solution: Defining

$$F(t) = e^{At}e^{Bt}, \tag{2.17}$$

where t is a real variable, we have

$$\frac{\mathrm{d}F}{\mathrm{d}t} = Ae^{At}e^{Bt} + e^{At}Be^{Bt} = \left(A + e^{At}Be^{-At}\right)F(t). \tag{2.18}$$

Since B commutes with the commutator $[A, B]$ by assumption, and since

$$[G(A), B] = [A, B]G'(A), \tag{2.19}$$

we have

$$[e^{At}, B] = t[A, B]e^{At}. \tag{2.20}$$

Hence,

$$e^{At}B = Be^{At} + t[A, B]e^{At}. \tag{2.21}$$

Multiplying both sides of the above equation by e^{-At}, we obtain

$$\frac{\mathrm{d}F}{\mathrm{d}t} = \left(A + B + t[A, B]\right)F(t). \tag{2.22}$$

Since A and B commute with $[A, B]$ by assumption, we can integrate the last differential equation as if these were numbers. This yields

$$F(t) = F(0)e^{(A+B)t + \frac{1}{2}[A,B]t^2}. \tag{2.23}$$

Setting $t = 0$ in Equation (2.17) we get $F(0) = I$, and hence

$$F(t) = e^{(A+B)t + \frac{1}{2}[A,B]t^2}. \tag{2.24}$$

Finally, setting $t = 1$, we get from Equations (2.17) and (2.24) Glauber's identity (2.16).

---

* Problem 2.4: Unitary operators

Prove that if H is Hermitian, $U = e^{iH}$ is unitary, i.e., $U^\dagger U = UU^\dagger = U^{-1}U = UU^{-1} = I$.

Solution: By definition,

$$U = e^{iH} = 1 + \frac{iH}{1!} - \frac{1}{2!}H^2 - \frac{i}{3!}H^3 \cdots \tag{2.25}$$

Hence,

$$U^\dagger = 1 - \frac{iH}{1!} - \frac{1}{2!}H^2 - \frac{i}{3!}H^3 \cdots$$
$$= 1 + \frac{-iH}{1!} + \frac{1}{2!}(-iH)^2 + \frac{1}{3!}(-iH)^3 \cdots = e^{-iH}. \tag{2.26}$$

Therefore,

$$U^\dagger U = e^{-iH} e^{iH}, \tag{2.27}$$

and using Glauber's identity (Problem 2.3), we get

$$U^\dagger U = e^{-iH} e^{iH} = e^{-iH+iH} = I, \tag{2.28}$$

which proves that U is unitary.

---

* Problem 2.5: Useful identity to perform unitary transformations

Prove the following identity [1]:

$$e^{\xi A} B e^{-\xi A} = B + \xi[A, B] + \frac{\xi^2}{2!}[A, [A, B]] + \frac{\xi^3}{3!}[A, [A, [A, B]]] + \cdots \tag{2.29}$$

This identity is very useful when performing unitary transformations on some Hamiltonians to find their energy spectrum (see Problem 2.12).

Solution: By definition,

$$e^{\xi \hat{A}} = \sum_{n}^{\infty} \frac{1}{n!} \xi^n A^n. \tag{2.30}$$

Hence,

$$e^{\xi A} B e^{-\xi A} = \left[\sum_{n=0}^{\infty} \frac{1}{n!} \xi^n A^n\right] B \left[\sum_{k=0}^{\infty} \frac{1}{k!} \xi^k A^k\right]. \tag{2.31}$$

Expanding, we get after regrouping the terms in increasing power of ξ,

$$e^{\xi A} B e^{-\xi A} = A^0 B A^0 + \xi(AB - BA) + \frac{\xi^2}{2!}(A^2 B - 2ABA + BA^2) + \cdots \tag{2.32}$$

i.e.,

$$e^{\xi A} B e^{-\xi A} = B + \xi[A, B] + \frac{\xi^2}{2!}(A[A, B] + [B, A]A) + \cdots$$

$$= B + \xi[A, B] + \frac{\xi^2}{2!}[A[A, B]] + \cdots \tag{2.33}$$

Higher-order terms can be worked out using lengthy but straightforward algebra and shown to be of the generic form given in Equation (2.29).

---

* Problem 2.6: The shift operator

If z, p_z are the position and the corresponding conjugate (momentum) operators for motion of a particle along the z-direction, their commutator satisfies

$$[z, p_z] = i\hbar. \tag{2.34}$$

According to the result of the previous problem, the operator $U = e^{\frac{i}{\hbar}\xi p_z}$ (with ξ real) is unitary because p_z is a Hermitian operator, and performing the operation $UF(z)U^{-1}$ on any operator $F(z)$ is a "unitary transformation." Show that

$$e^{\frac{i}{\hbar}\xi p_z} F(z) e^{-\frac{i}{\hbar}\xi p_z} = F(z+\xi). \qquad (2.35)$$

This is why the unitary operator $e^{\frac{i}{\hbar}\xi p_z}$ is called the shift operator.

Solution: Using the result of the previous problem, we get

$$e^{\frac{i}{\hbar}\xi p_z} z e^{-\frac{i}{\hbar}\xi p_z} = z + \frac{i\xi}{\hbar}[p_z, z] + \cdots , \qquad (2.36)$$

where all higher-order terms are zero. Hence,

$$e^{\frac{i}{\hbar}\xi p_z} z e^{-\frac{i}{\hbar}\xi p_z} = z + \xi. \qquad (2.37)$$

Using the results of Problem 2.1(c) leads to the desired result:

$$e^{\frac{i}{\hbar}\xi p_z} F(z) e^{-\frac{i}{\hbar}\xi p_z} = F\left(e^{\frac{i}{\hbar}\xi p_z} z e^{-\frac{i}{\hbar}\xi p_z}\right) = F(z+\xi). \qquad (2.38)$$

** Problem 2.7: Additional unitary operators

(a) Show that the operator $U = e^{-\lambda(a_z - a_z^\dagger)}$ is unitary, where λ is real and a_z and $a_z^\dagger$ are annihilation and creation operators that are Hermitian conjugates of each other. The dagger (†) stands for Hermitian conjugate.

(b) If A is an $M \times M$ matrix of the form $A = i\alpha[a]$, where α is a real parameter and $[a]$ is an $M \times M$ matrix with all its elements equal to unity, prove that $U = e^A$ is unitary and is given by

$$U = I + \frac{(e^{i\alpha M} - 1)}{M}[a], \qquad (2.39)$$

where I is the $M \times M$ identity matrix. This exercise will be used in Chapter 8 when studying scattering from a two-dimensional delta scatterer in a quantum wave guide formed in a two-dimensional electron gas.

Solution: (a) To prove that $U = e^{-\lambda(a_z - a_z^\dagger)}$ is unitary, we just need to prove that the operator in the exponent is of the form iH, where H is a Hermitian operator (see Problem 2.4). We rewrite $-\lambda(a_z - a_z^\dagger)$ as $i\mu(a_z - a_z^\dagger)$, where $\mu = i\lambda$ is purely imaginary. Next, we show that the operator $= H = \mu(a_z - a_z^\dagger)$ is Hermitian.

The Hermitian adjoint of H is

$$H^\dagger = (\mu a_z)^\dagger - (\mu a_z^\dagger)^\dagger = \mu^* a_z^\dagger - \mu^* a_z = \mu a_z - \mu a_z^\dagger = H, \qquad (2.40)$$

where we used the fact that, μ being purely imaginary, $\mu^* = -\mu$.

This completes the proof that the operator U is unitary.

(b) We start with the following definition of e^A:

$$e^A = I + A + \frac{A^2}{2!} + \frac{A^3}{3!} + \cdots \qquad (2.41)$$

For $A = i\alpha[a]$, we have

$$A^n = (i\alpha)^n [a]^n = (i\alpha)^n M^{n-1}[a], \qquad (2.42)$$

as can easily be shown by induction (remember that all the elements of $[a]$ are 1).

So,

$$e^{i\alpha[a]} = I + i(\alpha M)\frac{[a]}{M} + \frac{(i\alpha M)^2}{2!}\frac{[a]}{M} + \cdots \qquad (2.43)$$

Hence,

$$e^{i\alpha[a]} = I + \left[(i\alpha)M + \frac{1}{2!}(i\alpha M)^2 + \cdots \right]\frac{[a]}{M}. \qquad (2.44)$$

Therefore,

$$U = e^{i\alpha[a]} = I + \frac{(e^{i\alpha M} - 1)}{M}[a]. \qquad (2.45)$$

To prove that U is unitary, we must show that $U^\dagger U = I$.

$$U^\dagger = I + \frac{(e^{-i\alpha M} - 1)}{M}[a], \qquad (2.46)$$

so

$$U^\dagger U = \left(I + \frac{1}{M}(e^{-i\alpha M} - 1)[a] \right)\left(I + \frac{1}{M}(e^{i\alpha M} - 1)[a] \right). \qquad (2.47)$$

Therefore,

$$U^\dagger U = I + \frac{1}{M}\left(e^{i\alpha M} + e^{-i\alpha M} - 2 \right)[a]$$
$$+ \frac{1}{M^2}\left(e^{-i\alpha M} - 1 \right)\left(e^{i\alpha M} - 1 \right)M[a], \qquad (2.48)$$

or

$$U^\dagger U = I + \frac{1}{M}\left(2\cos(\alpha M) - 2 \right)[a] + \frac{2}{M}\left(1 - \cos(\alpha M) \right)[a] = I,$$

which proves the unitarity of U.

*** Problem 2.8: Virial theorem

Consider the Hamiltonian in one dimension

$$H = \frac{p_z{}^2}{2m^*} + V(z), \qquad (2.49)$$

where

$$p_z = \frac{\hbar}{i} \frac{d}{dz} \qquad (2.50)$$

and

$$V(z) = \lambda z^n, \qquad (2.51)$$

where λ is real and n is an integer.

First prove the following result:

$$[H, zp_z] = i\hbar - \left[2 \left(\frac{{p_z}^2}{2m^*} \right) + \lambda n z^n \right], \qquad (2.52)$$

where H is the Hamiltonian above.

Starting with the results of the previous step, prove that if $|\phi\rangle$ is an eigenstate of H, the following is true:

$$2\langle T \rangle = n\langle V \rangle, \qquad (2.53)$$

where $\langle \cdots \rangle$ stands for the expectation value in the state $|\phi\rangle$ and T is the kinetic energy operator expressed as

$$T = \frac{-\hbar^2}{2m^*} \frac{d^2}{dz^2}. \qquad (2.54)$$

Equation (2.53) is known as the virial theorem.

Solution: **Preliminary lemma:** If $|\phi\rangle$ is an eigenstate of H, then $\langle \phi \mid [H, A] \, \phi \rangle = 0$, for any quantum mechanical operator A.

The expectation value of any quantum mechanical operator represents the average value of the corresponding physical variable which we expect to measure in an experiment. Therefore, the expectation value of a quantum mechanical operator must be real (we cannot measure an imaginary quantity). Since the expectation value of any Hermitian operator is always real (see Appendix C), all legitimate quantum mechanical operators are Hermitian.

Since $|\phi\rangle$ is an eigenstate of H, the following relation holds: $H|\phi\rangle = E|\phi\rangle$, where E is the eigenenergy. Recalling that both H and A are Hermitian and that E is real, we obtain

$$\begin{aligned} \langle \phi \mid [H, A]\phi \rangle &= \int d^3\vec{r}\, \phi^*(\vec{r})(HA - AH)\phi(\vec{r}) \\ &= \int d^3\vec{r} \left\{ [AH\phi(\vec{r})]^* \, \phi(\vec{r}) - \phi^*(\vec{r})AH\phi(\vec{r}) \right\} \\ &= E \int d^3\vec{r} \left\{ \phi^*(\vec{r})A\phi(\vec{r}) - \phi^*(\vec{r})A\phi(\vec{r}) \right\} = 0. \end{aligned} \qquad (2.55)$$

This property will be used below.

Starting with the operator identity (see Appendix C)

$$[A, BC] = [A, B]C + B[A, C], \tag{2.56}$$

we get

$$[H, zp_z] = [H, z]p_z + z[H, p_z]. \tag{2.57}$$

Substituting the Hamiltonian given in Equation (2.49) in the preceding equation leads to

$$[H, zp_z] = \left[\left(\frac{p_z{}^2}{2m^*} + \lambda z^n \right), z \right] p_z + z \left[\left(\frac{p_z{}^2}{2m^*} + \lambda z^n \right), p_z \right]$$

$$= \left[\frac{p_z{}^2}{2m^*}, z \right] p_z + \lambda \left[z^n, z \right] p_z + \frac{z}{2m^*} \left[p_z{}^2, p_z \right] + z \left[\lambda z^n, p_z \right]. \tag{2.58}$$

The second and third commutators are equal to zero. Furthermore,

$$[p_z, z^n] = -i\hbar n z^{n-1}, \tag{2.59}$$

and

$$[z, p_z{}^2] = 2i\hbar p_z. \tag{2.60}$$

Using these results, we get

$$[H, zp_z] = \frac{1}{2m^*}(-2i\hbar p_z)p_z + \lambda z(i\hbar n z^{n-1})$$

$$= \frac{1}{2m^*}(-2i\hbar p_z)p_z + i\lambda \hbar n z^n$$

$$= i\hbar \left[-2 \left(\frac{p_z{}^2}{2m^*} \right) + \lambda n z^n \right]. \tag{2.61}$$

Since $|\phi\rangle$ is an eigenstate of H, we can apply the preliminary lemma derived earlier and get

$$\langle [H, zp_z] \rangle = 0 = i\hbar \left(-2\langle T \rangle + n\langle V \rangle \right). \tag{2.62}$$

In other words,

$$2\langle T \rangle = n\langle V \rangle. \tag{2.63}$$

** Problem 2.9: Generalized version of the virial theorem

For an arbitrary $V(z)$, show that

$$\langle T \rangle = \frac{1}{2} \left\langle z \frac{dV}{dz} \right\rangle, \tag{2.64}$$

where the average is taken over an eigenstate $|\phi\rangle$ of the Hamiltonian $H = T + V(z)$.

Solution: From the previous problem, we get

$$[H, zp_z] = \left[\frac{p_z{}^2}{2m^*} + V(z), z\right]p_z + z\left[\frac{p_z{}^2}{2m^*} + V(z), p_z\right]$$

$$= \frac{1}{2m^*}[p_z{}^2, z]p_z + z[V(z), p_z], \tag{2.65}$$

because z and $V(z)$ commute, as do p_z^2 and p_z. Making use of the identities (see Appendix C)

$$[z, p_z{}^2] = 2i\hbar p_z \tag{2.66}$$

and

$$[p_z, V(z)] = -i\hbar \frac{dV}{dz}, \tag{2.67}$$

we get

$$[H, zp_z] = \hbar\left[-2\left(\frac{p_z{}^2}{2m^*}\right) + z\frac{dV}{dz}\right]. \tag{2.68}$$

Hence, since $|\phi\rangle$ is an eigenstate of H,

$$\frac{\langle[H, zp_z]\rangle}{\hbar} = -2\left\langle\frac{p_z{}^2}{2m^*}\right\rangle + \left\langle z\frac{dV}{dz}\right\rangle. \tag{2.69}$$

Since the left-hand side is equal to zero, we finally get the desired result,

$$\langle T\rangle = \frac{1}{2}\left\langle z\frac{dV}{dz}\right\rangle. \tag{2.70}$$

---

* Problem 2.10: Sum rule

Evaluate the commutator

$$[p_z, [p_z, H]], \tag{2.71}$$

where p_z is the z-component of the momentum operator and $H = \frac{p^2}{2m^*} + V(\vec{r})$ is the Hamiltonian of the system.

Use this result to prove the following sum rule:

$$\sum_n (E_n - E_m)\,|\langle n|p_z|m\rangle|^2 = \frac{\hbar^2}{2}\left\langle m\left|\frac{\partial^2 V}{\partial z^2}\right|m\right\rangle, \tag{2.72}$$

where $|n\rangle$ and $|m\rangle$ are eigenvectors of H with eigenvalues E_n and E_m, respectively.

Solution: Starting with the relations

$$H|n\rangle = \left[\frac{p^2}{2m^*} + V(\vec{r})\right]|n\rangle = E_n|n\rangle, \tag{2.73}$$

$$[p_z, H] = -i\hbar\frac{\partial V}{\partial z}, \tag{2.74}$$

and

$$[p_z, [p_z, H]] = -\hbar^2 \frac{\partial^2 V}{\partial^2 z},$$ (2.75)

we get

$$-\hbar^2 \left\langle m \left| \frac{d^2 V}{d^2 z} \right| m \right\rangle = \sum_n \left(\langle m|p_z|n\rangle \langle n|[p_z, H]|m\rangle \right.$$
$$\left. - \langle m|[p_z, H]|n\rangle \langle n|p_z|m\rangle \right).$$ (2.76)

But

$$\langle n|[p_z, H]|m\rangle = \langle n|p_z H - H p_z|m\rangle$$
$$= \langle n|p_z E_m|m\rangle - \langle n|H p_z|m\rangle$$
$$= E_m \langle n|p_z|m\rangle - \langle Hn|p_z|m\rangle$$
$$= E_m \langle n|p_z|m\rangle - E_n \langle n|p_z|m\rangle$$
$$= (E_n - E_m)\langle n|p_z|m\rangle,$$ (2.77)

where we made use of the Hermiticity of H in the third line.

Similarly,

$$\langle m|[p_z, H]|n\rangle = (E_m - E_n)\langle m|p_z|n\rangle.$$ (2.78)

Because p_z is a Hermitian operator,

$$\langle m|p_z|n\rangle = \langle p_z m|n\rangle = \langle n|p_z|m\rangle^*,$$ (2.79)

where the asterisk denotes complex conjugate.

Using the last three relations in Equation (2.76), we get

$$\sum_n (E_n - E_m)|\langle n|p_z|m\rangle|^2 = \frac{\hbar^2}{2} \left\langle m \left| \frac{d^2 V}{d^2 z} \right| m \right\rangle.$$ (2.80)

*** Problem 2.11: Generalized sum rule [3]

Prove the following sum rule for any Hermitian operator F:

$$\sum_m (E_m - E_n)|\langle n|F|m\rangle|^2 = -\frac{1}{2} \langle n|[F, [F, H]]|n\rangle,$$ (2.81)

where $|n\rangle$ and $|m\rangle$ are eigenvectors of the operator H with eigenvalues E_n and E_m, respectively.

Solution: We have

$$\sum_m (E_m - E_n)|\langle n|F|m\rangle|^2 = \sum_m (E_m - E_n)\langle n|F|m\rangle \langle n|F|m\rangle^*.$$ (2.82)

Hence.

$$\sum_m (E_m - E_n)|\langle n|F|m\rangle|^2 = \sum_m (E_m - E_n)\langle n|F|m\rangle\langle m|F^\dagger|n\rangle, \qquad (2.83)$$

where $F^\dagger$ is the Hermitian conjugate of F. This last equation can be rewritten as follows (since H is always Hermitian, $H|m\rangle = E_m|m\rangle$ and $H|n\rangle = E_n|n\rangle$):

$$\sum_m (E_m - E_n)|\langle n|F|m\rangle|^2 = \sum_m \langle n|[FH - HF]|m\rangle\langle m|F^\dagger|n\rangle$$

$$= \sum_m \langle n|[F, H]|m\rangle\langle m|F^\dagger|n\rangle. \qquad (2.84)$$

Using the normalization condition

$$\sum_m |m\rangle\langle m| = 1, \qquad (2.85)$$

we finally get

$$\sum_m (E_m - E_n)|\langle n|F|m\rangle|^2 = \langle n|[F, H]F^\dagger|n\rangle. \qquad (2.86)$$

If F is a Hermitian operator, then, by definition, $\langle n|F|m\rangle^* = \langle m|F|n\rangle$, and hence the summation can be rewritten as

$$\sum_m (E_m - E_n)|\langle n|F|m\rangle|^2 = \sum_m (E_m - E_n)\langle n|F|m\rangle\langle m|F|n\rangle. \qquad (2.87)$$

Therefore,

$$\sum_m (E_m - E_n)|\langle n|F|m\rangle|^2 = -\sum_m \langle n|F|m\rangle\langle m|[FH - HF]|n\rangle$$

$$= -\sum_m \langle n|F|m\rangle\langle m|[F, H]|n\rangle$$

$$= -\langle n|F[F, H]|n\rangle = \langle n|F[H, F]|n\rangle. \qquad (2.88)$$

Combining Equations (2.86) and (2.88), and noting that $F^\dagger = F$ when F is Hermitian, we obtain the following sum rule for any Hermitian operator F:

$$\sum_m (E_m - E_n)|\langle n|F|m\rangle|^2 = -\frac{1}{2}\langle n|[F, [F, H]]|n\rangle. \qquad (2.89)$$

Since the inception of quantum mechanics, sum rules have been used extensively in various branches of physics, including atomic, molecular, solid state, and particle physics. To delve further into the subject, the reader should utilize the suggested reading section at the end of the chapter.

**** Problem 2.12: Polarizability of one-dimensional harmonic oscillator

Using the concept of unitary transformation, show that the polarizability α of a one-dimensional harmonic oscillator $\frac{1}{2}m^*\omega_0^2 z^2$ in a uniform external electric field is given by $\alpha = \frac{e^2}{m^*\omega_0}$.

Some of the basic properties of the one-dimensional harmonic oscillator are listed in Appendix B.

Solution: We first rewrite the Hamiltonian of a one-dimensional harmonic oscillator,

$$H_0 = \frac{p_z{}^2}{2m^*} + \frac{1}{2}m^*\omega_0^2 z^2, \tag{2.90}$$

by introducing the annihilation a_z and creation $a_z^\dagger$ operators:

$$a_z = \frac{1}{\sqrt{2}}\left(\beta_0 z + \frac{i}{\hbar\beta_0}p_z\right) \tag{2.91}$$

and

$$a_z^\dagger = \frac{1}{\sqrt{2}}\left(\beta_0 z - \frac{i}{\hbar\beta_0}p_z\right), \tag{2.92}$$

where

$$\beta_0 = \frac{m^*\omega_0^{\frac{1}{2}}}{\hbar}. \tag{2.93}$$

It is left to the reader to show that operators a_z, $a_z^\dagger$ satisfy the commutation rule

$$[a_z, a_z^\dagger] = 1. \tag{2.94}$$

Using the operators a_z and $a_z^\dagger$, H_0 can be rewritten as

$$H_0 = \hbar\omega_0\left(N_z + \frac{1}{2}\right), \tag{2.95}$$

where we have introduced the occupation number operator

$$N_z = a_z^\dagger a_z. \tag{2.96}$$

In the presence of an external (uniform) electric field, we must add the Stark interaction to the Hamiltonian H_0. The total Hamiltonian can then be written as

$$H = H_0 - qE_z. \tag{2.97}$$

Next, we perform a unitary transform on H_0 using the unitary operator $U = e^S$ with

$$S = -\lambda(a_z - a_z^\dagger), \tag{2.98}$$

λ being some real parameter to be determined later. Using this operator, we perform the unitary transform

$$H = e^S H_0 e^{-S}. \tag{2.99}$$

Making use of the identity (see Problem 2.5)

$$e^{\xi\hat{A}}\hat{B}e^{\xi\hat{A}} = \hat{B} + \xi[\hat{A},\hat{B}] + \frac{\xi^2}{2!}[\hat{A},[\hat{A},\hat{B}]] + \frac{\xi^3}{3!}[\hat{A},[\hat{A},[\hat{A},\hat{B}]]] + \cdots \qquad (2.100)$$

With $\xi = 1$, $\hat{B} = H_0$, and $\hat{A} = S$, the transformed Hamiltonian H becomes

$$e^S H_0 e^{-S} = H_0 + [S,H_0] + \frac{1}{2!}[S,[S,H_0]] + \cdots \qquad (2.101)$$

First, we calculate $[S,H_0]$:

$$[S,H_0] = [-\lambda(a_z - a_z^{\dagger}),H_0]. \qquad (2.102)$$

Making use of Equation (2.94), we obtain

$$[S,H_0] = -\lambda\hbar\omega_0(a_z + a_z^{\dagger}). \qquad (2.103)$$

Therefore,

$$[S,[S,H_0]] = 2\lambda^2\hbar\omega_0\left(a_z^{\dagger}a_z - a_z a_z^{\dagger}\right) = 2\lambda^2\hbar\omega_0\left[a_z^{\dagger},a_z\right] = 2\lambda^2\hbar\omega_0, \qquad (2.104)$$

where we once again used Equation (2.94).

Since the last commutator is a constant, we conclude that all the terms after the third one in the expansion (2.100) are identically zero.

Grouping the previous results, the transformed Hamiltonian is therefore

$$H = e^S H_0 e^{-S} = H_0 - \lambda\hbar\omega_0\left(a_z + a_z^{\dagger}\right) + \lambda^2\hbar\omega_0. \qquad (2.105)$$

Since

$$z = \frac{1}{\sqrt{2}\beta_0}(a_z + a_z^{\dagger}), \qquad (2.106)$$

we deduce that the second term in Equation (2.105) is identical to the Stark shift $-qEz$ if we choose λ such that

$$\lambda = \frac{1}{\sqrt{2m^*\hbar\omega_0}}\frac{qE}{\omega_0}. \qquad (2.107)$$

Equation (2.105) can then be rewritten as

$$H - \frac{q^2}{2m^*\omega_0^2}E^2 = H_0 - qEz. \qquad (2.108)$$

In other words, the eigenvalues of $H_0 - qEz$ are the same as the eigenvalues of $H - \frac{q^2}{2m^*\omega_0}E^2$.

Calling $|n\rangle$ the eigenstates of H_0, the eigenstates of $H_0 - qEz$ are given by $e^S|n\rangle$, with the corresponding eigenvalues

$$E_n = E_n^0 - \frac{q^2}{2m^*\omega_0^2}E^2, \qquad (2.109)$$

E_n^0 being the eigenvalues of H_0, given by

$$E_n^0 = \hbar\omega_0 \left(n + \frac{1}{2}\right). \tag{2.110}$$

By definition, the polarizability α of the harmonic oscillator in its ground state is such that

$$E_0 = E_0^0 - \frac{1}{2}\alpha E^2. \tag{2.111}$$

By comparing Equations (2.109) and (2.111), the polarizability α is therefore given by

$$\alpha = \frac{q^2}{m\omega_0^2}. \tag{2.112}$$

---

* Problem 2.13: Decomposition of general 2×2 matrix in terms of Pauli matrices

A trivial decomposition of any 2×2 matrix

$$M = \begin{pmatrix} m_{11} & m_{12} \\ m_{21} & m_{22} \end{pmatrix} \tag{2.113}$$

is obviously

$$M = m_{11} \begin{pmatrix} 1 & 0 \\ 0 & 0 \end{pmatrix} + m_{12} \begin{pmatrix} 0 & 1 \\ 0 & 0 \end{pmatrix} + m_{21} \begin{pmatrix} 0 & 0 \\ 1 & 0 \end{pmatrix} + m_{22} \begin{pmatrix} 0 & 0 \\ 0 & 1 \end{pmatrix}. \tag{2.114}$$

The four matrices on the right-hand side form a complete basis for all 2×2 matrices.

A not so obvious decomposition of any 2×2 complex matrix M involves the 2×2 Pauli matrices defined in Appendix D. For a more detailed description of the Pauli matrices, see Chapter 2 in Ref. [4].

Show that any 2×2 matrix M can be expressed as follows:

$$M = a_0 I + \vec{a} \cdot \vec{\sigma}, \tag{2.115}$$

where

$$a_0 = \frac{1}{2}\text{Tr}(M), \tag{2.116}$$

$$\vec{\sigma} = \sigma_x \hat{x} + \sigma_y \hat{y} + \sigma_z \hat{z}, \tag{2.117}$$

and

$$\vec{a} = \frac{1}{2}\text{Tr}(M\vec{\sigma}), \tag{2.118}$$

where $\vec{\sigma} = (\sigma_x, \sigma_y, \sigma_z)$, and Tr stands for the trace of the matrix.

Solution: Starting with the definitions of the Pauli matrices in Appendix D, we can easily show that any 2×2 matrix can be written as

$$M = \frac{m_{11} + m_{22}}{2}I + \frac{m_{11} - m_{22}}{2}\sigma_z + \frac{m_{12} + m_{21}}{2}\sigma_x + i\frac{m_{12} - m_{21}}{2}\sigma_y, \quad (2.119)$$

which is equivalent to

$$M = a_0 I + \vec{a} \cdot \vec{\sigma}, \quad (2.120)$$

where

$$a_0 = \frac{1}{2}\text{Tr}(M) \quad (2.121)$$

and

$$\vec{a} = \frac{1}{2}\text{Tr}(M\vec{\sigma}). \quad (2.122)$$

In other words, the four matrices $(I, \sigma_x, \sigma_y, \sigma_z)$ form a complete set of bases in the space of 2×2 complex matrices.

Clearly, M is Hermitian if a_0 and the three components of the vector $\vec{a}$ are real.

The matrix decomposition (2.120) is very useful in studying the properties of the operators associated with quantum gates operating on qubits [2, 4].

---

* Problem 2.14: Operator identity

Prove that if θ is real and if the matrix A is such that $A^2 = I$, the following identity holds:

$$e^{i\theta A} = \cos\theta I + i\sin\theta A. \quad (2.123)$$

This is the generalization to operators of the well-known Euler relation for complex numbers: $e^{i\phi} = \cos\phi + i\sin\phi$.

Solution: From the Taylor series expansion

$$e^x = \sum_{k=0}^{\infty} \frac{x^k}{k!} \quad (2.124)$$

and the definition of the function of an operator, we get

$$e^{i\theta A} = I + (i\theta)A + \frac{(i\theta)^2 A^2}{2!} + \frac{(i\theta)^3 A^3}{3!} + \frac{(i\theta)^4 A^4}{4!} + \cdots, \quad (2.125)$$

or

$$e^{i\theta A} = \left(1 - \frac{\theta^2}{2!} + \frac{\theta^4}{4!} - \cdots + (-1)^k\frac{\theta^{2k}}{(2k)!}\right)I$$
$$+ i\left(\theta - \frac{\theta^3}{3!} + \frac{\theta^5}{5!} - \cdots + (-1)^k\frac{i\theta^{2k+1}}{(2k+1)!}\right)A, \quad (2.126)$$

which is indeed Equation (2.123) if we use the the Taylor expansions

$$\sin x = \sum_{k=0}^{\infty} (-1)^k x^{2k+1}/(2k+1)! \tag{2.127}$$

and

$$\cos x = \sum_{k=0}^{\infty} (-1)^k x^{2k}/(2k)! \tag{2.128}$$

This identity is very useful when studying the action of quantum gates on quantum bits in the study of quantum computing [2, 4].

---

* Problem 2.15: Equality for Pauli spin matrices

Prove the following equality for the Pauli spin matrices:

$$(\vec{\sigma} \cdot \vec{a})(\vec{\sigma} \cdot \vec{b}) = i\vec{\sigma} \cdot (\vec{a} \times \vec{b}) + \vec{a} \cdot \vec{b} I, \tag{2.129}$$

where I is the 2×2 identity matrix and $\vec{a}$ and $\vec{b}$ are any arbitrary three-dimensional vectors in real space $\mathbb{R}^3$.

Solution:

$$(\vec{\sigma} \cdot \vec{a})(\vec{\sigma} \cdot \vec{b}) = \sum_{j,k} \sigma_j A_j \sigma_k B_k = \sum_{j,k} A_j B_k \left[i \sum_l \epsilon_{jkl} \sigma_l + \delta_{ij} I \right], \tag{2.130}$$

where δ_{ij} is the Kronecker delta and ϵ_{jkl} is equal to zero if any two indices are equal, and equal to $+1$ (-1) for a cyclic (non-cyclic) permutation of the indices (1,2,3).

Hence,

$$(\vec{\sigma} \cdot \vec{a})(\vec{\sigma} \cdot \vec{b}) = i \sum_l \sigma_l \left[\sum_{j,k} \epsilon_{jkl} A_j B_k \right] + \sum_j A_j B_j I$$

$$= i\vec{\sigma} \cdot (\vec{a} \times \vec{b}) + \vec{a} \cdot \vec{b} I. \tag{2.131}$$

The next problem makes use of the identity (2.129) and is very useful for the Bloch sphere concept, which is a fundamental tenet of the field of spin-based quantum computing [2, 4].

Preliminary:

The well-known operator for an electron's spin is

$$\vec{S}_{\text{op}} = \frac{\hbar}{2} \vec{\sigma}, \tag{2.132}$$

whose three components $(\sigma_x, \sigma_y, \sigma_z)$ are the Pauli spin matrices whose basic properties are described in Appendix D. A measurement of the spin component along an arbitrary direction characterized by a unit vector $\hat{n}$ will yield results given by the eigenvalues of the operator

$$\vec{S} \cdot \hat{n}, \tag{2.133}$$

and these eigenvalues are $\pm \frac{\hbar}{2}$, irrespective of the direction of the unit vector $\hat{n}$. This last statement is easily proved by starting with the identity derived in the previous problem.

If the vectors $\vec{a}$ and $\vec{b}$ are equal to a unit vector $\hat{n}$, then the identity (2.129) reduces to

$$(\vec{\sigma} \cdot \hat{n})^2 = I, \tag{2.134}$$

i.e., the square of any component of $\vec{\sigma}$ is equal to the 2×2 identity matrix. Hence, the eigenvalues of $\vec{\sigma} \cdot \hat{n}$ are ± 1, and therefore the eigenvalues of the operator $\vec{S} \cdot \hat{n}$ must be $\pm \hbar/2$, which proves the result we were after. In other words, the measurement of the spin angular momentum along any arbitrary axis always yields the values $\pm \hbar/2$.

** Problem 2.16: Eigenvectors of the $\vec{\sigma} \cdot \vec{n}$ operator

Derive the explicit analytical expressions for the eigenvectors of $\vec{\sigma} \cdot \hat{n}$ corresponding to the eigenvalues $+1$ and -1 [2, 4].

Solution: Consider the operators

$$\frac{1}{2}(I \pm \vec{\sigma} \cdot \hat{n}), \tag{2.135}$$

which are 2×2 matrices (I is the 2×2 identity matrix) acting on an arbitrary spinor or qubit $|\chi\rangle$. A "spinor" is a 2×1 column vector describing the spin orientation of a particle with spin. If we operate on that with the operator $(\vec{\sigma} \cdot \hat{n})$, we get

$$(\vec{\sigma} \cdot \hat{n}) \left[\frac{1}{2}(I \pm \vec{\sigma} \cdot \hat{n})|\chi\rangle \right] = \frac{1}{2} \vec{\sigma} \cdot \hat{n}|\chi\rangle \pm \frac{1}{2}(\vec{\sigma} \cdot \hat{n})^2 |\chi\rangle$$

$$= \pm \left[\frac{1}{2}(I \pm \vec{\sigma} \cdot \hat{n})|\chi\rangle \right]. \tag{2.136}$$

This means that, for any $|\chi\rangle$, $\frac{1}{2}(1 \pm \vec{\sigma} \cdot \hat{n})|\chi\rangle$ are eigenvectors of $\vec{\sigma} \cdot \hat{n}$ with eigenvalues ± 1. Making use of the identity

$$\frac{1}{2}(I \pm \vec{\sigma} \cdot \hat{n}) = \frac{1}{2} \left[I \pm \sigma_z n_z \pm \frac{1}{2}(\sigma_x + i\sigma_y)(n_x - in_y) \pm \frac{1}{2}(\sigma_x - i\sigma_y)(n_x + in_y) \right],$$

$$\tag{2.137}$$

where n_x, n_y, and n_z are the x-, y- and z-components of the vector $\hat{n}$, and using spherical coordinates with polar angle θ and azimuthal angle ϕ, so that

$$(n_x, n_y, n_z) = (\sin\theta \cos\phi, \sin\theta \sin\phi, \cos\theta), \tag{2.138}$$

we get

$$n_x \pm i n_y = \sin\theta e^{\pm i\phi}, \tag{2.139}$$

which leads to

$$\frac{1}{2}(I \pm \vec{\sigma}\cdot\hat{n}) = \frac{1}{2}\left[I \pm \cos\theta\sigma_z \pm \frac{1}{2}(\sin\theta e^{-i\phi}\sigma_+ \pm \sin\theta e^{i\phi}\sigma_-)\right], \tag{2.140}$$

where the operators σ_+ and σ_- are given by $\sigma_+ = \sigma_x + i\sigma_y$ and $\sigma_- = \sigma_x - i\sigma_y$, respectively.

Let us define two special spinors as follows:

$$|0\rangle = \begin{bmatrix} 1 \\ 0 \end{bmatrix}$$

$$|1\rangle = \begin{bmatrix} 0 \\ 1 \end{bmatrix}. \tag{2.141}$$

Acting with the operators in Equation (2.140) on the spinor $|0\rangle$, we get

$$\frac{1}{2}(I + \vec{\sigma}\cdot\hat{n})|0\rangle = \cos\frac{\theta}{2}\left[\cos\frac{\theta}{2}|0\rangle + \sin\frac{\theta}{2}e^{i\phi}|1\rangle\right] \tag{2.142}$$

and

$$\frac{1}{2}(I - \vec{\sigma}\cdot\hat{n})|0\rangle = \sin\frac{\theta}{2}\left[\sin\frac{\theta}{2}|0\rangle - \cos\frac{\theta}{2}e^{i\phi}|1\rangle\right]. \tag{2.143}$$

The last two spinors can be easily normalized by dividing the first by $\cos\frac{\theta}{2}$ and the second by $\sin\frac{\theta}{2}$. This leads to the spinors

$$|\xi_n^+\rangle = \cos\frac{\theta}{2}|0\rangle + \sin\frac{\theta}{2}e^{i\phi}|1\rangle \tag{2.144}$$

and

$$|\xi_n^-\rangle = \sin\frac{\theta}{2}|0\rangle - \cos\frac{\theta}{2}e^{i\phi}|1\rangle. \tag{2.145}$$

Since we had proved that any spinor $(1/2)(1 \pm \vec{\sigma}\cdot\hat{n})|\chi\rangle$ is an eigenvector of $\vec{\sigma}\cdot\hat{n}$ with eigenvalues ± 1, it is obvious that the spinors $|\xi_n^+\rangle$ and $|\xi_n^-\rangle$ are eigenspinors of the operator $(\vec{\sigma}\cdot\hat{n})$ with eigenvalues $+1$ and -1, respectively.

---

* Problem 2.17: Quantum mechanical operators for charge density and velocity

The quantum mechanical operators for charge density and velocity are $q\delta(\vec{r})$ and $-\frac{i\hbar\vec{\nabla}_r}{m^*}$, respectively, where q is the charge of the electron and m^* is its effective mass. Show that these operators do not commute.

Solution: Let the commutator $\left[-\frac{i\hbar\vec{\nabla}_r}{m^*}, q\delta(\vec{r})\right]$ operate on a wave function $\psi(\vec{r}, t)$. This will result in:

$$\left[-\frac{i\hbar\vec{\nabla}_r}{m^*}q\delta(\vec{r}) + q\delta(\vec{r})\frac{i\hbar\vec{\nabla}_r}{m^*}\right]\psi(\vec{r}, t) = -\frac{iq\hbar\vec{\nabla}_r\delta(\vec{r})}{m^*}\psi(\vec{r}, t) \neq 0. \qquad (2.146)$$

Hence, the two operators do not commute.

Physical significance of this result: Classically, the current density is the product of the velocity and the charge density. Therefore, one might assume that the current density operator should be the product of the charge density operator and the velocity operator, and that the order in which the product is taken is immaterial. That is clearly not true. The order matters since the two operators do not commute. Chapter 5 contains a set of problems on the properties of the current density operator.

** **Problem 2.18: Hermiticity of operators**

Show that neither $-\frac{i\hbar}{m^*}\frac{\mathrm{d}}{\mathrm{d}z}q\delta(z)$ nor the operator $q\delta(z)\frac{i\hbar}{m^*}\frac{\mathrm{d}}{\mathrm{d}z}$ is Hermitian, but their symmetric combination is.

That is why the current density operator (see Chapter 5) is written as the symmetric combination of these two operators.

Solution: In the following, the prime denotes the first derivative. Without loss of generality, we will consider a time-independent wave function.

First, consider the operator $\frac{i\hbar}{m^*}\frac{\mathrm{d}}{\mathrm{d}z}q\delta(z - z_0)$ and check if it is Hermitian. Therefore, consider the integral

$$\int_{-\infty}^{+\infty} \mathrm{d}z\Phi^*(z)\left[\frac{i\hbar}{m^*}\frac{\mathrm{d}}{\mathrm{d}z}q\delta(z - z_0)\right]\Psi(z)$$

$$= \frac{iq\hbar}{m^*}\int_{-\infty}^{+\infty}\mathrm{d}z\Phi^*(z)\left[\delta'(z - z_0)\Psi(z) + \delta(z - z_0)\Psi'(z)\right]. \qquad (2.147)$$

Integrating the first term on the right-hand side (R.H.S.) by parts, we get

$$\text{R.H.S.} = \frac{iq\hbar}{m^*}\left[\Phi^*(z)\Psi(z)\delta(z - z_0)\Big|_{-\infty}^{+\infty} - \int_{-\infty}^{+\infty}\mathrm{d}z\left[\Phi^*(z)\Psi(z)\right]'\delta(z - z_0)\right.$$

$$\left. + \int_{-\infty}^{+\infty}\mathrm{d}z\Phi^*(z)\Psi'(z)\delta(z - z_0)\right]. \qquad (2.148)$$

The first term on the right-hand side of the preceding equation must vanish since the wave function is zero at $\pm\infty$. Hence,

$$\text{R.H.S.} = -\frac{iq\hbar}{m^*} \left[\int_{-\infty}^{+\infty} dz \Phi^*(z) \Psi'(z) \delta(z - z_0) \right.$$
$$\left. + \int_{-\infty}^{+\infty} dz \Phi'^*(z) \Psi(z) \delta(z - z_0) - \int_{-\infty}^{+\infty} dz \Phi^*(z) \Psi'(z) \delta(z - z_0) \right].$$
$$(2.149)$$

Simplifying, we get

$$\text{R.H.S.} = -\frac{iq\hbar}{m^*} \Phi'^*(z_0) \Phi(z_0) = -\frac{iq\hbar}{m^*} \left[\frac{d}{dz} \Phi^* \right](z_0) \Psi(z_0). \qquad (2.150)$$

Next, consider the integral

$$\int_{-\infty}^{+\infty} dz \left[\frac{i\hbar}{m^*} \frac{d}{dz} q \delta(z - z_0) \Phi(z) \right]^* \Psi(z)$$
$$= -\frac{iq\hbar}{m^*} \left[\int_{-\infty}^{+\infty} dz \delta'(z - z_0) \Phi^*(z) \Psi(z) + \delta(z - z_0) \Psi'^*(z) \Phi(z) \right]. \qquad (2.151)$$

Once again, integrate the first term on the right-hand side by parts to yield

$$\text{R.H.S.} = -\frac{iq\hbar}{m^*} \left[\Phi^*(z) \Psi(z) \delta(z - z_0) \Big|_{-\infty}^{+\infty} \right.$$
$$\left. + \int_{-\infty}^{+\infty} dz \delta(z - z_0) \left[\Phi^*(z) \Psi(z) \right]' - \int_{-\infty}^{+\infty} dz \delta(z - z_0) \Phi'^*(z) \Psi(z) \right].$$
$$(2.152)$$

Hence, proceeding as above,

$$\text{R.H.S.} = \frac{iq\hbar}{m^*} \Phi^*(z_0) \left[\frac{d}{dz} \Psi \right](z_0). \qquad (2.153)$$

Comparing Equations (2.150) and (2.153), we see that

$$\int_{-\infty}^{+\infty} dz \Phi^*(z) \left[\frac{i\hbar}{m^*} \frac{d}{dz} q \delta(z - z_0) \right] \Psi(z)$$
$$\neq \int_{-\infty}^{+\infty} dz \left[\frac{i\hbar}{m^*} \frac{d}{dz} q \delta(z - z_0) \Phi(z) \right]^* \Psi(z). \qquad (2.154)$$

Therefore, the operator $-\frac{i\hbar}{m^*} \frac{d}{dz} q \delta(z)$ is not Hermitian.

Next, consider the operator $q \delta(z - z_0) \frac{i\hbar}{m^*} \frac{d}{dz}$ and check if it is Hermitian. As before, consider the integral

$$\int_{-\infty}^{+\infty} dz \Phi^*(z) \left[q \delta(z - z_0) \frac{i\hbar}{m^*} \frac{d}{dz} \right] \Psi(z)$$
$$= \frac{iq\hbar}{m^*} \int_{-\infty}^{+\infty} dz \Phi^*(z) \delta(z - z_0) \Psi'(z) = \frac{iq\hbar}{m^*} \Phi^*(z_0) \left[\frac{d}{dz} \Psi \right](z_0), \qquad (2.155)$$

and then consider the other integral

$$\int_{-\infty}^{+\infty} dz \left[q\delta(z - z_0) \frac{i\hbar}{m^*} \frac{d}{dz} \Phi(z) \right]^* \Psi(z)$$

$$= -\frac{iq\hbar}{m^*} \int_{-\infty}^{+\infty} dz \Phi^{*\prime}(z)\delta(z - z_0))\Psi(z) = -\frac{iq\hbar}{m^*} \Psi(z_0) \left[\frac{d}{dz} \Phi^* \right](z_0), \quad (2.156)$$

where we used the fact that $\delta^*(z - z_0) = \delta(z - z_0)$. Clearly,

$$\int_{-\infty}^{+\infty} dz \Psi^*(z) \left[\frac{i\hbar}{m^*} \frac{d}{dz} q\delta(z - z_0) \right] \Psi(z)$$

$$\neq \int_{-\infty}^{+\infty} dz \left[\frac{i\hbar}{m^*} \frac{d}{dz} q\delta(z - z_0) \right] \Psi(z)^* \Psi(z), \quad (2.157)$$

so that the operator $q\delta(z - z_0)\frac{i\hbar}{m^*}\frac{d}{dz}$ is not Hermitian either.

Finally, focus on the symmetric combination and consider the integral

$$\int_{-\infty}^{+\infty} dz \Phi^*(z) \left[\frac{i\hbar}{m^*} \frac{d}{dz} q\delta(z - z_0) + q\delta(z - z_0) \frac{i\hbar}{m^*} \frac{d}{dz} \right] \Psi(z)$$

$$= \frac{iq\hbar}{m^*} \int_{-\infty}^{+\infty} dz \Phi^*(z)[\delta'(z - z_0)\Psi(z) + 2\delta(z - z_0)\Psi'(z)]. \quad (2.158)$$

Integrate the first term on the right-hand side by parts to yield

$$\text{R.H.S.} = \frac{iq\hbar}{m^*} \left[\Phi^*(z)\Psi(z)\delta(z - z_0) \Big|_{-\infty}^{+\infty} - \int_{-\infty}^{+\infty} dz[\Phi^*(z)\Psi(z)]'\delta(z - z_0) \right.$$

$$\left. + \int_{-\infty}^{+\infty} dz 2\Phi^*(z)\Psi'(z)\delta(z - z_0) \right]. \quad (2.159)$$

Hence,

$$\text{R.H.S.} = \frac{iq\hbar}{m^*} \left[\Phi^*(z_0)\Psi'(z_0) - \Phi'^*(z_0)\Psi(z_0) \right]. \quad (2.160)$$

Next, consider the other integral

$$\int_{-\infty}^{+\infty} dz \left\{ \left[\frac{i\hbar}{m^*} \frac{d}{dz} q\delta(z - z_0) + q\delta(z - z_0) \frac{i\hbar}{m^*} \frac{d}{dz} \right] \Phi(z) \right\}^* \Psi(z)$$

$$= -\frac{iq\hbar}{m^*} \int_{-\infty}^{+\infty} dz[\delta'(z - z_0)\Phi^*(z) + 2\delta(z - z_0)\Phi'^*(z)]\Psi(z). \quad (2.161)$$

Integrate the first term on the right-hand side by parts to yield

$$\text{R.H.S.} = -\frac{iq\hbar}{m^*} \Phi^*(z)\Psi(z)\delta(z - z_0) \Big|_{-\infty}^{+\infty}$$

$$+ \frac{iq\hbar}{m^*} \int_{-\infty}^{+\infty} dz \left[\Phi^*(z)\Psi(z) \right]'\delta(z - z_0) - \frac{iq\hbar}{m^*} \int_{-\infty}^{+\infty} dz 2\Phi^*(z)\Psi(z)\delta(z - z_0),$$

$$(2.162)$$

which reduces to

$$\text{R.H.S.} = \frac{iq\hbar}{m^*}\left[\Phi^*(z_0)\Psi'(z_0) - \Phi'^*(z_0)\Psi(z_0)\right]. \tag{2.163}$$

Therefore, from Equations (2.160) and (2.163), we infer that

$$\int_{-\infty}^{+\infty} dz \Phi^*(z) \left[\frac{i\hbar}{m^*}\frac{\mathrm{d}}{\mathrm{d}z}q\delta(z-z_0) + q\delta(z-z_0)\frac{i\hbar}{m^*}\frac{\mathrm{d}}{\mathrm{d}z}\right]\Psi(z)$$

$$= \int_{-\infty}^{+\infty} dz \left\{\left[\frac{i\hbar}{m^*}\frac{\mathrm{d}}{\mathrm{d}z}q\delta(z-z_0) + q\delta(z-z_0)\frac{i\hbar}{m^*}\frac{\mathrm{d}}{\mathrm{d}z}\right]\Phi(z)\right\}^* \Psi(z). \tag{2.164}$$

This equality, by definition, implies Hermiticity. Thus, the symmetric combination is Hermitian.

---

* Problem 2.19: Sturm–Liouville equation

Any time-dependent wave function can be expanded in a complete orthonormal set and written as a weighted sum of orthonormal functions in the following way:

$$\Psi(\vec{r},t) = \sum_n R_n(t)\phi_n(\vec{r}), \tag{2.165}$$

where the $\phi_n(\vec{r})$ are orthonormal functions of space but not time. The weights or coefficients of expansion $R_n(t)$ depend on time, but not space.

The so-called density matrix is defined as $\rho_{nm}(t) = R_n^*(t)R_m(t) = \langle\phi_n|\rho|\phi_m\rangle$.

Show that the density matrix obeys the equation

$$i\hbar\frac{\partial\rho(t)}{\partial t} = [H(\vec{r},t),\rho(t)], \tag{2.166}$$

where $H(\vec{r},t)$ is the time-dependent Hamiltonian describing an electron whose wave function is $\Psi(\vec{r},t)$, and the square bracket denotes the commutator.

For a more thorough introduction to the density matrix, see Chapter 5 of Ref. [4].

Solution: Substituting the expansion for the wave function in the time-dependent Schrödinger equation

$$i\hbar\frac{\partial\Psi(\vec{r},t)}{\partial t} = H(\vec{r},t)\Psi(\vec{r},t), \tag{2.167}$$

we get

$$i\hbar\frac{\partial\sum_j R_j(t)\phi_j(\vec{r})}{\partial t} = H(\vec{r},t)\sum_j R_j(t)\phi_j(\vec{r}), \tag{2.168}$$

or

$$i\hbar\sum_j \frac{\partial R_j(t)}{\partial t}\phi_j(\vec{r}) = \sum_j R_j(t)H(\vec{r},t)\phi_j(\vec{r}), \tag{2.169}$$

where in deriving the last equality we used the fact that the Hamiltonian does not operate on the coefficients $R_j(t)$.

Multiplying Equation (2.169) by the "bra" $\langle \phi_m |$ and integrating over all space, we obtain

$$i\hbar \sum_j \frac{\partial R_j(t)}{\partial t} \langle \phi_m | \phi_j \rangle = \sum_j R_j(t) \langle \phi_m | H(\vec{r},t) | \phi_j \rangle = \sum_j R_j(t) H_{m,j}(t), \quad (2.170)$$

where $H_{m,j} = \langle \phi_m | H(\vec{r},t) | \phi_j \rangle$.

Using the fact that $\langle \phi_m | \phi_j \rangle = \delta_{m,j}$, we get

$$i\hbar \sum_j \frac{\partial R_j(t)}{\partial t} \delta_{m,j} = i\hbar \frac{\partial R_m(t)}{\partial t} = \sum_j R_j(t) H_{m,j}(t). \quad (2.171)$$

Taking the complex conjugate of both sides, we get

$$-i\hbar \sum_j \frac{\partial R_n^*(t)}{\partial t} = \sum_j R_j^*(t) H_{n,j}^*(t) = \sum_j R_j^*(t) H_{j,n}(t), \quad (2.172)$$

where we have used the Hermiticity of the Hamiltonian to derive the last equality, i.e., $H_{n,j}^* = H_{j,n}$. Next,

$$i\hbar \frac{[\partial R_n^*(t) R_m(t)]}{\partial t} = i\hbar R_n^*(t) \frac{\partial R_m(t)}{\partial t} + i\hbar R_m(t) \frac{\partial R_n^*(t)}{\partial t}$$

$$= \sum_j [R_n^*(t) R_j(t) H_{m,j}(t) - R_j^*(t) R_m(t) H_{j,n}(t)], \quad (2.173)$$

where we have used Equations (2.171) and (2.172). Then, using the definition of the density matrix, we get

$$i\hbar \frac{\partial \rho_{m,n}(t)}{\partial t} = \sum_j [\rho_{j,n} H_{m,j}(t) - \rho_{m,j}(t) H_{j,n}(t)], \quad (2.174)$$

or

$$i\hbar \frac{\partial \rho(t)}{\partial t} = [H(t)\rho(t) - \rho(t)H(t)] = [H(t), \rho(t)], \quad (2.175)$$

which is the Sturm–Liouville equation describing the time evolution of the density matrix.

---

* Problem 2.20: Ehrenfest's theorem [5, 6]

Show that the expectation value of any time-dependent operator $A(t)$ obeys the equation

$$i\hbar \frac{\partial \langle A(t) \rangle}{\partial t} = \langle [A(t), H(t)] \rangle + i\hbar \left\langle \frac{\partial A(t)}{\partial t} \right\rangle. \quad (2.176)$$

Solution: The expectation value of an operator is, by definition,

$$\langle A \rangle = \langle \phi | A | \phi \rangle / \langle \phi | \phi \rangle. \tag{2.177}$$

Assuming that the wave function is normalized,

$$\langle A \rangle = \int \mathrm{d}^3 \vec{r} \, \phi^*(\vec{r}, t) A \phi(\vec{r}, t). \tag{2.178}$$

Differentiating this last expression with respect to time and remembering that A is time dependent, we get

$$i\hbar \frac{\partial \langle A \rangle}{\partial t} = \int \mathrm{d}^3 \vec{r} \left[i\hbar \frac{\partial \phi^*}{\partial t} A \phi + \phi^* A i\hbar \frac{\partial \phi}{\partial t} + i\hbar \phi^* \frac{\partial A}{\partial t} \phi \right]. \tag{2.179}$$

Using the Schrödinger equation

$$i\hbar \frac{\partial \phi}{\partial t} = H\phi \tag{2.180}$$

and its complex conjugate

$$-i\hbar \frac{\partial \phi^*}{\partial t} = H\phi^* \tag{2.181}$$

in Equation (2.179), we get

$$i\hbar \frac{\partial \langle A \rangle}{\partial t} = \int \mathrm{d}^3 \vec{r} \left[-(H\phi)^* A\phi + \phi^* AH\phi + i\hbar \phi^* \frac{\partial A}{\partial t} \phi \right]. \tag{2.182}$$

Since H is Hermitian, the following relation holds:

$$\int \mathrm{d}^3 \vec{r} (H\Phi)^* A\phi = \int \mathrm{d}^3 \vec{r} \phi^* H A\phi. \tag{2.183}$$

Using this result in Equation (2.182), we obtain

$$i\hbar \frac{\partial \langle A \rangle}{\partial t} = \int \mathrm{d}^3 \vec{r} \left[AH - HA + i\hbar \frac{\partial A}{\partial t} \right] \phi = \langle [A, H] \rangle + i\hbar \left\langle \frac{\partial A}{\partial t} \right\rangle. \tag{2.184}$$

This last equation is referred to as *Ehrenfest's theorem*.

---

* Problem 2.21: Application of Ehrenfest's theorem: The one-dimensional harmonic oscillator

The Hamiltonian of a particle of mass m^* in a one-dimensional parabolic potential (the so-called one-dimensional simple harmonic oscillator potential) is

$$H = -\frac{\hbar^2}{2m^*} \frac{\mathrm{d}^2}{\mathrm{d}z^2} + \frac{1}{2} m^* \omega^2 z^2, \tag{2.185}$$

where ω is the curvature of the parabolic potential (see Appendix C).

Use Ehrenfest's theorem to show explicitly that the velocity of the particle is given by

$$v = \frac{\mathrm{d}\langle z \rangle}{\mathrm{d}t} = \frac{\langle p \rangle}{m^*}. \tag{2.186}$$

You do not need to know the electron's wave function to solve this problem.

Solution: Ehrenfest's theorem states that the time evolution of the expectation value of any operator A obeys the relation

$$i\hbar \frac{\partial \langle A \rangle}{\partial t} = \langle [A, H] \rangle + i\hbar \left\langle \frac{\partial A}{\partial t} \right\rangle, \tag{2.187}$$

where the angular brackets denote expectation value and the square bracket denotes the commutator, i.e., $[A, H] = AH - HA$. Note that the second term on the right-hand side vanishes if A is a time-independent operator.

We first have to find the commutator $[z, H] = zH - Hz$. Let this commutator be Λ. Then, the following relation must be satisfied:

$$(zH - Hz)\phi(z) = \Lambda\phi(z). \tag{2.188}$$

Using the Hamiltonian in Equation (2.185), the left-hand side (L.H.S.) of Equation (2.188) becomes

$$\begin{aligned}
\text{L.H.S.} &= \left[-z\frac{\hbar^2}{2m^*}\frac{\mathrm{d}^2}{\mathrm{d}z^2} + \frac{1}{2}m^*\omega^2 z^3 + \frac{\hbar^2}{2m^*}\frac{\mathrm{d}^2}{\mathrm{d}z^2}z - \frac{1}{2}m^*\omega^2 z^3 \right] \phi(z) \\
&= -z\frac{\hbar^2}{2m^*}\frac{\mathrm{d}^2\phi(z)}{\mathrm{d}z^2} + \frac{\hbar^2}{2m^*}\frac{\mathrm{d}}{\mathrm{d}z}\left[\phi(z) + z\frac{\mathrm{d}\phi(z)}{\mathrm{d}z} \right] \\
&= -z\frac{\hbar^2}{2m^*}\frac{\mathrm{d}^2\phi(z)}{\mathrm{d}z^2} + \frac{\hbar^2}{2m^*}\frac{\mathrm{d}\phi(z)}{\mathrm{d}z} + \frac{\hbar^2}{2m^*}\frac{\mathrm{d}\phi(z)}{\mathrm{d}z} + z\frac{\hbar^2}{2m^*}\frac{\mathrm{d}^2\phi(z)}{\mathrm{d}z^2} \\
&= \frac{\hbar^2}{m^*}\frac{\mathrm{d}\phi(z)}{\mathrm{d}z} \\
&= \Lambda\phi(z). \tag{2.189}
\end{aligned}$$

Therefore, the commutator $[z, H] = \frac{\hbar^2}{m^*}\frac{\mathrm{d}}{\mathrm{d}z}$. Also note that since z and t are independent variables, i.e., the operator z is not time dependent, we have $\left\langle \frac{\partial z}{\partial t} \right\rangle = 0$. Therefore, using the last result in the Ehrenfest theorem, we get

$$i\hbar \frac{\partial \langle z \rangle}{\partial t} = \langle [z, H] \rangle + 0 = \left\langle \frac{\hbar^2}{m^*}\frac{\partial}{\partial z} \right\rangle. \tag{2.190}$$

Hence,

$$\frac{\partial \langle z \rangle}{\partial t} = \frac{\mathrm{d}\langle z \rangle}{\mathrm{d}t} = v = \frac{1}{i\hbar}\left\langle \frac{\hbar^2}{m^*}\frac{\partial}{\partial z} \right\rangle = \frac{\langle -i\hbar\frac{\partial}{\partial z} \rangle}{m^*} = \frac{\langle p_z \rangle}{m^*}. \tag{2.191}$$

---

* Problem 2.22: Application of Ehrenfest's theorem: Electron in a uniform electric field

The Hamiltonian for an electron in a uniform electric field is given by

$$H = -\frac{\hbar^2}{2m^*}\frac{\mathrm{d}^2}{\mathrm{d}z^2} - \alpha z. \qquad (2.192)$$

Use Ehrenfest's theorem to find $\frac{\mathrm{d}\langle z\rangle}{\mathrm{d}t}$ and $\frac{\mathrm{d}\langle p_z\rangle}{\mathrm{d}t}$.

Solution: Applying Ehrenfest's theorem, we get

$$i\hbar\frac{\mathrm{d}\langle z\rangle}{\mathrm{d}t} = \langle zH - Hz\rangle \qquad (2.193)$$

and

$$i\hbar\frac{\mathrm{d}\langle p_z\rangle}{\mathrm{d}t} = \langle p_zH - Hp_z\rangle. \qquad (2.194)$$

Now,

$$[zH - Hz]\phi(z) = \left[-z\frac{\hbar^2}{2m^*}\frac{\mathrm{d}^2}{\mathrm{d}z^2} - \alpha z^2 + \frac{\hbar^2}{2m^*}\frac{\mathrm{d}^2}{\mathrm{d}z^2}z + \alpha z^2\right]\phi(z)$$

$$= -z\frac{\hbar^2}{2m^*}\frac{\mathrm{d}^2\phi(z)}{\mathrm{d}z^2} + z\frac{\hbar^2}{2m^*}\frac{\mathrm{d}^2\phi(z)}{\mathrm{d}z^2} + \frac{\hbar^2}{m^*}\frac{\mathrm{d}\phi(z)}{\mathrm{d}z} = \frac{\hbar^2}{m^*}\frac{\mathrm{d}\phi(z)}{\mathrm{d}z}.$$
$$(2.195)$$

This implies

$$\frac{\langle zH - Hz\rangle}{i\hbar} = \frac{-i\hbar\langle\frac{\mathrm{d}}{\mathrm{d}z}\rangle}{m^*} = \frac{\langle p_z\rangle}{m^*} \qquad (2.196)$$

and

$$\frac{\langle dz\rangle}{\mathrm{d}t} = \frac{\langle p_z\rangle}{m^*}. \qquad (2.197)$$

To calculate the expectation value of the momentum operator, you will, of course, have to know the wave function of the electron in the electric field, but that is a different matter.

Also,

$$[p_zH - Hp_z]\phi(z) = \left[-i\frac{\hbar^3}{2m^*}\frac{\mathrm{d}^3}{\mathrm{d}z^3} - i\hbar\alpha - i\hbar\alpha z\frac{\mathrm{d}}{\mathrm{d}z} + i\frac{\hbar^3}{2m^*}\frac{\mathrm{d}^3}{\mathrm{d}z^3} + i\hbar\alpha z\frac{\mathrm{d}}{\mathrm{d}z}\right]\phi(z)$$

$$= -i\hbar\alpha\phi(z). \qquad (2.198)$$

Therefore,

$$\frac{\langle p_zH - Hp_z\rangle}{i\hbar} = \frac{\langle dp_z\rangle}{\mathrm{d}t} = -\alpha. \qquad (2.199)$$

* Problem 2.23: Ehrenfest's theorem from the Sturm–Liouville equation

(a) Show that the average value of an operator M in a state ϕ can be calculated as follows:

$$\langle M \rangle = \text{Tr}(\rho M), \tag{2.200}$$

where Tr stands for the trace operator.

(b) Using the previous result, the equation for the time evolution of the density matrix in the Sturm–Liouville equation, and the time evolution of the expectation value of an operator given by the Ehrenfest theorem, calculate the time evolution of the expectation value of an operator M and rederive Ehrenfest's theorem.

Solution: (a) Using $\Psi(\vec{r}, t) = \sum_n R_n(t)\phi_n(\vec{r})$, we get

$$\langle \Psi|M|\Psi \rangle = \sum_m \sum_n R_m^*(t) R_n(t) \langle \phi_m|M|\phi_n \rangle, \tag{2.201}$$

i.e.,

$$\langle \Psi|M|\Psi \rangle = \sum_{m,n} \rho_{mn} M_{mn}, \tag{2.202}$$

which we rewrite as

$$\langle \Psi|M|\Psi \rangle = \sum_n \langle \phi_n|\rho \left(\sum_m |\phi_m\rangle\langle\phi_m| \right) M|\phi_n\rangle. \tag{2.203}$$

Using the closure relation $\sum_m |\phi_m\rangle\langle\phi_m| = I$, we finally get

$$\langle \Psi|M|\Psi \rangle = \text{Tr}(\rho M). \tag{2.204}$$

(b) Taking the time derivative of $\text{Tr}(\rho M)$, we get

$$\begin{aligned}
\frac{d\langle M \rangle}{dt} &= \frac{d}{dt}\text{Tr}(\rho M) \\
&= \text{Tr}\left(\frac{d\rho}{dt}M\right) \text{Tr}\left(\rho\frac{dM}{dt}\right).
\end{aligned} \tag{2.205}$$

Using the Sturm–Liouville equation, we get

$$\begin{aligned}
\frac{d\langle M \rangle}{dt} &= -\frac{i}{\hbar}\text{Tr}\,[[H,\rho]M] + \left\langle \frac{dM}{dt} \right\rangle \\
&= -\frac{i}{\hbar}\text{Tr}\,[H\rho M - \rho HM] + \left\langle \frac{dM}{dt} \right\rangle \\
&\quad -\frac{i}{\hbar}\text{Tr}\,[\rho MH - \rho HM] + \left\langle \frac{dM}{dt} \right\rangle.
\end{aligned} \tag{2.206}$$

Using the properties of the trace, we obtain $\mathrm{Tr}(AB) = \mathrm{Tr}(BA)$,

$$\frac{\mathrm{d}\langle M\rangle}{\mathrm{d}t} = -\frac{i}{\hbar}\mathrm{Tr}\left[\rho[M,H]\right] + \left\langle\frac{\mathrm{d}M}{\mathrm{d}t}\right\rangle, \tag{2.207}$$

or

$$\frac{\mathrm{d}\langle M\rangle}{\mathrm{d}t} = -\frac{i}{\hbar}\langle[M,H]\rangle + \left\langle\frac{\mathrm{d}M}{\mathrm{d}t}\right\rangle, \tag{2.208}$$

which is Ehrenfest's theorem proved in a different way.

---

* Problem 2.24: The Cayley approximation to a unitary operator

If the Hamiltonian of a system is time independent and its wave function at time $t = 0$ is known, the solution to the time-dependent Schrödinger equation can be calculated as follows:

$$\psi(\vec{r},t) = U\psi(\vec{r},0) = \mathrm{e}^{-\frac{i}{\hbar}Ht}\psi(\vec{r},0). \tag{2.209}$$

As shown in Problem 2.4, the operator U is unitary.

In a numerical approach to the Schrödinger equation, the wave function at later times is typically calculated using small time steps δt and the following iteration procedure:

$$\psi(\vec{r},t+\delta t) = U\psi(\vec{r},t) = \mathrm{e}^{-\frac{i}{\hbar}H\delta t}\psi(\vec{r},t). \tag{2.210}$$

Numerically, the following (Cayley) approximation is used for the operator on the right-hand side of Equation (2.210):

$$\mathrm{e}^{-i\frac{\delta tH}{\hbar}} = \left(1 - \frac{i\delta tH}{2\hbar}\right)\Big/\left(1 + \frac{i\delta tH}{2\hbar}\right). \tag{2.211}$$

If $\delta t/\hbar$ is selected to be a small quantity, show that the Cayley approximation is unitary to order $(\delta t)^2$.

Solution: Performing a Taylor expansion of the operator $1/\left(1 + \frac{i\delta tH}{2\hbar}\right)$, the right-hand side of Equation (2.211) becomes

$$\mathrm{R.H.S.} = \left(1 - \frac{i\delta tH}{2\hbar}\right)\Big/\left(1 - \frac{i}{2}\frac{\delta tH}{\hbar} + \left(\frac{i}{2}\right)^2\frac{(\delta t)^2H^2}{\hbar^2} - \cdots\right). \tag{2.212}$$

Expanding this last equation up to order $(\delta t)^2$, we get:

$$\mathrm{R.H.S.} = 1 - i\frac{\delta tH}{\hbar} - \frac{1}{2}\frac{(\delta t)^2H^2}{\hbar^2} - \cdots \tag{2.213}$$

This is identical to the Taylor expansion of $\mathrm{e}^{-i\frac{\delta tH}{\hbar}}$ to order δ^2. Indeed,

$$\mathrm{e}^{-i\frac{\delta tH}{\hbar}} = 1 - i\frac{\delta tH}{\hbar} - \frac{1}{2}\frac{(\delta t)^2H^2}{\hbar} + O[(\delta t)^3]. \tag{2.214}$$

Application: The Cayley approximation described above is used to develop the Crank–Nicholson scheme to solve the time-dependent Schrödinger equation. This scheme ensures the conservation of probability over the simulation domain, i.e., the integration of the probability density associated over the time-varying wave function stays constant as a function of time. In other words, the wave function stays normalized as a function of time if it was properly normalized at time $t = 0$. The Crank–Nicholson scheme is studied in detail in Problem 13.6, where the algorithm to solve the one-dimensional time-dependent Schrödinger equation is described in full detail for the case of reflecting and absorbing boundary conditions at the ends of the simulation domain.

Suggested problems

- Using the commutation relations $[x, p_x] = i\hbar$, $[y, p_y] = i\hbar$, and $[z, p_z] = i\hbar$, show that $[L_x, L_y] = i\hbar L_z$, where L_x, L_y, and L_z are the components of the angular momentum $\vec{L} = \vec{r} \times \vec{p}$, the cross product of the position and momentum of a particle.

- Prove that the exponential operator containing the first spatial derivative plays the role of a displacement operator, i.e.:

$$(e^{\alpha \frac{\partial}{\partial z}}) f(z) = f(z + \alpha). \tag{2.215}$$

- Let A be a Hermitian operator with a non-degenerate spectrum, i.e., $A|n\rangle = a_n|n\rangle$ with all the a_n being distinct. Prove that any arbitrary function of the operator can be expressed as

$$F(A) = \sum_n F(a_n) \prod_{n \neq n'} \frac{a_{n'} I - A}{a_{n'} - a_n}, \tag{2.216}$$

where I is the unit operator. This expansion is called *Sylvester's formula*.

- Apply Sylvester's formula to calculate e^{bA} for the cases where the 2×2 matrix A has the two distinct eigenvalues λ_1 and λ_2, and b is a complex number.

- Apply the results of the previous problem to $F(\sigma_z) = e^{i\theta\sigma_z}$ and show that Sylvester's formula agrees with Equation (2.123).

- Using $\vec{F} = \vec{r}$ in Problem 2.11 and the Hamiltonian $H = \frac{p^2}{2m^*} + V(\vec{r})$, derive the Thomas–Reiche–Kuhn sum rule:

$$\sum_n (E_n - E_m)|\langle n|\vec{r}|m\rangle|^2 = \frac{3\hbar^2}{2m^*}. \tag{2.217}$$

- Using $F = e^{i\vec{k}\cdot\vec{r}}$ in Problem 2.11 and the Hamiltonian $H = \frac{p^2}{2m^*} + V(\vec{r})$, derive the Bethe–Bloch sum rule:

$$\sum_n (E_n - E_m)|\langle n|e^{i\vec{k}\cdot\vec{r}}|m\rangle|^2 = \frac{\hbar^2 k^2}{2m^*}. \tag{2.218}$$

- Show that any 2×2 Hermitian operator can be written as a linear combination of the 2×2 unit matrix and the Pauli matrices.

- Prove that the two eigenspinors $|\xi_n^+\rangle$ and $|\xi_n^-\rangle$ are indeed orthogonal, i.e., in Dirac's notation,

$$\langle \xi_n^+ | \xi_n^- \rangle = 0. \tag{2.219}$$

- If $|\phi(\vec{r})\rangle = \phi_1(\vec{r})|0\rangle + \phi_2(\vec{r})|1\rangle$ is a normalized state of a spin-1/2 particle, calculate the probabilities that a measurement of $(\hbar/2)\sigma_n$ will give $\pm 1/2\hbar$.

- If the wave function of a spin-1/2 electron is not completely known but it is known to be the eigenstate $|0\rangle$ of σ_z with probability $|c_1|^2$ and eigenstate $|1\rangle$ of σ_z with probability $|c_2|^2$, (with $|c_1|^2 + |c_2|^2 = 1$), what are the probabilities that a measurement of the spin component in the $\hat{n}$ direction will be $+1$ and -1 (in units of $\hbar/2$). Check that the sum of these probabilities is unity.

- Using the fact that

$$[\sigma_x, \sigma_y] = 2i\sigma_z, \tag{2.220}$$

calculate the commutator $[R_x(\theta_1), R_y(\theta_2)]$ where the rotation matrices are defined as follows:

$$R_x(\theta_1) = e^{-i\frac{\theta_1}{2}\sigma_x} \tag{2.221}$$

and

$$R_y(\theta_2) = e^{-i\frac{\theta_2}{2}\sigma_y}. \tag{2.222}$$

These operators are very useful to make rotations on the Bloch sphere [2, 4].

- If the state of a spin is given by the qubit

$$|\Phi\rangle = N[3|0\rangle - i|1\rangle], \tag{2.223}$$

where $|0\rangle$ and $|1\rangle$ are the normalized eigenstates of the σ_z Pauli matrix (see Appendix D), find the normalization constant N of the qubit.

For an ensemble of spins prepared in the qubit state above, calculate the average value and standard deviation when measuring the component $S_y = \frac{\hbar}{2}\sigma_y$.

- For any analytical function $f(x)$ with a Taylor expansion, prove that the following equality holds:

$$f(\theta\vec{\sigma} \cdot \vec{n}) = \left[\frac{f(\theta) + f(-\theta)}{2}\right] I + \left[\frac{f(\theta) - f(-\theta)}{2}\right] \vec{\sigma} \cdot \vec{n}, \tag{2.224}$$

where I is the 2×2 identity matrix. This is a generalization of the identity proven in Problem 2.14.

- Using the properties of the Pauli spin matrices given in Appendix D, calculate $e^{i\alpha\sigma_n}$ with $\sigma_n = \lambda\sigma_x + \mu\sigma_y$ and $\lambda^2 + \mu^2 = 1$, when α, λ, and μ are real.

- All that is known about a spin-1/2 particle is that it is in a state in which $S_z = \frac{\hbar}{2}\sigma_z$ has the values $\pm\frac{\hbar}{2}$ with probabilities $|c_1|^2$ and $|c_2|^2$, respectively, with $|c_1|^2 + |c_2|^2 = 1$. What are the probabilities to measure $\pm\frac{\hbar}{2}$ for the components of $S_n = \frac{\hbar}{2}\vec{\sigma} \cdot \vec{n}$?

- For a particle of mass m^* subjected to a constant force F,

  (1) Show that $<p^2> - <p>^2$ is independent of time.

  (2) Starting with the Schrödinger equation in momentumm space, find a relation between $\frac{\partial}{\partial t}|\langle p|\phi(p)\rangle|^2$ and $\frac{\partial}{\partial p}|\langle p|\phi(p)\rangle|^2$.

  (3) Integrate the equation obtained in the previous step. Give a physical interpration to the result of this integration.

- Consider a free particle with mass m^* in one dimension.

  (1) By applying Ehrenfest's theorem, show that $\langle z\rangle(t)$ is a linear function of time when $\langle p\rangle(t)$ is a constant.

  (2) Derive the equations of motion for $\langle z^2\rangle$ and $\langle zp_z+p_zz\rangle$ and integrate them.

  (3) Show that the following relation holds:

$$[\Delta z(t)]^2 = [\Delta z(t=0)]^2 + \frac{t^2}{m^{*2}}(\Delta p)^2(t=0). \tag{2.225}$$

References

[1] Cohen-Tannoudji, C., Diu, B., and Laloe, F. (2000) *Quantum Mechanics*, Complement B_{II}, Hermann, Paris.

[2] Nielsen, M. A. and Chuang, I. L. (2000) *Quantum Computation and Quantum Information*, Chapter 2, Cambridge University Press, New York.

[3] Wang, S. (1999) Generalization of the Thomas–Reiche–Kuhn and the Bethe sum rules. *Phys. Rev. A* 60, pp. 262–266.

[4] Bandyopadhyay, S. and Cahay, M. (2015) *Introduction to Spintronics*, 2nd edition, CRC Press, Boca Raton, FL.

[5] Kroemer, H., (1994) *Quantum Mechanics for Engineering, Materials Science, and Applied Physics*, Section 7.6.4, Prentice Hall, Englewood Cliffs, NJ.

[6] Ohanian, H.C. (1990) *Principles of Quantum Mechanics*, Section 5.2, Prentice Hall, Upper Saddle River, NJ.

Suggested Reading

- Belloni, M. and Robinett, R. W. (2008) Quantum mechanical sum rules for two model systems. *American Journal of Physics* 76(9), pp. 814–821.

- Epstein, J. H. and Epstein, S. T. (1961) Some applications of hypervirial theorems to the calculation of average values. *American Journal of Physics* 30, pp. 266–268.

- Hadjimichael, E., Currie, W., and Fallieros, S. (1997) The Thomas–Reiche–Kuhn sum rule and the rigid rotator. *American Journal of Physics* 65, pp. 335–341.

- Fernandez, F. M. (2002) The Thomas–Reiche–Kuhn sum rule for the rigid rotator. *Int. J. Math. Ed. Sci. Tech.* 33, pp. 636–640.

- Jackiw, R. (1967) Quantum mechanical sum rules. *Physical Review* 157, pp. 1220–1225.

- Beertsch, G. and Ekardt, W. (1985) Application of sum rules to the response of small metal particles. *Physical Review B* 32, pp. 7659–7663.

Chapter 3: Bound States

The following set of problems deals with one-dimensional bound state calculations, which can be performed analytically or via the numerical solution of a transcendental equation [1]. These problems give some insight into more complicated three-dimensional bound state problems whose solutions typically require numerically intensive approaches.

** Problem 3.1: Bound state in a one-dimensional attractive delta scatterer

A "bound state" of a potential is, as the name suggests, a state such that if an electron is in that state, then it is bound to the potential and does not stray too far away from it. Classically, this would mean that the total energy of the particle (kinetic + potential) is negative. A quantum mechanical definition of a bound state is that the wave function associated with that state vanishes at infinite distance from the potential so that the probability of finding an electron infinitely far from the potential is zero.

Find the bound state energy and corresponding normalized eigenfunction in a one-dimensional attractive delta potential (this potential is representative of a strongly screened ionized impurity scatterer in a solid) expressed as

$$V(z) = -\Gamma \delta(z), \tag{3.1}$$

with $\Gamma > 0$.

Hint:

(a) First, integrate the Schrödinger equation in a small interval around $z = 0$ and show that the derivative of the eigenfunction $\phi(z)$ has a discontinuity at $z = 0$. Determine the value of the discontinuity in terms of Γ, m^*, and $\phi(0)$. Assume a constant effective mass throughout.

(b) Eigenfunctions describing bound states of a localized potential, i.e., states whose wave functions decay to zero at infinite distances from the potential, must be of the form

$$\phi(z) = A_- e^{\kappa z} + B_- e^{-\kappa z} \tag{3.2}$$

for $z < 0$ and

$$\phi(z) = A_+ e^{\kappa z} + B_+ e^{-\kappa z} \tag{3.3}$$

for $z > 0$. The quantity κ is a positive real constant. Express it as a function of the particle's total energy E and m^*. Remember that for a bound state $E < 0$ (the classical definition of a bound state).

Problem Solving in Quantum Mechanics: From Basics to Real-World Applications for Materials Scientists, Applied Physicists, and Devices Engineers, First Edition.
Marc Cahay and Supriyo Bandyopadhyay.
© 2017 John Wiley & Sons Ltd. Published 2017 by John Wiley & Sons Ltd.

What is the general expression for the transmission matrix M defined as follows:

$$\begin{pmatrix} A_+ \\ B_+ \end{pmatrix} = M \begin{pmatrix} A_- \\ B_- \end{pmatrix}? \tag{3.4}$$

Find the explicit expression for the matrix M that will ensure that the bound state wave function is square integrable. Use this result to derive the expression for the bound state energy and its corresponding eigenfunction.

(c) Calculate the width Δz (variance) of the wave function associated with the bound state.

Solution:
(a) Starting with the one-dimensional Schrödinger equation

$$-\frac{\hbar^2}{2m^*}\phi''(z) - \Gamma\delta(z)\phi(z) = E_\phi(z), \tag{3.5}$$

where the superscript $''$ denotes the second derivative with respect to z, and integrating it on both sides of the delta potential located at $z = 0$ from $z = -\epsilon$ to $z = +\epsilon$ leads to:

$$-\frac{\hbar^2}{2m}[\phi'(\varepsilon) - \phi'(\varepsilon)] - \Gamma\phi(0) = E\int_{-\varepsilon}^{\varepsilon}\phi(z)\,dz, \tag{3.6}$$

where the superscript $'$ denotes the first derivative with respect to z.

We note from Equation (3.5) that, since $\phi''(z)$ has a delta singularity at $z = 0$, $\phi'(z)$ has a jump discontinuity at $z = 0$. Hence, $\phi(z)$ is continuous. With $\varepsilon \to 0$, Equation (3.6) becomes

$$\phi'(+0) - \phi'(-0) = -\frac{2m^*\Gamma}{\hbar^2}\phi(0). \tag{3.7}$$

(b) For $z \neq 0$ and $E < 0$, we rewrite the Schrödinger equation as

$$\phi''(z) = -\frac{2mE}{\hbar^2}\phi(z) = -\kappa^2\phi(z), \tag{3.8}$$

with

$$\kappa = \left(\frac{2m^*E}{\hbar^2}\right)^{1/2}, \tag{3.9}$$

where we have taken the positive root.

With solutions of the form

$$\phi(z) = A_-e^{\kappa z} + B_-e^{-\kappa z}, \; z < 0; \quad \phi(z) = A_+e^{\kappa z} + B_+e^{-\kappa z}, \; z > 0, \tag{3.10}$$

the continuity of $\phi(z)$ at $z = 0$ requires

$$A_- + B_- = A_+ + B_+ = \phi(0), \tag{3.11}$$

while Equation (3.7) leads to

$$\kappa \left(A_+ - B_+ - A_- + B_- \right) = -\frac{2m^*\Gamma}{\hbar^2}\phi(0).$$ (3.12)

Defining the parameter

$$\Lambda = \frac{m^*\Gamma}{\hbar^2\kappa},$$ (3.13)

we obtain, from Equations (3.11) and (3.12),

$$\begin{pmatrix} A_+ \\ B_+ \end{pmatrix} = \begin{pmatrix} 1 - \Lambda & -\Lambda \\ \Lambda & 1 + \Lambda \end{pmatrix} \begin{pmatrix} A_- \\ B_- \end{pmatrix}.$$ (3.14)

Thus the transmission matrix M is given by:

$$M = \begin{pmatrix} 1 - \Lambda & -\Lambda \\ \Lambda & 1 + \Lambda \end{pmatrix}.$$ (3.15)

We search for bound states by insisting that the wave function $\phi(z)$ given by Equation (3.10) vanishes at $z = \pm\infty$. This requires

$$B_- = A_+ = 0.$$ (3.16)

This is possible only if $1 - \Lambda = 0$ and $A_- = B_+$. Thus

$$M = \begin{pmatrix} 0 & -1 \\ 1 & 2 \end{pmatrix}.$$ (3.17)

Therefore, there is only one bound state since there is only one allowed value of Λ.

The corresponding wave function is given by (see Equation (3.10))

$$\phi_0(z) = \sqrt{\kappa_0}e^{-\kappa_0|z|},$$ (3.18)

where κ_0 is found by setting $\Lambda = 1$ in Equation (3.13) and solving for κ_0, which yields

$$\kappa_0 = \frac{m^*\Gamma}{\hbar^2}.$$ (3.19)

The factor $\sqrt{\kappa_0}$ is needed in Equation (3.18) for normalization of the wave function, i.e., for the probability density integrated over all space to be unity. Indeed,

$$\int_{-\infty}^{\infty} \phi^2(z)dz = \kappa_0 \int_{-\infty}^{\infty} e^{-2\kappa_0|z|}dz = 2\kappa_0 \int_{0}^{\infty} e^{-2\kappa_0 z}dz = 1.$$ (3.20)

The energy of the bound state is given by

$$E_0 = \left\langle \phi_0(z) \left| -\frac{\hbar^2}{2m^*}\frac{d^2}{dz^2} \right| \phi_0(z) \right\rangle = -\frac{\hbar^2\kappa_0^2}{2m^*} = -\frac{m^*\Gamma^2}{2\hbar^2}.$$ (3.21)

Note that the bound state energy is *negative*. Hence, the quantum mechanical definition of a bound state, i.e., a state whose wave function vanishes at $\pm\infty$, is *completely equivalent* to the classical definition that it is a state with negative total energy.

(c) By symmetry, $\langle z \rangle = 0$. We obtain easily

$$\langle z^2 \rangle = \int_{-\infty}^{\infty} \phi^2(z) z^2 \mathrm{d}z = \frac{1}{2\kappa_0^2}. \tag{3.22}$$

Hence, $\Delta z = [\langle z^2 \rangle - \langle z \rangle^2]^{1/2} = \frac{1}{\sqrt{2}\kappa_0}$.

** Problem 3.2: One-dimensional Schrödinger equation in momentum representation

Derive the one-dimensional Schrödinger equation for a particle moving in a one-dimensional attractive delta scatterer $-\Gamma\delta(z)$, where Γ is positive. Solve the resulting equation for the bound state and show that it agrees with the results of the previous problem.

Solution: The wave function $\bar{\phi}(p)$ in momentum space is obtained from a Fourier transform of the wave function $\phi(z)$ in the spatial representation:

$$\phi(z) = (2\pi\hbar)^{-1/2} \int_{-\infty}^{+\infty} \bar{\phi}(p) \exp\left(\frac{ipz}{\hbar}\right) \mathrm{d}p. \tag{3.23}$$

By definition, the one-dimensional delta function can be expressed as

$$\delta(z) = (2\pi\hbar)^{-1} \int_{-\infty}^{+\infty} \exp\left(\frac{ipz}{\hbar}\right) \mathrm{d}p. \tag{3.24}$$

Using the last two equations, we get (using the concept of "convolution")

$$\delta(z)\phi(z) = (2\pi\hbar)^{-3/2} \int_{-\infty}^{+\infty} \mathrm{d}p \int_{-\infty}^{+\infty} \mathrm{d}p' \bar{\phi}(p - p') \exp\left(\frac{ipz}{\hbar}\right). \tag{3.25}$$

Using the preceding results, the one-dimensional Schrödinger equation becomes:

$$\frac{p^2}{2m^*}\phi(z) - \Gamma\delta(z)\phi(z) - E\phi(z) \tag{3.26}$$

$$= (2\pi\hbar)^{-1/2} \int_{-\infty}^{+\infty} \mathrm{d}p \left[\left(\frac{p^2}{2m^*} - E\right)\bar{\phi}(p)\right.$$

$$\left. - \frac{\Gamma}{2\pi\hbar} \int_{-\infty}^{+\infty} \mathrm{d}p' \bar{\phi}(p - p')\right] \exp\left(\frac{ipz}{\hbar}\right) = 0.$$

For the last equality to hold, the integrand in the last equation, i.e., the quantity within the square brackets, must be identically zero. This leads to the Schrödinger equation in momentum space:

$$\left(\frac{p^2}{2m^*} - E\right)\bar{\phi}(p) - \frac{\Gamma}{2\pi\hbar}\int_{-\infty}^{+\infty}\bar{\phi}(p-p')dp' = 0. \tag{3.27}$$

The integral in Equation (3.27) can be written as

$$\int_{-\infty}^{+\infty}\bar{\phi}(p-p')dp' = -\int_{-\infty}^{+\infty}\bar{\phi}(p-p')d(p-p') = -\int_{-\infty}^{+\infty}\bar{\phi}(p)dp = -C, \tag{3.28}$$

where, obviously, C is a constant independent of p since the integration is carried out over the variable p.

Equation (3.27) therefore has the solution

$$\bar{\phi}(p) = -\frac{\Gamma C}{2\pi\hbar}\frac{1}{\left[\frac{p^2}{2m^*} - E\right]}. \tag{3.29}$$

The integral in Equation (3.28) converges only for *bound states* where $\bar{\phi}(p)$ goes to zero at $p = \pm\infty$, i.e., for $E < 0$.

For $E < 0$, introducing p_0 such that

$$E = -\frac{p_0^2}{2m^*}, \tag{3.30}$$

we get

$$\bar{\phi}(p) = -\frac{m^*\Gamma C}{2\pi\hbar}\frac{1}{p^2 + p_0^2}. \tag{3.31}$$

Using the last result in Equation (3.28), we get

$$-C = \int_{-\infty}^{+\infty}\bar{\phi}(p)dp = -\frac{m^*\Gamma C}{2\pi\hbar}\int_{-\infty}^{+\infty}\frac{dp}{p^2 + p_0^2}$$
$$= -\frac{m^*\Gamma C}{hp_0}\times\tan^{-1}\left(\frac{p}{p_0}\right)\Big|_{-\infty}^{\infty}$$
$$= -\frac{m^*\Gamma C}{\hbar p_0}. \tag{3.32}$$

The solution of the last equation is obviously $p_0 = m^*\Gamma/\hbar$, leading to the following energy for the bound state:

$$E_0 = -\frac{p_0^2}{2m^*} = -\frac{m^*\Gamma^2}{2\hbar^2}, \tag{3.33}$$

which agrees with the result of Problem 3.1.

***** Problem 3.3: The one-dimensional double delta potential

Find the bound state energies of a particle of mass m^* moving in the one-dimensional potential with a double minimum potential energy profile approximated as two attractive delta scatterers separated by a distance a:

$$V(z) = -\Gamma[\delta(z) + \delta(z-a)], \ \Gamma > 0. \tag{3.34}$$

Solution: The one-dimensional time-independent Schrödinger equation for this problem is given by:

$$\phi''(z) + \mu\delta(z)\phi(z) + \mu\delta(z-a)\phi(z) = \kappa^2\phi(z), \tag{3.35}$$

where we have used the shorthand notation

$$\mu = \frac{2m^*\Gamma}{\hbar^2}, \tag{3.36}$$

and $\kappa^2 = \frac{-2m^*E}{\hbar^2} = \frac{2m^*|E|}{\hbar^2}$, for $E < 0$. As before, a single prime denotes the first derivative and a double prime denotes the second derivative.

The wave function $\phi(z)$ must be continuous, and its spatial derivative at $z = 0$ and $z = a$ must satisfy the following conditions:

$$\phi'(0_+) - \phi'(0_-) = -\mu\phi(0) \tag{3.37}$$

and

$$\phi'(a_+) - \phi'(a_-) = -\mu\phi(a). \tag{3.38}$$

The solutions of the Schrödinger equation corresponding to bound states are of the form:

$$\phi(z) = Ae^{\kappa z}, \ z < 0, \tag{3.39}$$
$$\phi(z) = Be^{\kappa z} + Ce^{-\kappa z}, \ 0 < z < a, \tag{3.40}$$

and

$$\phi(z) = De^{-\kappa z}, \ z > a. \tag{3.41}$$

The requirements that $\phi(z)$ be continuous at $z = 0$ and $z = a$, along with the boundary conditions in Equations (3.37) and (3.38), lead to the following set of four equations for the four unknowns A, B, C, and D:

$$A = B + C = \phi(0), \tag{3.42}$$
$$Be^{\kappa a} + Ce^{-\kappa a} = De^{-\kappa a} = \phi(a), \tag{3.43}$$
$$\kappa(B - C - A) = -\mu\phi(0), \tag{3.44}$$
$$-De^{-\kappa a} - B\kappa e^{\kappa a} + C\kappa e^{-\kappa a} = -\mu\phi(a). \tag{3.45}$$

We rewrite these last equations as follows:

$$A - B - C = 0, \tag{3.46}$$
$$(\mu - \kappa)A + \kappa B - \kappa C = 0, \tag{3.47}$$
$$Be^{\kappa a} + Ce^{-\kappa a} - De^{-\kappa a} = 0, \tag{3.48}$$
$$-B\kappa e^{\kappa a} + C\kappa e^{-\kappa a} + (\mu - \kappa)De^{-\kappa a} = 0. \tag{3.49}$$

These four coupled equations can be written in a matrix form:

$$
\begin{bmatrix}
1 & -1 & -1 & 0 \\
\mu - \kappa & \kappa & -\kappa & 0 \\
0 & e^{\kappa a} & e^{-\kappa a} & -e^{-\kappa a} \\
0 & -\kappa e^{\kappa a} & \kappa e^{-\kappa a} & (\mu - \kappa)e^{-\kappa a}
\end{bmatrix}
\begin{bmatrix}
A \\
B \\
C \\
D
\end{bmatrix} = 0.
\tag{3.50}
$$

Non-trivial solutions for A, B, C, and D exist only if the following determinant is equal to zero:

$$
\det
\begin{bmatrix}
1 & -1 & -1 & 0 \\
\mu - \kappa & \kappa & -\kappa & 0 \\
0 & e^{\kappa a} & e^{-\kappa a} & -e^{-\kappa a} \\
0 & -\kappa e^{\kappa a} & \kappa e^{-\kappa a} & (\mu - \kappa)e^{-\kappa a}
\end{bmatrix} = 0.
\tag{3.51}
$$

Writing out the determinant and setting it equal to zero leads to the following relation:

$$
\mu^2 e^{-2\kappa a} - (\mu - 2\kappa)^2 = 0,
\tag{3.52}
$$

which must be solved for κ to find the energies of the bound states.

We rewrite Equation (3.52) as

$$
e^{-\xi} = \pm \left(1 - \frac{\xi}{\xi_0} \right)
\tag{3.53}
$$

by introducing the new variables $\xi = \kappa a$ and $\xi_0 = \frac{\mu a}{2} = \frac{m^* a \Gamma}{\hbar^2}$. The allowed values of ξ, and hence the energies of the bound state E, correspond to the intersections of the exponential curve $\eta = e^{-\xi}$ and the straight lines

$$
\eta = \pm \left(1 - \frac{\xi}{\xi_0} \right).
\tag{3.54}
$$

The two curves, $\eta = e^{-\xi}$ and $\eta = -1 + \frac{\xi}{\xi_0}$, will always have an intersection no matter what the strength Γ of the delta potentials is. We call the value of ξ at the intersection ξ_S. The curve $\eta = e^{-\xi}$ intersects the straight line $\eta = 1 - \frac{\xi}{\xi_0}$ at $\xi = 0$. This intersection does not represent a bound state since then $E = 0$.

A second intersection is possible if and only if

$$
\frac{d}{d\xi} e^{-\xi} \bigg|_{\xi=0} < \frac{d}{d\xi} \left(1 - \frac{\xi}{\xi_0} \right) = -\frac{1}{\xi_0},
\tag{3.55}
$$

i.e., if $\xi_0 > 1$. We designate this new solution ξ_A. Next, we study in more detail the bound states associated with ξ_S and ξ_A. We note that the straight lines intercept the η axis at ξ_0. Therefore, since $e^{-\xi} > 0$, we must have $\xi_S > \xi_0$ while $\xi_A < \xi_0$.

We first consider the solution associated with ξ_S. Introducing the new variable $s = \frac{\xi_S}{2\xi_0}$, Equations (3.46)–(3.49) can be rewritten as

$$
A - B - C = 0,
\tag{3.56}
$$
$$
A(1 - s) + s(B - C) = 0,
\tag{3.57}
$$
$$
B e^{\xi_S} + e^{-\xi_S}(C - D) = 0,
\tag{3.58}
$$

and
$$-Bse^{\xi_S} + Cse^{-\xi_S} + (1-s)De^{-\xi_S} = 0. \tag{3.59}$$

Once again, we can write the last four equations in a matrix form, set the determinant equal to zero for non-trivial solutions of A, B, C, and D, and find that the zero determinant condition mandates

$$e^{-\xi_S} = -1 + 2s. \tag{3.60}$$

From Equations (3.58) and (3.59), we obtain

$$C - D = -Bse^{2\xi_S} \tag{3.61}$$

and

$$D + (C - D)s = Bse^{2\xi_S}. \tag{3.62}$$

Combining Equations (3.61) and (3.62) leads to

$$D = 2sBe^{2\xi_S} \tag{3.63}$$

and

$$C = (2s-1)Be^{2\xi_S} = Be^{\xi_S}. \tag{3.64}$$

Substituting Equation (3.64) into Equation (3.56), we get

$$A(1-s) + sB(1 - e^{\xi_S}) = 0, \tag{3.65}$$

or

$$Ae^{-\xi_S} - 2sB = 0. \tag{3.66}$$

Equations (3.63) and (3.66) imply

$$D = Ae^{\xi_S}. \tag{3.67}$$

Equations (3.64) and (3.67) lead to the conclusion that the eigenfunction $\phi_S(z)$ associated with ξ_S is symmetric, i.e.,

$$\phi_S(z) = \phi_S(z - a). \tag{3.68}$$

In fact, substituting Equations (3.64) and (3.67) in Equations (3.40)–(3.41) and defining $\kappa_S = \frac{\xi_S}{a}$, we find that the eigenfunctions in different domains of z are given by:

$$
\begin{aligned}
\phi(z) &= Ae^{-\kappa_S|z|} & (z < 0), \\
\phi(z) &= Be^{\kappa_S z} + Be^{\kappa_S(z-a)} & (0 < z < a), \\
\phi(z) &= Ae^{-\kappa_S(z-a)} & (z > a).
\end{aligned}
\tag{3.69}
$$

The other bound state exists if $\xi_0 \left(= \frac{m^* a\Gamma}{\hbar^2} \right) > 1$. For this solution, which we called ξ_A, we have

$$e^{-\xi_A} = 1 - 2s', \tag{3.70}$$

where $s' = \frac{\xi_A}{2\xi_0}$. In this case, we obtain equations similar to Equations (3.56)–(3.59) with ξ_S and s replaced by ξ_A and s', respectively. We now obtain

$$C = -Be^{\xi_A} \tag{3.71}$$

and

$$D = -Ae^{\xi_A}. \tag{3.72}$$

These last two relations imply that the corresponding eigenfunction $\phi_A(z)$ is anti-symmetric, i.e.,

$$\phi_A(z) = -\phi_A(a - z). \tag{3.73}$$

We note that there is always at least one symmetric bound state with energy

$$E_S = -\frac{\hbar^2 {\kappa_S}^2}{2m^*}. \tag{3.74}$$

The antisymmetric bound state, when it exists, is less bound than the symmetric state because $\kappa_S > \kappa_A$.

Application: The results of the previous problem can serve as a simple theory of the binding of two nuclei to form an ionized molecule containing a single electron, such as H_2^+. The attraction due to the formation of the symmetric bound state is opposed by the repulsion of the two ions. The energy associated with the latter is given by $\frac{e^2}{4\pi\epsilon a}$, where ϵ is the dielectric constant of the medium. Hence, the energy of the molecular ion H_2^+ is given by

$$W = \frac{mV_0^2}{2\hbar^2} + \frac{e^2}{4\pi\epsilon a} - \frac{\hbar^2 k_S^2}{2m} = \frac{mV_0^2}{2\hbar^2} + \frac{e^2}{4\pi\epsilon a} - \frac{\hbar^2 \xi_S^2}{2ma^2}. \tag{3.75}$$

Since ξ_S is the solution of $e^{-\xi} = -1 + \frac{\xi}{\xi_0}$ and ξ_0 depends on a, so does ξ_S.

The equilibrium distance a for H_2^+ is obtained when W is a minimum and binding occurs only if $W < 0$.

*** Problem 3.4: Bound state of a one-dimensional scatterer at a heterointerface

We consider the problem of a one-dimensional delta scatterer located at the interface between two dissimilar materials, as shown in Figure 3.1. Find an analytical expression for the energy of the bound state E^* of the delta scatterer and show that this bound state disappears when the height of the potential step ΔE_c is equal to four times the magnitude E_0 of the energy of the bound state of the delta scatterer corresponding to $\Delta E_c = 0$. In Problem 3.1, E_0 was shown to be given by

$$|E_0| = \frac{m^* \Gamma^2}{2\hbar^2}, \tag{3.76}$$

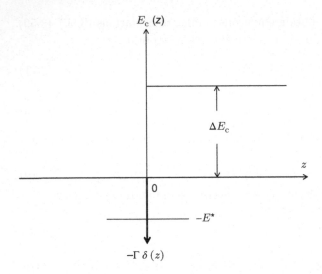

Figure 3.1: Illustration of the bound state energy $-E^*$ of a one-dimensional delta scatterer located at a heterointerface. This energy depends on the magnitude of the potential jump ΔE_c.

where Γ is the strength of the delta scatterer. Assume a constant effective mass m^* throughout.

Solution: The Schrödinger equation for this problem is given by

$$-\frac{\hbar^2}{2m^*}\phi''(z) - \Gamma\delta(z)\phi(z) + \Delta E_c\theta(z)\phi(z) = E\phi(z), \tag{3.77}$$

where Γ is the strength of the delta scatterer and $\theta(z)$ is the unit step function (also known as the Heaviside function).

We seek the bound state solution of this equation, i.e., a solution with negative energy E.

For $z < 0$, we have

$$\phi(z) = Ae^{\kappa z}, \tag{3.78}$$

and for $z > 0$,

$$\phi(z) = Be^{-\kappa' z}, \tag{3.79}$$

with

$$\kappa = \frac{1}{\hbar}\sqrt{2m^*|E|} \text{ and } \kappa' = \frac{1}{\hbar}\sqrt{2m^*(|E| + \Delta E_c)}. \tag{3.80}$$

At $z = 0$, the following conditions must be satisfied:

$$\phi'(0_+) - \phi'(0_-) = -\frac{2m^*\Gamma}{\hbar}\phi(0) \tag{3.81}$$

and

$$\phi(0_+) = \phi(0_-). \tag{3.82}$$

This leads to the following two equations for the two unknowns A and B:

$$A = B \tag{3.83}$$

and

$$\kappa A + \kappa' B = \frac{2m^*\Gamma}{\hbar^2} B. \tag{3.84}$$

Using Equation (3.83), the latter equation becomes

$$\kappa + \kappa' = \frac{2m^*\Gamma}{\hbar^2}. \tag{3.85}$$

Substituting in the expressions for κ and κ', we get:

$$\sqrt{2m|E|} + \sqrt{2m(|E| + \Delta E_{\mathrm{c}})} = \frac{2m\Gamma}{\hbar}. \tag{3.86}$$

The magnitude of the bound state energy corresponding to $\Delta E_{\mathrm{c}} = 0$ is given by $E_0 = \frac{m^*\Gamma^2}{2\hbar^2}$ and, squaring Equation (3.86) on both sides, we get

$$2|E| + \Delta E_{\mathrm{c}} + 2\sqrt{|E|(|E| + \Delta E_{\mathrm{c}})} = 4E_0. \tag{3.87}$$

Solving for $|E|$ gives the magnitude of the bound state energy E^*:

$$E^* = E_0 \left(1 - \frac{\Delta E_{\mathrm{c}}}{4E_0} \right). \tag{3.88}$$

This last equation shows that the step potential reduces the magnitude of the bound state energy of the delta scatterer. Furthermore, the bound state disappears when $\Delta E_{\mathrm{c}} = 4E_0$.

** Problem 3.5: One-dimensional particle in a box

This is the quintessential problem of bound states in a quantum confined structure. Consider a one-dimensional potential well with infinitely high barriers located between $z = 0$ and $z = W$, as shown in Figure 3.2. Find the allowed energies of an electron confined within the well and their corresponding wave functions.

Solution: Since the potential is not changing with time, we will solve the time-independent Schrödinger equation

$$-\frac{\hbar^2}{2m^*}\frac{\mathrm{d}^2}{\mathrm{d}z^2}\phi(z) + E_{\mathrm{c}}(z)\phi(z) = E\phi(z), \tag{3.89}$$

where $E_{\mathrm{c}}(z)$ is assumed to be constant within the interval $[0, W]$. Since potential is always undefined to the extent of an arbitrary constant, we can, without loss of generality, set the potential equal to zero.

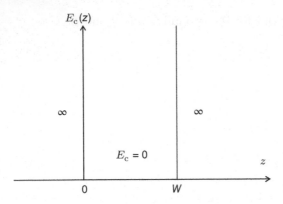

Figure 3.2: The one-dimensional particle in a box problem. $E_c(z) = 0$ inside the box in the range $[0, W]$ and is ∞ otherwise.

The first point to note is that the electron can never penetrate an infinite barrier and hence the probability of finding the electron outside the well is exactly zero. Since the probability is the squared magnitude of the wave function, we can deduce that $\phi(z < 0) = \phi(z > W) = 0$. Furthermore, since the wave function must be continuous in space, we obtain the boundary conditions $\phi(z = 0) = \phi(z = W) = 0$, i.e., the wave function vanishes at the boundaries.

We can write Equation (3.89) as

$$\frac{\mathrm{d}^2}{\mathrm{d}z^2}\phi(z) + k^2\phi(z) = 0, \tag{3.90}$$

where $k = \sqrt{2m^*E}/\hbar$. The solution of the above second-order differential equation is

$$\phi(z) = A\cos(kz) + B\sin(kz). \tag{3.91}$$

Since $\phi(z = 0) = 0$, $A = 0$; hence, $\phi(z) = B\sin(kz)$.

Now, since $\phi(z = W) = 0$, the following condition must be satisfied:

$$B\sin(kW) = 0. \tag{3.92}$$

The above equation can be satisfied if and only if either $B = 0$ or $kW = n\pi$, where n is an integer. Making $B = 0$ will make the wave function vanish everywhere, and therefore that is not a non-trivial solution. Thus, we must have $k = \frac{n\pi}{W}$, and the possible wave functions for the particle in a box are:

$$\phi_n(z) = B\sin\left(\frac{n\pi z}{W}\right). \tag{3.93}$$

We can find B from the normalization condition, i.e.,

$$\int_0^W \phi^*(z)\phi(z)\mathrm{d}z = 1, \tag{3.94}$$

which yields $B = \sqrt{\frac{2}{W}}$. Therefore, the particle in a box wave functions are

$$\phi_n(z) = \sqrt{\frac{2}{W}} \sin\left(\frac{n\pi z}{W}\right).$$ (3.95)

The kinetic energy associated with each of the allowed wave functions (i.e. for various values of the integer n) is found by evaluating the expectation value of the kinetic energy operator with the corresponding wave function, but there is a much simpler way. Since $k = \sqrt{2m^*E}/\hbar$ and $k = \frac{n\pi}{W}$, we get directly that

$$E_n = n^2 \frac{\hbar^2 \pi^2}{2m^* W^2}.$$ (3.96)

---

* Problem 3.6: Lowest bound state energies of an impurity located in a one-dimensional quantum well [2, 3].

We consider the problem of a one-dimensional attractive delta scatterer (i.e., a scatterer with a negative potential) located in the middle of a one-dimensional box of size W, as shown in Figure 3.3. It is assumed that $E_c(z)$ is 0 in the interval $0 < z < W$ and ∞ outside the box. The effective mass m^* of the electron is assumed to be constant throughout.

Find the two lowest bound states of this problem, one with negative energy and one with positive energy. The former can be considered as a modification of Problem 3.1 in which the energy of the delta scatterer is modified because the impurity is

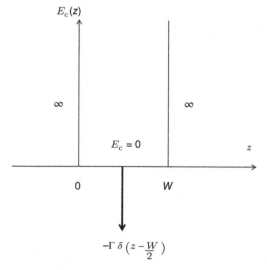

Figure 3.3: Bound state problem for a one-dimensional attractive delta scatterer in the middle of a box of size W with infinite barriers, i.e., $E_c(z) = \infty$ outside the box.

*located in a finite box. The latter is obtained as a modification of the ground state
of a particle in a box due to the presence of the attractive delta scatterer.*

Solution:
Case a: Bound state with positive energy: The solution of the Schrödinger
equation for a particle in a one-dimensional box is one of the simplest problems in
quantum mechanics. As shown in the previous problem, the wave functions of the
eigenstates are given by

$$\phi_n(z) = \sqrt{\frac{2}{W}} \sin\left(\frac{n\pi z}{W}\right),\tag{3.97}$$

where n is an integer ($n = 1, 2, 3, \ldots$) and W is the width of the box. The corresponding eigenenergies are given by

$$E_n = n^2 \frac{\hbar^2 \pi^2}{2m^* W^2}.\tag{3.98}$$

The presence of the delta scatterer will lower the energy of the ground state E_1
below the value given by Equation (3.96).

To find the new bound state with positive energy, we write the solutions of the
Schrödinger equation inside the box as follows:

$$\phi_1 = A\sin(kz) \text{ for } z < W/2,\tag{3.99}$$

and

$$\phi_2 = B\sin[k(W-z)] \text{ for } z > W/2,\tag{3.100}$$

where $k = \frac{1}{\hbar}\sqrt{2m^*E}$, E being the energy of the new bound state with positive
energy. Note that since E is positive, k is *real*.

At $z = W/2$, the wave function must be continuous. Hence, we must have
$\phi_1(W/2) = \phi_2(W/2)$, leading to $A = B$.

Inside the well, the Schrödinger equation is given by

$$-\frac{\hbar^2}{2m}\phi''(z) - \Gamma\delta\left(z - \frac{W}{2}\right)\phi(z) = E\phi(z).\tag{3.101}$$

Integrating this equation from from $W/2 - \epsilon$ to $W/2 + \epsilon$, with $\epsilon \to 0$, we get

$$\phi'\left(\frac{W}{2}^+\right) - \phi'\left(\frac{W}{2}^-\right) = -\frac{2m^*\Gamma}{\hbar^2}\phi\left(\frac{W}{2}\right).\tag{3.102}$$

Using the expressions of the wave functions to the left and right of the delta scatterer
given above, this last equation becomes

$$-kB\cos(kW/2) - kA\cos(kW/2) = -\frac{2m^*\Gamma}{\hbar^2}A\sin(kW/2).\tag{3.103}$$

Since $A = B$, we get

$$\cos\left(\frac{kW}{2}\right) = \frac{m^*\Gamma}{k\hbar^2}\sin\left(\frac{kW}{2}\right). \tag{3.104}$$

The bound state energies for $E > 0$ are therefore solutions of the transcendental equation

$$\tan\left(\frac{kW}{2}\right) = \frac{\frac{kW}{2}}{\frac{k_0 W}{2}}, \tag{3.105}$$

where we have introduced the quantity

$$k_0 = \frac{m^*\Gamma}{\hbar^2}. \tag{3.106}$$

The solutions of this transcendental equation must be obtained graphically (see the list of suggested problems). Since the trigonometric function tan has multiple branches, there are many bound states as long as the straight line given by the right-hand side of Equation (3.105) intersects those branches. There is always an intersection with the lowest branch for $k = 0$, which corresponds to a bound state energy with zero energy. This special case is examined further in Problem 3.11.

Case b: Bound state with negative energy: We seek solutions to the bound state problem with $E = -|E| < 0$. In this case, since E is negative, k would be imaginary and hence the wave functions are not described by sine functions.

For $z < \frac{W}{2}$, we must have

$$\phi(z) = Ae^{\kappa z} + Be^{-\kappa z}. \tag{3.107}$$

For $z > \frac{W}{2}$, we must have

$$\phi(z) = Ce^{\kappa z} + De^{-\kappa z}, \tag{3.108}$$

with $\kappa = \frac{1}{\hbar}\sqrt{2m|E|}$.

Continuity of $\phi(z)$ at $z = \frac{W}{2}$ requires

$$Ae^{\frac{\kappa W}{2}} + Be^{\frac{-\kappa W}{2}} = Ce^{\frac{\kappa W}{2}} + De^{\frac{-\kappa W}{2}}. \tag{3.109}$$

Equation (3.102) implies

$$\left(\kappa Ce^{\frac{\kappa W}{2}} - \kappa De^{\frac{-\kappa W}{2}}\right) - \left(\kappa Ae^{\frac{\kappa W}{2}} - \kappa Be^{\frac{-\kappa W}{2}}\right)$$
$$= \frac{-2m\Gamma}{\hbar^2}\left(Ae^{\frac{\kappa W}{2}} + Be^{\frac{-\kappa W}{2}}\right). \tag{3.110}$$

In addition, we must have

$$\phi(z = 0) = A + B = 0 \tag{3.111}$$

and

$$\phi(z = W) = Ce^{\kappa W} + De^{-\kappa W} = 0. \tag{3.112}$$

These last four equations give us four equations with four unknowns (A, B, C, D).

We eliminate B and C by rewriting Equations (3.111) and (3.112) as follows:

$$B = -A \tag{3.113}$$

and

$$D = -Ce^{2\kappa W}. \tag{3.114}$$

Plugging these last relations into Equation (3.109) leads to

$$A = C\left(\frac{1 - e^{\kappa W}}{1 - e^{-\kappa W}}\right), \tag{3.115}$$

and Equation (3.110) becomes

$$\kappa C e^{\frac{\kappa W}{2}} + \kappa C e^{\frac{3\kappa W}{2}} = A\left[\kappa\left(e^{\frac{\kappa W}{2}} + e^{\frac{-\kappa W}{2}}\right) - \frac{2m\Gamma}{\hbar^2}\left(e^{\frac{\kappa W}{2}} - e^{\frac{-\kappa W}{2}}\right)\right]. \tag{3.116}$$

Finally, using Equation (3.115), we get

$$\kappa e^{\frac{\kappa W}{2}} C\left(1 + e^{\kappa W}\right) = C\left(\frac{1 - e^{\kappa W}}{1 - e^{-\kappa W}}\right) e^{\frac{\kappa W}{2}}$$
$$\times \left[\kappa\left(1 + e^{-\kappa W}\right) - \frac{2m\Gamma}{\hbar^2}\left(1 - e^{-\kappa W}\right)\right], \tag{3.117}$$

which simplifies to

$$\kappa\left(1 + e^{\kappa W}\right) = \left(e^{-\kappa W} - 1\right)\left(\kappa - \frac{2m\Gamma}{\hbar^2}\right). \tag{3.118}$$

A few extra steps finally lead to the transcendental equation for the bound state with negative energy:

$$\kappa\left[1 + \frac{1}{\tanh\left(\frac{\kappa W}{2}\right)}\right] = \frac{2m^*\Gamma}{\hbar^2}. \tag{3.119}$$

For large values of W, the solution of this equation is $\kappa = \frac{m^*\Gamma}{\hbar^2}$, implying that the magnitude of the energy of an attractive delta scatterer is equal to $\frac{m^*\Gamma^2}{2\hbar^2}$, as derived in Problem 3.1.

An explicit solution of the transcendental Equation (3.119) is considered in a suggested problem at the end of the chapter.

** Problem 3.7: Bound states of a finite quantum well or a one-dimensional square well potential of finite depth [4–7]

This problem is frequently dealt with in most entry-level classes in quantum mechanics and is covered here for the sake of a comparison with a solution to the same problem obtained using the transfer matrix technique to be discussed in Chapter 7. The purpose is to show that the transfer matrix approach leads to a less cumbersome derivation of the final results.

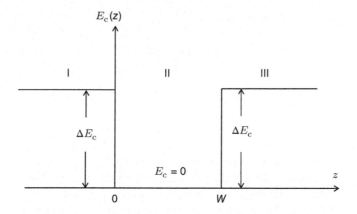

Figure 3.4: A one-dimensional potential well with $E_c(z)$ equal to zero in the interval $[0, W]$ and ΔE_c outside.

Find the bound state energies of the one-dimensional potential well of depth ΔE_c and width W as shown in Figure 3.4. Use the bottom of the well as the zero of energy. Assume a constant effective mass m^* of the electron throughout.

Solution: We seek solutions of the Schrödinger equation for all z with $E < \Delta E_c$. The solution of the Schrödinger equation in region I ($z < 0$) which vanishes at $-\infty$ (i.e., corresponds to a bound state) is given by

$$\phi_I(z) = Ae^{-\kappa z}, \qquad (3.120)$$

with $\kappa = \frac{1}{\hbar}\sqrt{2m^*(\Delta E_c - E)}$. In region II ($z < W$), it is

$$\phi_{II}(z) = B\sin(k_0 z) + C\cos(k_0 z), \qquad (3.121)$$

with $k_0 = \frac{1}{\hbar}\sqrt{2m^*E}$, and, in region III, it is

$$\phi_{III}(z) = De^{-\kappa z}. \qquad (3.122)$$

Matching ϕ and $\frac{d\phi}{dz}$ at $z = 0$ and $z = W$, we obtain the four equations

$$A = C,$$
$$A\kappa = k_0 B,$$
$$B\sin(k_0 W) + C\cos(k_0 W) = De^{-\kappa W},$$
$$Bk_0\cos(k_0 W) - k_0 C\sin(k_0 W) = -D\kappa e^{-\kappa W}. \qquad (3.123)$$

Using the last two equations, we get

$$B\sin(k_0 W) + C\cos(k_0 W) = -B\frac{k_0}{\kappa}\cos(k_0 W) + \frac{k_0}{\kappa}C\sin(k_0 W). \qquad (3.124)$$

Expressing C as a function of B using the preceding equations, we get

$$B\sin(k_0 W) + 2\left(\frac{k_0}{\kappa}\right)B\cos(k_0 W) = \left(\frac{k_0}{\kappa}\right)^2\sin(k_0 W)B. \qquad (3.125)$$

Multiplying both sides by $\frac{\kappa}{k_0 B}$, we finally obtain

$$2\cos(k_0 W) + \left[\frac{\kappa}{k_0} - \frac{k_0}{\kappa}\right]\sin(k_0 W) = 0, \tag{3.126}$$

or

$$2\cot(k_0 W) = \left(\frac{k_0}{\kappa} - \frac{\kappa}{k_0}\right). \tag{3.127}$$

This is a transcendental equation that must be solved numerically for E. The bound state energies are the solutions of this equation which satisfy the condition $E < \Delta E_c$. The problem of finding the bound state of a symmetric square well is often solved by considering solutions which are either odd or even with respect to the center of the well, since, according to the results of Problem 3.5, this problem has non-degenerate solutions with even and odd parity. The transcendental Equation (3.127) covers both types of solution. The equivalence of the two approaches has been analyzed extensively in the literature (see list of references).

** **Problem 3.8: Approximate solution for the lowest bound state of a finite one-dimensional quantum well, i.e., a well with finite barrier heights [8]**

In the previous problem, we showed that finding the bound state energy levels in a one-dimensional square well potential of finite depth requires the solution of a transcendental equation. Hereafter, a simple physical argument is given to obtain the approximate ground state energy of a finite square well starting with the well-known results for the one-dimensional particle in a box.

First, calculate the penetration depth of the wave function in one of the barriers on the side of the well as follows:

$$\delta = N^2 \int_0^\infty dz z e^{-2\kappa z}, \tag{3.128}$$

where N is the normalization factor required to normalize the wave function. Calculate N assuming that the well width is negligible and use this result to find δ. The latter is an estimate of the penetration depth of the wave function in the regions outside the well.

Next, calculate the energy of the ground state by using the particle in a box result and replacing the well width W by $W + 2\delta$, where 2δ accounts for the decay of the wave function on either side of the well.

Solution: Neglecting the thickness of the quantum well, the normalization coefficient is found from

$$1 = N^2 \int_{-\infty}^0 e^{2\kappa z} dz + N^2 \int_0^\infty e^{-2\kappa z} dz = 2N^2 \int_0^\infty e^{-2\kappa z} dz = \frac{N^2}{\kappa}, \tag{3.129}$$

leading to $N^2 = \kappa$, with $\kappa = \frac{1}{\hbar}\sqrt{2m^*(\Delta E_c - E)}$. This leads to $\delta = \frac{N^2}{4\kappa^2} = \frac{1}{4\kappa}$.

Using the expressions for the energy levels of a particle in a box, the energy levels in the well with finite barrier heights are approximated as

$$E_n^* = \frac{n^2 \hbar^2 \pi^2}{2m^*(W + 2\delta)^2} = \frac{n^2 \hbar^2 \pi^2}{2m^* \left(W + \frac{1}{2\kappa}\right)^2}, \tag{3.130}$$

with $\kappa = \frac{1}{\hbar}\sqrt{2m^*(\Delta E_{\mathrm{c}} - E_n^*)}$. The last equation is an equation for E_n^* that must be solved iteratively.

---

* Problem 3.9: Features of the wave function of an eigenstate in a finite one-dimensional quantum well

Someone solved the Schrödinger equation in the one-dimensional potential well shown in Figure 3.5 and plotted the wave function of the fifth excited state (which is the sixth eigenstate) as shown. Explain why this solution cannot be correct even qualitatively. Think of as many reasons as you can.

Solution: There are many errors:

- The amplitude of the wave function is larger to the right, implying that there is a higher probability of finding the electron to the right than to the left. However, the potential is deeper to the left and hence the electron will tend to be localized to the left and the amplitude should decrease as we go from left to right.

- The wave function has more wiggles (or higher-frequency oscillations) to the right. This means that the second derivative of the wave function is larger on the right. Since the kinetic energy is proportional to the second derivative, it appears that the electron will have higher kinetic energy (and hence lower potential energy) on the right since the total energy (kinetic + potential) is

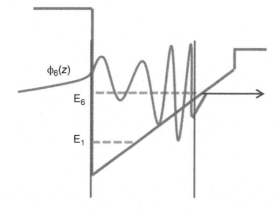

Figure 3.5: Erroneously calculated spatial variation of the wave function of an excited state in a one-dimensional heterostructure.

constant in an eigenstate. However, the potential energy is higher on the right. Hence the wave function should have had more wiggles on the left and fewer on the right.

- The penetration of the wave function into the left barrier is more than that into the right barrier. This is not possible given that the left barrier is taller than the right barrier.

- The first derivative of the wave function is discontinuous at the right, which is not permitted.

- The wave function has seven nodes. The number of nodes of the nth eigenstate should be $n - 1$, which means that the wave function of the sixth eigenstate should have five nodes and not seven.

** Problem 3.10: Zero-energy bound states in a quantum well [9]

Show that the Schrödinger equation for a particle in a box of width W containing a repulsive delta scatterer $\Gamma\delta(z - z_0)$ allows a bound state with zero energy. Derive the relation between the parameters Γ, W, and z_0 for that bound state to exist. Determine the expression of the normalized wave function associated with this bound state.

Solution: A zero-energy state corresponds to a wave function satisfying the Schrödinger equation

$$-\frac{\hbar^2}{2m^*}\phi''(z) - \Gamma\delta(z - z_0)\phi(z) = 0. \tag{3.131}$$

A solution of that zero energy state corresponds to a wave function of the form

$$\phi(z) = \alpha z \text{ for } z < z_0 \tag{3.132}$$

and

$$\phi(z) = \beta(W - z) \text{ for } z > z_0, \tag{3.133}$$

where α and β can both be taken positive.

At $z = z_0$, the wave function must be continuous, and integrating Equation (3.131) from $z = z_0 - \epsilon$ to $z = z_0 + \epsilon$ ($\epsilon \to 0$) leads to

$$-\frac{\hbar^2}{2m}\left[\phi'(z_0{}^+) - \phi'(z_0{}^-)\right] = -2k_0\phi(z_0), \tag{3.134}$$

where we have introduced the quantity $k_0 = \frac{m^*\Gamma}{\hbar^2}$.

This last relation, together with the continuity of the wave function at z_0, leads to the following equations for α and β:

$$\alpha z_0 - \beta(W - z_0) = 0 \tag{3.135}$$

and

$$\alpha(1 - 2k_0 z_0) + \beta = 0. \tag{3.136}$$

We write these coupled equations in matrix form:

$$\begin{bmatrix} z_0 & z_0 - W \\ 1 - 2k_0 z_0 & 1 \end{bmatrix} \begin{bmatrix} \alpha \\ \beta \end{bmatrix} = 0. \tag{3.137}$$

For non-trivial solutions of α and β, the determinant of the 2×2 matrix above should vanish, which leads to the following solution for k_0:

$$k_0 = \frac{W}{2 z_0 (W - z_0)}. \tag{3.138}$$

This establishes the relation between the strength Γ of the delta scatterer and the parameters z_0 and W for the existence of the zero-energy bound state.

The normalization of the associated wave function requires

$$\int_0^{z_0} \alpha^2 z^2 \mathrm{d}z + \int_{z_0}^{W} \beta^2 (W - z)^2 \mathrm{d}z = 1. \tag{3.139}$$

Using Equation (3.135) to relate β to α in the above equation, we get

$$\alpha^2 \left[\int_0^{z_0} z^2 \mathrm{d}z + z_0^2 \int_{z_0}^{W} \mathrm{d}z \right] = 1, \tag{3.140}$$

which yields

$$\alpha = \sqrt{\frac{3}{W}} \frac{1}{z_0},$$

$$\beta = \sqrt{\frac{3}{W}} \frac{1}{W - z_0}. \tag{3.141}$$

The results of this last problem can be used to establish a correspondence between Poisson and Schrödinger equations [10].

***

*** Problem 3.11: Bound states of a quantum well in a semi-infinite space

A particle of mass m^* moves in the one-dimensional conduction band energy profile shown in Figure 3.6. Find the energy of the bound states and their corresponding wave functions. Compare your results to those for a potential well of the form

$$E_c(z) = -\Delta E_c \tag{3.142}$$

for $z < |W|$ and $E_c(z) = 0$ otherwise. Determine the number of bound states as a function of the well parameters ΔE_c and W.

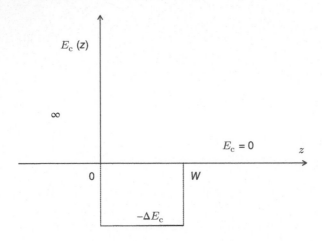

Figure 3.6: A one-dimensional potential well in a semi-infinite space.

Solution: The Schrödinger equation for this problem is

$$\phi''(z) = -\alpha^2\phi(z) \text{ for } 0 < z < W, \tag{3.143}$$

$$\phi''(z) = \beta^2\phi(z) \text{ for } z > W, \text{ and} \tag{3.144}$$

$$\phi(z) = 0 \text{ for } z < 0, \tag{3.145}$$

where the following quantities were introduced:

$$\alpha = \left[\frac{2m^*}{\hbar^2}(E + \Delta E_c)\right]^{1/2}, \tag{3.146}$$

$$\beta = \left(-\frac{2m^*E}{\hbar^2}\right)^{1/2}, \tag{3.147}$$

which are both real for bound states.

Since $\phi(z = 0) = 0$, bound state solutions with $-\Delta E_c < E < 0$ are of the form

$$\phi(z) = A\sin(\alpha z) \text{ for } 0 \leq z \leq W \tag{3.148}$$

and

$$\phi(z) = Be^{-\beta z} \text{ for } z \geq W. \tag{3.149}$$

Enforcing the continuity of $\phi(z)$ and $\phi'(z)$ at $z = W$, we get the following two equations:

$$B = Ae^{\beta W}\sin(\alpha W) \tag{3.150}$$

and

$$\beta B = -Ae^{\beta W}\alpha\cos(\alpha W). \tag{3.151}$$

Introducing the variables $\xi = \alpha W$ and $\eta = \beta W$, we can combine and write the last two equations as

$$
\begin{bmatrix} e^{\eta} \sin \xi & -1 \\ \alpha e^{\eta} \cos \xi & \beta \end{bmatrix} \begin{bmatrix} A \\ B \end{bmatrix} = 0.
\tag{3.152}
$$

The requirement that non-trivial solutions exist for A and B dictates that the determinant of the 2×2 matrix vanishes, which results in the following relation:

$$
\xi \cot \xi = -\eta.
\tag{3.153}
$$

In addition, using Equations (3.146) and (3.147), it is found that the two variables ξ and η must also satisfy the relation

$$
\xi^2 + \eta^2 = 2m^* W^2 \Delta E_c / \hbar^2.
\tag{3.154}
$$

The bound state solutions can be obtained by solving the last two equations simultaneously for η and ξ. One approach for accomplishing this is to plot the last two equations in the (ξ, η) plane.

The intersection(s) of the circle given by Equation (3.154) and the different branches associated with Equation (3.153) correspond to the bound state solutions. The larger the value of $W^2 \Delta E_c$, the larger will be the radius of the circle, leading to more possible bound states.

If $\Delta E_c < \frac{\pi^2 \hbar^2}{8m^* W^2}$, there is no bound state.

If $\frac{\pi^2 \hbar^2}{8m^* W^2} < \Delta E_c < \frac{\pi^2 \hbar^2}{2m^* W^2}$, there will be one bound state only.

It is easy to show that the number of bound states is the largest integer less than

$$
\frac{2}{\pi} \left(\frac{2m^* W^2 \Delta E_c}{\hbar^2} \right)^{1/2}.
\tag{3.155}
$$

The solutions discussed above are identical to the odd solutions of the potential energy profile given by $E_c(z) = -\Delta E_c$ for $-W < z < W$ and $E_c(z) = 0$ otherwise. Indeed, for this problem, the odd solutions satisfy the same equations as in this problem for $z > 0$ with $\phi(0) = 0$.

---

* Problem 3.12: Coupled finite wells

Consider two identical square wells of finite depth separated by a distance $2z_0$, as shown in Figure 3.7. The wells are well separated, but still allow some overlap between the wave functions of the ground states in each well. These are called "coupled wells." The depth of each well is V_0 and the width of each well is W. Let the ground state in each isolated well have an energy E_0. Find the energies of the ground and first excited states in the coupled well system.

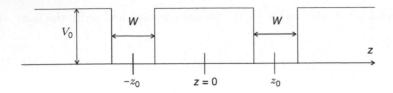

Figure 3.7: Two coupled finite square wells, each of width W.

Solution: The time-independent Schrödinger equation describing the coupled well system is

$$-\frac{\hbar^2}{2m}\frac{d^2\psi(z)}{dz^2} + V(z)\psi(z) = E\psi(z). \qquad (3.156)$$

The potential $V(z)$ can be written as the sum of the potentials associated with each well:

$$V(z) = V_1(z) + V_2(z). \qquad (3.157)$$

Since the wells are well separated, the coupling is weak and the wave function of the lowest energy state can be written as a linear superposition of the ground state wave functions in each isolated well:

$$\psi(z) = C_1\phi_1(z) + C_2\phi_2(z), \qquad (3.158)$$

where the ϕ_i are the ground state wave functions in isolated wells 1 and 2.

Substituting Equation (3.158) in Equation (3.156), we get

$$C_1\left[-\frac{\hbar^2}{2m}\frac{d^2\phi_1(z)}{dz^2} + V_1(z)\phi_1(z)\right] + C_2 V_2(z)\phi_2(z)$$
$$+ C_2\left[-\frac{\hbar^2}{2m}\frac{d^2\phi_2(z)}{dz^2} + V_2(z)\phi_2(z)\right] + C_1 V_1(z)\phi_1(z)$$
$$= E\left[C_1\phi_1(z) + C_2\phi_2(z)\right]. \qquad (3.159)$$

Note that the term within the first square bracket on the left-hand side is nothing but $E_1\phi_1(z)$, while the term within the second square bracket is $E_2\phi_2(z)$, where E_1 and E_2 are the ground states in the two isolated wells.

Therefore, the last equation can be written as

$$C_1 E_1\phi_1(z) + C_2 V_2(z)\phi_2(z) + C_2 E_2\phi_2(z) + C_1 V_1(z)\phi_1(z)$$
$$= E\left[C_1\phi_1(z) + C_2\phi_2(z)\right]. \qquad (3.160)$$

If we multiply the last equation throughout by $\phi_1^*(z)$ and integrate over all z from $-\infty$ to $+\infty$, we will get

$$C_1 E_1 + C_1 A + C_2 E_2 B + C_2 D = EC_1 + EC_2 B, \qquad (3.161)$$

where we used the fact that the wave function in normalized, i.e., $\int_{-\infty}^{\infty} \phi_1^*(z)$ $\phi_1(z)\mathrm{d}z = 1$, and

$$A = \int_{-\infty}^{\infty} \phi_1^*(z)V_2(z)\phi_1(z)\mathrm{d}z,$$

$$B = \int_{-\infty}^{\infty} \phi_1^*(z)\phi_2(z)\mathrm{d}z,$$

$$D = \int_{-\infty}^{\infty} \phi_1^*(z)V_1(z)\phi_2(z)\mathrm{d}z. \tag{3.162}$$

It is easy to see that if we multiply Equation (3.160) throughout by $\phi_2^*(z)$ instead of $\phi_1^*(z)$ and integrate over all z from $-\infty$ to $+\infty$, we will get

$$C_2 E_2 + C_2 A' + C_1 E_1 B' + C_1 D' = EC_2 + EC_1 B', \tag{3.163}$$

where

$$A' = \int_{-\infty}^{\infty} \phi_2^*(z)V_1(z)\phi_2(z)\mathrm{d}z,$$

$$B' = \int_{-\infty}^{\infty} \phi_2^*(z)\phi_1(z)\mathrm{d}z,$$

$$D' = \int_{-\infty}^{\infty} \phi_2^*(z)V_2(z)\phi_1(z)\mathrm{d}z. \tag{3.164}$$

Since the two wells are *identical*, $A = A'$, $B = B'$, and $D = D'$. Furthermore, $E_1 = E_2 = E_0$.

We can rewrite Equations (3.161) and (3.163) in matrix form:

$$\begin{bmatrix} E_0 - E + A & B(E_0 - E) + D \\ B(E_0 - E) + D & E_0 - E + A \end{bmatrix} \begin{bmatrix} C_1 \\ C_2 \end{bmatrix} = 0. \tag{3.165}$$

For non-trivial solutions of C_1 and C_2, the determinant of the 2×2 matrix must vanish; hence, we obtain

$$[E_0 - E + A]^2 = [B(E - E_0) + D]^2, \tag{3.166}$$

or

$$E = \begin{cases} E_0 + \frac{A-D}{1-B} & \text{(ground state)} \\ E_0 + \frac{A-D}{1-B} & \text{(first excited state).} \end{cases} \tag{3.167}$$

We can calculate the quantities A, B, and D. The wave functions in the isolated wells can be written as (check that they are normalized)

$$\phi_1(z) = \begin{cases} 2\sqrt{\kappa}e^{\kappa(z+z_0)} & (z < -z_0) \\ 2\sqrt{\kappa}e^{-\kappa(z+z_0)} & (z > -z_0) \end{cases}$$

$$\phi_2(z) = \begin{cases} 2\sqrt{\kappa}e^{\kappa(z-z_0)} & (z < z_0) \\ 2\sqrt{\kappa}e^{-\kappa(z-z_0)} & (z > z_0). \end{cases} \tag{3.168}$$

The quantity κ can be approximately related to E_0 (see Problem 3.9) as

$$W + \kappa/2 = \frac{\pi\hbar}{\sqrt{2mE_0}}. \tag{3.169}$$

With these wave functions, the integrals in Equation (3.162) can be evaluated; this is left as an exercise for the reader.

---

* Problem 3.13: Solution to the Schrödinger equation in a triangular well

Calculate the energy levels of an electron in a one-dimensional triangular well due to a constant electric field present for $z > 0$ (see Figure 3.8). Assume that $E_c(z) = \infty$ for $z < 0$.

Solution: We start with the one-dimensional Schrödinger equation

$$-\frac{\hbar^2}{2m^*}\ddot{\phi}(z) + E_c(z)\phi(z) = E\phi(z), \tag{3.170}$$

with $E_c(z) = qE_{el}z$, where E_{el} is the constant electric field for $z > 0$ and q is the magnitude of the charge of the electron.

We first rewrite the Schrödinger equation as

$$\ddot{\phi}(z) - \frac{2m^*}{\hbar^2}(qE_{el}z - E)\phi(z) = 0. \tag{3.171}$$

Next, we introduce the new variable ξ such that

$$C\xi = \frac{2m^*}{\hbar^2}(qE_{el}z - E), \tag{3.172}$$

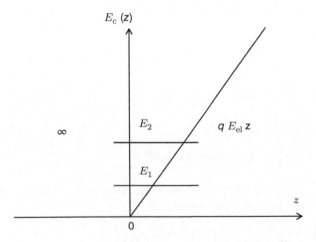

Figure 3.8: A triangular quantum well formed by a constant electric field E_{el} in the region $z > 0$. $E_c(z) = \infty$ for negative z. The quantity q is the magnitude of the charge of the electron.

where C is an undetermined constant. Making use of this substitution, the Schrödinger equation becomes

$$\frac{d^2}{d\xi^2}\phi(\xi) - C^3 \left(\frac{2m^*qE_{el}}{\hbar^2}\right)^{-2} \xi\phi(\xi) = 0. \tag{3.173}$$

If we select

$$C = \left(\frac{2m^*qE_{el}}{\hbar^2}\right)^{2/3}, \tag{3.174}$$

the Schrödinger equation reduces to

$$\frac{d^2}{d\xi^2}\phi(\xi) - \xi\phi(\xi) = 0. \tag{3.175}$$

The solution of this equation which does not blow up at infinity is proportional to the Airy function Ai($-\xi$) [11]. If we further require that the wave function must be zero at $z = 0$, then a solution to the Schrödinger equation exists only if Ai($\xi(z = 0)$) = 0. Because of our choice of C, we have

$$\xi = \left(\frac{2m^*qE_{el}}{\hbar^2}\right)^{1/3} \left(z - \frac{E}{qE_{el}}\right). \tag{3.176}$$

The first zero of Ai($-\xi$) is for $-\xi = -2.41$, which leads to the following relation for the energy of the ground state in the triangular well:

$$E_1 = 2.34 \left(\frac{\hbar^2 q^2 E_{el}^2}{2m^*}\right)^{1/3}. \tag{3.177}$$

The next two energy levels are found using the next two zeroes of the Airy function:

$$E_2 = 4.09 \left(\frac{\hbar^2 q^2 E_{el}^2}{2m^*}\right)^{1/3} \tag{3.178}$$

and

$$E_3 = 5.52 \left(\frac{\hbar^2 q^2 E_{el}^2}{2m^*}\right)^{1/3}. \tag{3.179}$$

Matlab code to generate the lowest energy levels in the triangular well is given in Appendix G.

---

* Problem 3.14: Degeneracy in a two-dimensional electron gas

Consider a rectangular two-dimensional electron gas as shown in Figure 3.9. What should be the relation between L_x and and L_y to guarantee that no two energy eigenvalues are degenerate? Assume hard-wall boundary conditions, i.e., the potential barriers confining the two-dimensional electron gas on all sides are infinitely high.

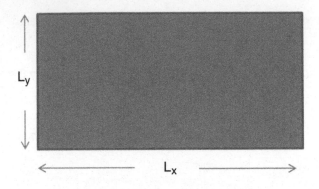

Figure 3.9: The two-dimensional particle in a box. $E_c(x,y) = 0$ in a box of size $L_x \times L_y$ and equal to ∞ outside the box.

Solution: The energy eigenstates are $E_{m,n} = \frac{\hbar^2}{2m^*} \left[\left(\frac{m\pi}{L_x} \right)^2 + \left(\frac{n\pi}{L_y} \right)^2 \right]$. Therefore, degeneracies occur when

$$\left(\frac{m\pi}{L_x} \right)^2 + \left(\frac{n\pi}{L_y} \right)^2 = \left(\frac{m'\pi}{L_x} \right)^2 + \left(\frac{n'\pi}{L_y} \right)^2. \tag{3.180}$$

We recast the last equation as follows:

$$\frac{L_x}{L_y} = \sqrt{\frac{(m+m')(m-m')}{(n+n')(n-n')}} = \sqrt{\frac{MM'}{NN'}}, \tag{3.181}$$

i.e., $L_x^2/L_y^2 = P/Q$, where P and Q are integers.

Therefore, to eliminate all degeneracies, we have to ensure that the ratio L_x^2/L_y^2 cannot be expressed as a rational number. For instance, one can choose $L_x^2/L_y^2 = \sqrt{2}$ (which is an irrational number), leading to $L_x/L_y = 2^{1/4}$.

** Problem 3.15: Subband population in a cylindrical quantum wire

Consider a cylindrical InSb nanowire (or quantum wire) of diameter 50 nm. The effective mass of electrons in InSb (m*) is 0.0145 times the free electron mass of 9.1×10^{-31} kg. Find the energies of the three lowest eigenstates (also known as subbands) in the conduction band of the nanowire. What fraction of the electrons in the wire occupies the lowest subband at room temperature if the Fermi level is $100\,\mu eV$ above the lowest subband?

Solution: We have to first solve the time-independent Schrödinger equation in cylindrical coordinates:

$$-\frac{\hbar^2}{2m^*}\left[\frac{1}{r}\frac{\partial}{\partial r}\left(r\frac{\partial \psi(r,\theta,z)}{\partial r}\right)+\frac{1}{r^2}\frac{\partial^2 \psi(r,\theta,z)}{\partial \theta^2}+\frac{\partial^2 \psi(r,\theta,z)}{\partial z^2}\right]$$

$$+V(r,\theta,z)\psi(r,\theta,z)=E\psi(r,\theta,z), \tag{3.182}$$

where $\psi(r,\theta,z)$ is the wave function, r is the radial coordinate, θ is the polar coordinate, and z is the axial coordinate (along the nanowire axis).

The solution of this equation subject to the condition $V(r,\theta,z)=0$ for $r\leq D/2$ and $V(r,\theta,z)=\infty$ for $r>D/2$ (D is the wire diameter) is [12]

$$\psi_{m,n,k_z}(r,\theta,z)=NJ_m(k_{m,n}r)e^{im\theta}e^{ik_z z}, \tag{3.183}$$

where J_m is the Bessel function of the mth order, N is a normalization constant, and k_z is the wave vector along the nanowire axis. The energy eigenstates are given by

$$E_{m,n}=\frac{\hbar^2}{2m^*}\left(k_{m,n}^{\;2}+k_z^{\;2}\right). \tag{3.184}$$

The boundary condition dictates that the wave function vanishes at $r=D/2$. This means that we will have to set $k_{0,n}D/2$ equal to the zeros of the Bessel function for $m=0$ to find the allowed values of $k_{0,n}$. For $D=50\,\mathrm{nm}$, we get

$$k_{0,1}=9.6\times 10^7\,\mathrm{m}^{-1}, \tag{3.185}$$
$$k_{0,2}=2.2\times 10^8\,\mathrm{m}^{-1}, \tag{3.186}$$

and

$$k_{0,3}=3.5\times 10^8\,\mathrm{m}^{-1}. \tag{3.187}$$

The subband energy bottoms are found by setting $k_z=0$ in Equation (3.184). This yields the energies of the three lowest subbands as

$$E_{0,1}=\frac{\hbar^2 k_{0,1}^{\;2}}{2m^*}=0.9\,k_\mathrm{B}T, \tag{3.188}$$
$$E_{0,2}=\frac{\hbar^2 k_{0,2}^{\;2}}{2m^*}=4.8\,k_\mathrm{B}T, \tag{3.189}$$

and

$$E_{0,3}=\frac{\hbar^2 k_{0,3}^{\;2}}{2m^*}=12.1\,k_\mathrm{B}T, \tag{3.190}$$

where $k_\mathrm{B}T$ is the thermal energy at room temperature, i.e., 4.186×10^{-21} joules.

The probability of an electron occupying any energy state (at equilibrium) is given by the Fermi–Dirac occupation probability $1/[e^{(E-E_\mathrm{F})/k_\mathrm{B}T}+1]$, where E_F is the Fermi energy, which is $100\,\mu$eV or $3.8 \times 10^{-3}\,k_\mathrm{B}T$ above the lowest subband energy.

The probabilities of finding electrons in the first three energy levels are determined by the Fermi level placement, which yields

$$E_{0,1} - E_\mathrm{F} = -3.8 \times 10^{-3}\,k_\mathrm{B}T, \tag{3.191}$$
$$E_{0,2} - E_\mathrm{F} = 3.8962\,k_\mathrm{B}T, \tag{3.192}$$

and

$$E_{0,3} - E_\mathrm{F} = 11.1962\,k_\mathrm{B}T. \tag{3.193}$$

The corresponding occupation probabilities are given by

$$p(E_{0,1}) = \frac{1}{e^{-3.8\times 10^{-3}} + 1} = 0.5, \tag{3.194}$$

$$p(E_{0,2}) = \frac{1}{e^{3.896} + 1} = 0.02, \tag{3.195}$$

and

$$p(E_{0,3}) = \frac{1}{e^{11.196} + 1} = 1.37 \times 10^{-5}. \tag{3.196}$$

We are interested only in the electrons in the conduction band and seek to find the relative subband populations in the conduction band alone. Therefore, the fraction of electrons in the lowest subband is given by (neglecting the contributions of the fourth and higher subbands):

$$f(E_{0,1}) = \frac{p(E_{0,1})}{p(E_{0,1}) + p(E_{0,2}) + p(E_{0,3})} = 0.96. \tag{3.197}$$

Therefore, 96% of the electrons in the InSb nanowire occupy the lowest subband at room temperature.

Suggested problems

- Consider an electron of effective mass m^* trapped in an infinite one-dimensional potential well, i.e., $E_\mathrm{c}(z) = \infty$ for $|z| \geq W$ and $E_\mathrm{c}(z) = 0$ for $|z| \leq W$.

  (1) Derive the expressions for the energy eigenstates in momentum space. Consider the cases of even and odd solutions separately.

  (2) Using the previous results, derive analytical expressions for the momentum probability distributions for both even and odd eigenstates. Make sure these expressions are normalized.

  (3) Determine the location of the maxima in the momentum probability distribution functions of the even and odd eigenstates. Show that the values of the momenta associated with these maxima are solutions of transcendental equations. Write the transcendental equations for both even and odd eigenstates.

- Using the results of Problem 3.5, find the lowest bound state with positive energy inside a quantum well of width $100\,\text{Å}$ containing an attractive delta scatterer in its center whose strength is such that $\frac{k_0 W}{2} = \frac{m^* W \Gamma}{2\hbar^2} = \pi/2$. How does that energy compare with the ground state energy of the well without the delta scatterer? Plot the probability densities of the ground state of the particle in a box, with and without the delta scatterer.

- Starting with the results of Problem 3.5, show that the lowest symmetric bound state, i.e., with no node in the wave function, with non-zero positive Γ exists only if the following condition is satisfied for the strength of the attractive delta scatterer:

$$\frac{m^* \Gamma^2}{\hbar^2} > \frac{8}{\pi^2} E_1, \qquad (3.198)$$

  where E_1 is the ground state energy of the particle in a box problem.

- Write Matlab code to calculate the lowest positive bound state energy level of a box containing an attractive one-dimensional delta scatterer of strength $\Gamma = 5\,\text{eV-Å}$ located in the middle of a well as a function of the well width W, for W varying from 50 to $200\,\text{Å}$.

- Repeat Problem 3.5 if the attractive delta scatterer is located at an arbitrary z_0 value between 0 and W. Derive the transcendental equation that must be solved to find the lowest bound states with negative and positive energy.

- Consider a repulsive one-dimensional delta scatterer with (positive) strength Γ_2 located at a distance L from an attractive scatterer $-\Gamma_1 \delta(z)$ (positive Γ_1). Study how the bound state energy of the attractive scatterer is affected by the presence of the repulsive scatterer. Write Matlab code to study the bound state energy dependence on the distance L and the strength Γ_2 of the repulsive scatterer. Does a bound state exist for $\Gamma_2 = -\Gamma_1$ independent of the length L?

- Use the results of Problem 3.7 to calculate the energies of the two lowest bound states in a quantum well of depth $0.3\,\text{eV}$ and width $50\,\text{Å}$. Assume a constant effective mass $m^* = 0.067 m_0$ throughout. Compare the values with the result obtained by solving the transcendental Equation (3.130) derived in Problem 3.9.

- Consider the particle in a box problem shown in Figure 3.10 and assume that the electron has a constant effective mass.

  (1) Write the Schrödinger equation in regions I and II. Consider the case where the total energy of the electron is either below or above the height of the step ΔE_c.

  (2) What boundary conditions must be satified at $z = 0$, $z = L/2$, and $z = L$?

  (3) Using the above results, write down the set of simultaneous equations that must be solved to find the coefficients in the solutions to the Schrödinger equation (i.e. the expressions for the wave functions) in regions I and II.

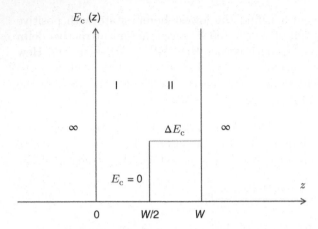

Figure 3.10: The bound state energy $-E^*$ of a one-dimensional delta scatter located at a heterointerface depends on the magnitude of the potential jump ΔE_c.

References

[1] Kroemer, H. (1994) *Quantum Mechanics for Engineering, Materials Science, and Applied Physics*, Chapter 2, Prentice Hall, Englewood Cliffs, NJ.

[2] Lapidus, I. R. (1982) One-dimensional hydrogen atom in an infinite square well. *American Journal of Physics* 50, pp. 563–564.

[3] Gettys, W. E. (1972) Quantum theory of a square well plus δ-function potential. *American Journal of Physics* 41, pp. 670–677.

[4] Bloch, J.-F. (2001) A new approach to bound states in potential wells. *American Journal of Physics* 69, pp. 1177–1181.

[5] Sprung, D. W. L., Wu, H., and Martorell, J. (1992) A new look at the square well potential. *Eur. J. Phys.* 13, pp. 21–25.

[6] Cameron Reed, B. (1990) A single equation for finite rectangular well energy eigenvalues. *American Journal of Physics* 58, pp. 503–504.

[7] Pitkanen, P. H. (1955) Rectangular potential well problem in quantum mechanics. *American Journal of Physics* 23, pp. 111–113.

[8] Garrett, S. (1979) Bound state energies of a particle in a finite square well: a simple approximation. *American Journal of Physics* 47, pp. 195–196.

[9] Gilbert, L. P., Belloni, M., Doncheski, M. A., and Robinett, R. W., Playing Quantum Physics Jeopardy with zero-energy states. *American Journal of Physics* 74, pp. 1035–1036.

[10] González, G. (2012) Relation between Poisson and Schrödinger equation. *American Journal of Physics*, 80(8), pp. 715–719.

[11] Vallée, O. and Soares, M. (2004) *Airy Functions and Applications To Physics*, Imperial College Press, London.

[12] Sercel, P. C. and and Vahala, K. J. (1991) Polarization dependence of optical absorption and emission in quantum wires. *Physical Review B* 44, 5681.

Suggested Reading

- Sedrakian, D. M. and Khachatrian, A. Zh. (2003) Determination of bound state energies for a one-dimensional potential field. *Physica E* 19, pp. 309–315.

- Mavromatis, H. A. (2000) Comment on "A single equation for finite rectangular well energy problems," *American Journal of Physics* 68, pp. 1151–1152.

- Barker, B. I., Rayborn, G. H., Ioup, J. W., and Ioup, G. E. (1991) Approximating the finite square well in an infinite well: energies and eigenfunctions. *American Journal of Physics* 59, pp. 1038–1042.

- Sprung, D. W. L., Wu, H., and Martorell, J. (1996) Poles, bound states, and resonances illustrated by the square well potential. *American Journal of Physics* 64, pp. 136–144.

- Buell, W. F. and Shadwick, B. A. (1994) Potentials and bound states. *American Journal of Physics* 63, pp. 256–258.

- Ganguly, A., Kuru, S., Negro, J., and Nieto, L. M. (2006) A study of the bound states for square wells with position-dependent mass. *Phys. Lett. A* 360, pp. 228–233.

- Johannin-Gilles, A., Abjean, R., Guern, Y., and Leriche, M. (1973) Maximum of the probability distribution of momenta in an infinite square well potential. *American Journal of Physics* 42, pp. 701–702.

This chapter starts with three different proofs of the generalized Heisenberg uncertainty relations, followed by illustrations of their application to the study of some bound state and scattering problems, including diffraction from a slit in a screen and quantum mechanical tunneling through a potential barrier.

** Problem 4.1: The Heisenberg uncertainty principle

For any two Hermitian operators that do not commute, the standard deviations of the observables A and B obey the following generalized Heisenberg uncertainty relation:

$$\Delta A \Delta B \geq \frac{1}{2} |\langle [A, B] \rangle|, \tag{4.1}$$

where $\Delta A = \sqrt{\langle A^2 \rangle - \langle A \rangle^2}$ and $\Delta B = \sqrt{\langle B^2 \rangle - \langle B \rangle^2}$ are the standard deviations of A and B, respectively; the averages (or expectation values) are taken over the state $|\phi\rangle$ of the system, and [A,B] is the commutator of the two operators A and B.

Solution: We give three separate proofs of the inequality in (4.1).

(a) First proof [1]

This proof is based on the following lemma:

Preliminary: Show that for any operator A, $\langle A^\dagger A \rangle \geq 0$.

Indeed, $\langle A^\dagger A \rangle = \int \phi^* A^\dagger A \phi \, d^3\vec{r} = \int (A\phi)^* A\phi \, d^3\vec{r} = \int |A\phi|^2 d^3\vec{r} \geq 0$.

Next, consider the operator $A + i\lambda B$, where λ is a real number. Based on the previous result, we can conclude that the function

$$\begin{aligned} f(\lambda) &= \langle (A + i\lambda B)^\dagger (A + i\lambda B) \rangle \\ &= \langle A^\dagger A \rangle + i\lambda \langle A^\dagger B \rangle - i\lambda \langle B^\dagger A \rangle + \lambda^2 \langle B^\dagger B \rangle \\ &\geq 0. \end{aligned} \tag{4.2}$$

Since A and B are Hermitian operators, $A^\dagger = A$ and $B^\dagger = B$. Hence,

$$f(\lambda) = \langle A^2 \rangle + \lambda^2 \langle B^2 \rangle + i\lambda \langle [A, B] \rangle \geq 0. \tag{4.3}$$

By differentiating the above function with respect to λ and setting the result equal to zero, we can show that the minimum of $f(\lambda)$ occurs for

$$\lambda_{\min} = -\frac{i}{2} \frac{\langle [A, B] \rangle}{\langle B^2 \rangle}. \tag{4.4}$$

Problem Solving in Quantum Mechanics: From Basics to Real-World Applications for Materials Scientists, Applied Physicists, and Devices Engineers, First Edition.
Marc Cahay and Supriyo Bandyopadhyay.

Note that the expectation value of the commutator of Hermitian operators, such as $\langle [A, B] \rangle$, is typically imaginary, and hence $\lambda_{\min}$ is real.

Substituting this value into Equation (4.3) leads to

$$f(\lambda_{\min}) = \langle A^2 \rangle - \frac{1}{4} \frac{(\langle [A, B] \rangle)^2}{\langle B^2 \rangle} + \frac{1}{2} \frac{(\langle [A, B] \rangle)^2}{\langle B^2 \rangle} \geq 0. \tag{4.5}$$

Hence,

$$\langle A^2 \rangle \langle B^2 \rangle \geq -\frac{1}{4} \langle [A, B] \rangle^2. \tag{4.6}$$

Note again that the right-hand side of the above inequality is a positive quantity since $\langle [A, B] \rangle$ is imaginary.

Next, we define the new Hermitian operators

$$\delta A = A - \langle A \rangle I \tag{4.7}$$

and

$$\delta B = B - \langle B \rangle I, \tag{4.8}$$

where I is the identity operator. Then, since the average or expectation value of an operator is a number and its product with the identity matrix commutes with any operator, we get that

$$[\delta A, \delta B] = AB - BA = [A, B]. \tag{4.9}$$

Hence,

$$\langle [\delta A, \delta B] \rangle = \langle [A, B] \rangle. \tag{4.10}$$

From Equation (4.6), we get

$$\langle (\delta A)^2 \rangle \langle (\delta B)^2 \rangle \geq -\frac{1}{4} \langle [\delta A, \delta B] \rangle^2. \tag{4.11}$$

It is easy to show that $\langle (\delta A)^2 \rangle = (\Delta A)^2$ and $\langle (\delta B)^2 \rangle = (\Delta B)^2$. As a result,

$$(\Delta A)^2 (\Delta B)^2 \geq -\frac{1}{4} \langle [A, B] \rangle^2. \tag{4.12}$$

Finally, since $\langle A, B \rangle$ is imaginary, its square is a negative quantity and hence we get

$$(\Delta A)^2 (\Delta B)^2 \geq \frac{1}{4} |\langle [A, B] \rangle|^2, \tag{4.13}$$

or

$$\Delta A \Delta B \geq \frac{1}{2} |\langle [A, B] \rangle|. \tag{4.14}$$

(b) Second proof [2]

For any two Hermitian operators A, B associated with physical observables, the average value of the product AB in a quantum state $|\phi\rangle$ will be some complex number $\langle\phi|AB|\phi\rangle = x + iy$, where x, y are real.

Since AB is Hermitian, $\langle\phi|BA|\phi\rangle = x - iy$ and the following equalities hold:

$$\langle\phi|[A,B]|\phi\rangle = \langle\phi|AB|\phi\rangle - \langle\phi|BA|\phi\rangle = (x+iy) - (x-iy) = 2iy \qquad (4.15)$$

and

$$\langle\phi|\{A,B\}|\phi\rangle = \langle\phi|AB|\phi\rangle + \langle\phi|BA|\phi\rangle = (x+iy) + (x-iy) = 2x, \qquad (4.16)$$

where the square brackets denote commutator and the curly brackets denote anti-commutator.

Hence,

$$|\langle\phi|[A,B]|\phi\rangle|^2 + |\langle\phi|\{A,B\}|\phi\rangle|^2 = 4(x^2 + y^2)$$
$$= 4\,|\langle\phi|AB|\phi\rangle|^2 = 4\,|\langle\phi|BA|\phi\rangle|^2, \qquad (4.17)$$

where we have used the fact that $\langle\phi|AB|\phi\rangle = x + iy$ and $\langle\phi|BA|\phi\rangle = x - iy$.

Using the Cauchy–Schwartz inequality

$$|\langle v|w\rangle|^2 \leq \langle v|v\rangle\langle w|w\rangle, \qquad (4.18)$$

with $|v\rangle = A|\phi\rangle$ and $|w\rangle = B|\phi\rangle$, we get

$$|\langle\phi|AB|\phi\rangle|^2 \leq \langle\phi|A^2|\phi\rangle\langle\phi|B^2|\phi\rangle, \qquad (4.19)$$

which leads (using Equation (4.17)) to

$$|\langle\phi|[A,B]|\phi\rangle|^2 \leq 4\langle\phi|A^2|\phi\rangle\langle\phi|B^2|\phi\rangle. \qquad (4.20)$$

Using the definition of the standard deviation associated with the measurement of an observable M, and defining two new operators C and D such that

$$C = A - \langle A\rangle, \; D = B - \langle B\rangle, \qquad (4.21)$$

we get

$$\langle C\rangle = 0, \; \Delta C = \sqrt{\langle C^2\rangle} = \Delta A = \sqrt{\langle A^2\rangle - \langle A\rangle^2} \qquad (4.22)$$

and

$$\langle D\rangle = 0, \; \Delta D = \sqrt{\langle B^2\rangle} = \Delta B = \sqrt{\langle B^2\rangle - \langle B\rangle^2}. \qquad (4.23)$$

Furthermore, $[A,B] = [C,D]$ and therefore the inequality (4.20) can be rewritten as

$$\Delta C \Delta D \geq \frac{|\langle\phi|[C,D]|\phi\rangle|}{2}. \qquad (4.24)$$

This is the generalized form of the Heisenberg uncertainty principle.

(c) Third proof

The next derivation is based on the following lemma [3].

Preliminary: For any Hermitian operator A and any quantum state $|\phi\rangle$, the following identity holds:

$$A|\phi\rangle = \langle A\rangle|\phi\rangle + \Delta A|\phi_\perp\rangle, \tag{4.25}$$

where $|\phi\rangle$, $|\phi_\perp\rangle$ are normalized vectors, $\langle\phi_\perp|\phi\rangle = 0$, i.e., $|\phi_\perp\rangle$ is orthogonal to $|\phi\rangle$. Also, $\langle A\rangle = \langle\phi|A|\phi\rangle$ and $\Delta A = \sqrt{\langle A^2\rangle - \langle A\rangle^2}$.

It is always possible to make a decomposition $A|\phi\rangle = \alpha|\phi\rangle + \beta|\phi_\perp\rangle$.

Then $\langle\phi|A|\phi\rangle = \langle\phi|(\alpha|\phi\rangle + \beta|\phi_\perp\rangle))\rangle$, leading to $\alpha = \langle A\rangle$ and $\langle\phi|A^\dagger A|\phi\rangle = \langle A\phi|A\phi\rangle = \langle(\alpha\langle\phi| + \beta\langle\phi_\perp|)|(\alpha|\phi\rangle + \beta|\phi_\perp\rangle))\rangle$, leading to $\beta = \Delta A$.

This last relation can now be used to give an alternate derivation of the Heisenberg uncertainty relations. Consider two Hermitian operators A and B; then the following relations hold:

$$A|\phi\rangle = \langle A\rangle|\phi\rangle + \Delta A|\phi_{\perp,A}\rangle \tag{4.26}$$

and

$$B|\phi\rangle = \langle B\rangle|\phi\rangle + \Delta B|\phi_{\perp,B}\rangle, \tag{4.27}$$

where $\langle\phi_{\perp,A}|\phi\rangle = 0$ and $\langle\phi_{\perp,B}|\phi\rangle = 0$ because of the orthogonality.

Multiplying the Hermitian conjugate of Equation (4.27) by Equation (4.26), and using the Hermiticity of B, we get

$$\langle\phi|BA|\phi\rangle = \langle B\rangle\langle A\rangle + \Delta B\Delta A\langle\phi_{\perp,B}|\phi_{\perp,A}\rangle. \tag{4.28}$$

Similarly,

$$\langle\phi|AB|\phi\rangle = \langle A\rangle\langle B\rangle + \Delta A\Delta B\langle\phi_{\perp,A}|\phi_{\perp,B}\rangle. \tag{4.29}$$

Substracting Equation (4.28) from Equation (4.29) leads to

$$\langle[A,B]\rangle = 2i\Delta A\Delta B\,\mathrm{Im}(\langle\phi_{\perp,B}|\phi_{\perp,A}\rangle), \tag{4.30}$$

where Im stands for the imaginary part.

Since the two vectors are normalized, a simple application of the Cauchy–Schwartz inequality leads to

$$|\mathrm{Im}\langle\phi_{\perp,A}|\phi_{\perp,B}\rangle| \leq 1. \tag{4.31}$$

Together with Equation (4.30), this leads to

$$\Delta A\Delta B \geq \frac{1}{2}|\langle[A,B]\rangle|, \tag{4.32}$$

which is the standard form of the generalized Heisenberg uncertainty principle.

Interpretation of the Heisenberg uncertainty principle: From a practical standpoint, the meaning of the inequality (4.32) is the following:

(a) We must first prepare a physical system a large number of times (or ensemble) in state $|\phi\rangle$.

(b) Then, perform measurements of the observable A on a fraction of the ensemble and of the observable B on the rest of the ensemble.

(c) When a meaningful number of measurements have been performed so that a standard deviation of these measurements can be calculated, the experimental results for the standard deviation ΔA of the measurements of the observable A and the standard deviation ΔB of the measurements of the observable B will satisfy the inequality

$$\Delta A \Delta B \geq \frac{|\langle \phi | [A, B] | \phi \rangle|}{2}. \tag{4.33}$$

---

* Problem 4.2: Heisenberg uncertainty principle and diffraction patterns

Consider a beam of particles of momentum p incident normally on a screen with a slit of width a, as shown in Figure 4.1. The particles are detected on a second screen parallel to the first one at a distance D from it. For what value of the slit width a will the width of the most intense fringe in the pattern observed on the second screen be minimum?

Solution: A diffraction pattern is observed on the screen, as shown in Figure 4.1. Because of the width of the slit, we have an uncertainty $\Delta y = a$ on the position of the electrons passing through the slit. Using Heisenberg's uncertainty relation, this gives rise to an uncertainty Δp_y in the y-component of the linear momentum after having passed the slit. From Figure 4.1, we have $\Delta p_y / p = \tan \theta$.

The position of the minima in the diffraction pattern are given by Bragg's diffraction law:

$$a \sin \theta = (2n + 1)\frac{\lambda}{2}, \tag{4.34}$$

where $\lambda = h/p$ from de Broglie's relation.

Setting $n = 0$ in the previous equation gives the position of the first minimum in the diffraction pattern. Referring back to Figure 4.1, the total width of the first maximum is therefore given by $W = 2d + a$. But $\tan \theta = d/D$, hence $W = a + 2D \tan \theta$. Since θ is small, $\tan \theta \sim \sin \theta$, hence $W = a + 2D\frac{\lambda}{2a} = a + \frac{D\lambda}{a}$. Therefore, W is minimum when $a = \sqrt{D\lambda}$, i.e., when the slit opening is equal to the geometrical mean of the wavelength of the incident particle and the distance between the two screens.

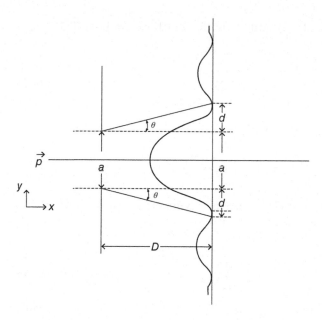

Figure 4.1: Diffraction pattern observed on a screen at a distance D from a parallel screen with an opening a due to a particle incident from the left with momentum $\vec{p}$ perpendicular to the plane of both screens.

** Problem 4.3: Heisenberg uncertainty principle for the one-dimensional particle in a box

Starting with the eigenstates and corresponding eigenvalues for the simple problem of a particle in a one-dimensional quantum box of size W (see Problem 3.5),

(a) Calculate the average values $\langle z \rangle$ and $\langle p \rangle$ (where p is the momentum in the z-direction) for an electron prepared in each of the eigenstates of the particle in a box.

(b) Calculate the standard deviations Δz, Δp for an electron prepared in each of the eigenstates of the particle in a box.

(c) Show that for large values of the quantum number n characterizing the eigenstates, the standard deviation Δz reduces to its classical value.

(d) Calculate $\Delta z \Delta p$ for each of the eigenstates found in part (a) and show that the Heisenberg uncertainty relation is satisfied.

Solution: The solution to the Schrödinger equation for a particle in a one-dimensional box is one of the simplest problems in quantum mechanics (see Problem 3.5). The eigenfunctions are

$$\phi_n(z) = \sqrt{\frac{2}{W}} \sin\left(\frac{n\pi z}{W}\right), \qquad (4.35)$$

where n is an integer ($n = 1, 2, 3, \dots$) and W is the width of the box. The corresponding eigenvalues are given by

$$E_n = n^2 \frac{\hbar^2 \pi^2}{2m^* W^2}. \tag{4.36}$$

(a) Using these results, the following average values are found:

$$\langle z \rangle = \int_0^W z |\phi_n(z)|^2 \mathrm{d}z = \frac{W}{2}. \tag{4.37}$$

This is to be expected since all eigenstates produce symmetrical probability distributions $|\phi_n|^2$ with respect to the center of the box.

Furthermore,

$$\langle p \rangle = \int_0^W \mathrm{d}z \phi_n(z) \frac{\hbar}{i} \frac{\mathrm{d}}{\mathrm{d}z} \phi_n(z) = 0. \tag{4.38}$$

This is also expected since a particle traveling back and forth between the two walls of the box will have the same magnitude of its momentum but opposite signs when traveling from left to right and right to left. Therefore, the average momentum of the particle trapped inside the box will be zero. Alternately, one can view the electron wave inside the box as a standing wave because of the infinite potential barriers at the boundaries of the box, and a standing wave has zero momentum.

(b) The standard deviations are given by

$$\langle \Delta z \rangle^2 = \langle z^2 \rangle - \langle z \rangle^2 = \int_0^W \mathrm{d}z z^2 \phi_n^2(z) - \left[\frac{W}{2}\right]^2. \tag{4.39}$$

So

$$\langle \Delta z \rangle^2 = \frac{2}{W} \int_0^W \mathrm{d}z z^2 \sin^2\left(\frac{n\pi z}{W}\right) - \frac{W^2}{4}. \tag{4.40}$$

Therefore,

$$\langle \Delta z \rangle^2 = \frac{W^2}{12} \left[1 - \frac{6}{(n\pi)^2}\right]. \tag{4.41}$$

Similarly,

$$\langle \Delta p \rangle^2 = \langle p^2 \rangle - \langle p \rangle^2 = \langle p^2 \rangle. \tag{4.42}$$

Hence,

$$\langle \Delta p \rangle^2 = \int_0^W \mathrm{d}z \phi_n(z) p^2 \phi_n(z) \mathrm{d}z. \tag{4.43}$$

But $p^2 \phi_n = 2m^* E_n \phi_n(z)$, so

$$\langle \Delta p \rangle^2 = 2m^* E_n \int_0^W \mathrm{d}z |\phi_n|^2 = 2m^* \left(\frac{\hbar^2 n^2 \pi^2}{2m^* W^2}\right) = \frac{\hbar^2 n^2 \pi^2}{W^2}. \tag{4.44}$$

(c) As n increases, we have, from part (b),

$$\lim_{n\to\infty} \langle \Delta z \rangle^2 = \frac{W^2}{12}.$$ (4.45)

We compare this to its classical limit, which is equal to

$$\Delta z^2 = \frac{1}{W} \int_0^W \left(z - \frac{W}{2} \right)^2 dz = \frac{W^2}{12}.$$ (4.46)

So, the quantum mechanical result does indeed converge to its classical value for eigenstates with large quantum numbers.

(d) Using the previous results, we get

$$\Delta z \Delta p = \frac{\hbar}{2} \frac{n\pi}{\sqrt{3}} \left[1 - \frac{6}{n^2\pi^2} \right]^{\frac{1}{2}}.$$ (4.47)

Hence, the value of $\Delta z \Delta p$ (in units of $\frac{\hbar}{2}$) is equal to 1.136, 2.271, 5.254, for $n = 1, 2$, and 3, respectively. All these values are larger than 1, in agreement with the Heisenberg uncertainty principle for position and momentum. Matlab code giving the values of the product $\Delta z \Delta p$ as a function of the quantum number n is given in Appendix G.

*** Problem 4.4: The Heisenberg uncertainty principle for an approximate ground state wave function of the particle in a box

Suppose that a particle in a one-dimensional well (with infinitely high potential barriers) of width L has a wave function of the form

$$\phi(z) = N(z^2 - Lz).$$ (4.48)

(a) What is the value of the normalization constant?

(b) What are the expectation values of the position, momentum, and kinetic energy of the particle?

(c) What is the uncertainty in the position and momentum of the particle?

(d) Using the results of the two previous steps, calculate the product $\Delta z \Delta p$ and compare your result to the one obtained for the exact ground state of the particle in a box, as derived in Problem 4.3.

Solution:
(a) Because the wave function $\psi(z) = N(z^2 - Lz)$ has no nodes except at $z = 0$ and $z = L$, it is an approximation to the ground state wave function in a well with infinite barriers. For the latter to be normalized, the normalization coefficient must

be determined starting with the condition

$$\int_0^L \psi^*(z)\psi(z)dz = \int_0^L |\psi(z)|^2 dz = 1. \tag{4.49}$$

This leads to the equation

$$N^2 \int_0^L (z^2 - Lz)^2 dz = N^2 \left(\frac{1}{5}L^5 - \frac{1}{2}L^5 + \frac{1}{3}L^5\right) = 1. \tag{4.50}$$

Hence,

$$N = \sqrt{\frac{30}{L^5}}. \tag{4.51}$$

(b) The expectation value of the position is given by:

$$\langle z \rangle = \int_0^L z|\psi(z)|^2 dz = N^2 \int_0^L z(z^2 - Lz)^2 dz. \tag{4.52}$$

This leads to

$$\langle z \rangle = N^2 \left(\frac{1}{6}L^6 - \frac{2}{5}L^6 + \frac{1}{4}L^6\right) = \frac{30}{L^5} \times \frac{1}{60}L^6 = \frac{1}{2}L. \tag{4.53}$$

This last result is to be expected since the wave function in Equation (4.48) is symmetric with respect to $z = L/2$.

The expection value of the momentum is given by

$$\langle p \rangle = \int_0^L \psi^* \left(-\frac{\hbar}{i}\frac{d\psi(z)}{dz}\right) dz = \frac{\hbar N^2}{i} \int_0^L (z^2 - Lz)(2z - L)dz. \tag{4.54}$$

Carrying out the integration leads to

$$\langle p \rangle = \frac{\hbar N^2}{i} \left(\frac{1}{2}L^4 - L^4 + \frac{1}{2}L^4\right) = 0. \tag{4.55}$$

This is to be expected since the integrand in Equation (4.54) is antisymmetric with respect to the center of the well.

The expectation value of the kinetic energy of the particle is given by:

$$\langle E \rangle = \int_0^L \psi^*(z) \left(-\frac{\hbar^2}{2m}\frac{d^2}{dz^2}\right) \psi(z)dz. \tag{4.56}$$

This gives

$$\langle E \rangle = -\frac{\hbar^2 N^2}{m} \int_0^L (z^2 - Lz)dz = -\frac{\hbar^2 N^2}{m} \left(-\frac{1}{6}L^3\right) = \frac{5\hbar^2}{L^2 m}. \tag{4.57}$$

(c) The uncertainty Δz in the position of the particle is given by

$$\Delta z = \sqrt{\langle z^2 \rangle - \langle z \rangle^2}. \tag{4.58}$$

We first calculate

$$\overline{z^2} = \int_0^L z^2 |\psi(z)|^2 \mathrm{d}z = N^2 \int_0^L (z^6 - 2z^5 L + z^4 L^2)\mathrm{d}z. \tag{4.59}$$

This leads to

$$\langle z^2 \rangle = N^2 \times \frac{L^7}{105} = \frac{30}{L^5} \times \frac{L^7}{105} = \frac{2L^2}{7}. \tag{4.60}$$

Hence,

$$\Delta z = \sqrt{\langle z^2 \rangle - \langle z \rangle^2} = \sqrt{\frac{2L^2}{7} - \left(\frac{1}{2}L\right)^2} = 0.1898L. \tag{4.61}$$

Similarly, the uncertainty in the momentum of the particle Δp is given by

$$\Delta p = \sqrt{\langle p^2 \rangle - \langle p \rangle^2} = \sqrt{\langle p^2 \rangle} = \sqrt{2m\langle E \rangle}. \tag{4.62}$$

Using Equation (4.57), we get

$$\langle p^2 \rangle - 2m\langle E \rangle = \frac{10\hbar^2}{L^2}. \tag{4.63}$$

Hence,

$$\Delta p = \sqrt{\langle p^2 \rangle} = \sqrt{\frac{10\hbar^2}{L^2}} = 3.16\frac{\hbar}{L}. \tag{4.64}$$

(d) Using the above result, the uncertainty product $\Delta z \Delta p$ is given by

$$\Delta z \Delta p = 1.2\frac{\hbar}{2}, \tag{4.65}$$

which is greater than $\frac{\hbar}{2}$, in agreement with Heisenberg's uncertainty relations for position and momentum.

We also note that the product $\Delta z \Delta p$ in Equation (4.65) is larger than its value for the exact ground state of the particle in a box, equal to 1.136 $\frac{\hbar}{2}$, as shown in Problem 4.3.

---

* **Problem 4.5: The Heisenberg uncertainty principle for the one-dimensional attractive delta scatterer**

Starting with the results of Problem 3.2,

(a) *Find the probability density of the momentum in the bound state. Determine the value of p for which the probability density in momentum space is maximum.*

Calculate the variance Δp, then the product $\Delta z \Delta p$, and show that it is in agreement with Heisenberg's uncertainty principle.

(b) Starting with the wave function of the bound state in momentum space (see Problem 3.2), calculate the variance Δz and Δp and show that the product of these two quantities agrees with the results of part (a).

Solution: Using the results of Problem 3.2, the probability distribution is obtained via a Fourier transform:

$$v(p) = (2\pi\hbar)^{-1/2} \int_{-\infty}^{\infty} \phi(z) e^{-\frac{ipz}{\hbar}} dz, \qquad (4.66)$$

which results in

$$v(p) = \left(\frac{2}{\pi\hbar}\right)^{1/2} \frac{\kappa_0^{3/2}}{\kappa_0^2 + \left(\frac{p}{\hbar}\right)^2}. \qquad (4.67)$$

The probability density for the momentum to be between p and $p + dp$ is therefore

$$|v(p)|^2 dp = \frac{2}{\pi} \frac{(\hbar\kappa_0)^3 dp}{[(\hbar\kappa_0)^2 + p^2]^2}. \qquad (4.68)$$

The maximum of $|v(p)|^2$ occurs when $p = 0$ and, by symmetry, $\langle p \rangle = 0$.

Also, $\langle p^2 \rangle$ is given by

$$\langle p^2 \rangle = \int_{-\infty}^{\infty} |v(p)|^2 p^2 dp = \frac{2}{\pi}(\hbar\kappa_0)^2 \int_{-\infty}^{\infty} \frac{u^2 du}{(1+u^2)^2} = (\hbar\kappa_0)^2. \qquad (4.69)$$

Since $\int_{-\infty}^{\infty} \frac{u^2 du}{(1+u^2)^2} = \pi/2$, we get

$$\Delta p = \hbar\kappa_0. \qquad (4.70)$$

Therefore, $\Delta p \Delta z = \frac{\hbar}{\sqrt{2}} \geq \frac{\hbar}{2}$, in agreement with that the Heisenberg uncertainty principle.

**** Problem 4.6: The one-dimensional ionized hydrogen molecule

Determine the probability density in momentum space for the even (symmetric) and odd (antisymmetric) bound states of the one-dimensional ionized hydrogen molecule considered in Problem 3.3.

Calculate the standard deviations Δz and Δp for the symmetric bound state. Derive the analytical expression for the product $\Delta z \Delta p$ [4, 5].

Solution: First, we present an alternative approach to derive the analytical expressions for the wave functions associated with the bound state of the one-dimensional ionized hydrogen molecule (see Problem 3.3) described by the Hamiltonian

$$H\phi(z) \equiv \frac{-\hbar^2}{2m^*} \frac{\mathrm{d}^2\phi(z)}{\mathrm{d}z^2} + V(z)\phi(z) = E\phi(z), \tag{4.71}$$

where the atomic potential $V(z)$ is composed of two delta functions of equal strength $-\Gamma$ (the minus sign for an attractive potential), one located at $z = -a$ and the other at $z = a$:

$$V(z) = -\Gamma\delta(z - a) - \Gamma\delta(z + a). \tag{4.72}$$

Solutions to Equation (4.71) are of the form

$$\phi_{\mathrm{I}}(z) = Be^{\kappa z} \quad \text{for } z < -a, \tag{4.73}$$

$$\phi_{\mathrm{II}}(z) = Ce^{\kappa z} + De^{-\kappa z} \quad \text{for } -a < z < a, \tag{4.74}$$

and

$$\phi_{\mathrm{III}}(z) = Fe^{\kappa z} \quad \text{for } z > a, \tag{4.75}$$

where $\kappa = \frac{1}{\hbar}\sqrt{-2m^*E}$, with E being the (negative) energy of the bound states.

According to the results of Problem 1.4, since $V(z)$ is symmetric, the Schrödinger equation admits symmetric (even) and antisymmetric (odd) solutions. Therefore, in the general solution given by Equations (4.73)–(4.75), the following constraints must apply to the parameters (B, C, D, F):

$$B = F \quad \text{and} \quad C = D \quad \text{(even solution)},$$
$$B = -F \quad \text{and} \quad C = -D \quad \text{(odd solution)}.$$

To determine the parameters B and C we must impose the continuity of the wave functions at $z = \pm a$, and the following conditions must also hold:

$$\frac{\mathrm{d}\phi(\pm a + \epsilon)}{\mathrm{d}z} - \frac{\mathrm{d}\phi(\pm a - \epsilon)}{\mathrm{d}z} = -2\alpha\phi(\pm a), \tag{4.76}$$

where $\epsilon \to 0$ and $\alpha = m^*\Gamma/\hbar^2$.

Energy eigenvalues for odd solutions: For the odd solution, continuity of ϕ at $z = \pm a$ and Equation (4.76) lead to the following four equations:

$$Be^{-\kappa a} = C(e^{-\kappa a} - e^{\kappa a}), \tag{4.77}$$

$$\kappa C(e^{-\kappa a} + e^{\kappa a}) - \kappa Be^{-\kappa a} = -2\alpha Be^{-\kappa a}, \tag{4.78}$$

$$Be^{-\kappa a} = C(e^{-\kappa a} - e^{\kappa a}), \tag{4.79}$$

and

$$-\kappa C(e^{-\kappa a} + e^{\kappa a}) = 2\alpha Be^{-\kappa a}. \tag{4.80}$$

Equations (4.79)–(4.80) are the same as Equations (4.77)–(4.78). We are therefore left with two equations for the two unknowns B and C. Solving for C in Equation (4.77) and plugging the result into Equation (4.78) leads to the transcendental equation

$$\kappa \coth(\kappa a) + \kappa = 2\alpha. \tag{4.81}$$

Multiplying both sides by a and defining $\gamma = \kappa a$, we obtain the following transcendental equation to determine the energy of the odd bound state:

$$\beta = \gamma\left[1 + \coth(\gamma)\right] = 2\alpha a. \tag{4.82}$$

A similar derivation leads to the following transcendental equation for the solution of the even bound state:

$$\beta = \gamma\left(1 + \tanh(\gamma)\right) = 2\alpha a. \tag{4.83}$$

Finding the eigenfunctions and their normalization coefficients: For the odd bound state to be normalized, the following equality must hold:

$$1 = \int_{-\infty}^{-a} B^2 e^{\frac{2\gamma}{a}z} dz + \int_{-a}^{a} C^2 \left(e^{\frac{\gamma}{a}z} - e^{-\frac{\gamma}{a}z}\right)^2 dz + \int_{a}^{\infty} (-B)^2 e^{-\frac{2\gamma}{a}z} dz. \tag{4.84}$$

Performing the integration leads to

$$1 = \frac{C^2 a}{\gamma}\left[e^{2\gamma} - e^{-2\gamma}\right] - 4C^2 a + \frac{a}{\gamma}B^2 e^{-2\gamma}. \tag{4.85}$$

With the use of Equation (4.79), this last equality leads to the following explicit expressions for the parameters B and C:

$$B = \left(\frac{\gamma}{2a}\right)^{1/2} \left(e^{2\gamma} - 2\gamma - 1\right)^{-1/2} \left(e^{-\gamma} - e^{\gamma}\right) e^{\gamma} \tag{4.86}$$

and

$$C = \left(\frac{\gamma}{2a}\right)^{1/2} \left(e^{2\gamma} - 2\gamma - 1\right)^{-1/2}. \tag{4.87}$$

Plugging the values of B and C into Equations (4.73)–(4.75) leads to the following explicit expression for the wave function associated with the odd bound state:

$$\phi_{\text{odd}} = \epsilon(z)\left(\frac{2\gamma}{a}\right)^{1/2} \left(e^{2\gamma} - 2\gamma - 1\right)^{-1/2} \sinh(\gamma)e^{-\gamma\left(\frac{|z|-a}{a}\right)} \quad (|z| > a) \tag{4.88}$$

$$= \left(\frac{2\gamma}{a}\right)^{1/2} \left(e^{2\gamma} - 2\gamma - 1\right)^{-1/2} \sinh\left(\frac{\gamma z}{a}\right) \quad (|z| < a), \tag{4.89}$$

where $\epsilon(z) = -1$ for $z < 0$ and $+1$ for $z > 0$.

A similar procedure leads to the following expression for the wave function associated with the even bound state:

$$\phi_{\text{even}} = \left(\frac{2\gamma}{a}\right)^{1/2} \left(e^{2\gamma} + 2\gamma + 1\right)^{-1/2} \cosh(\gamma)e^{-\gamma\left(\frac{|z|-a}{a}\right)} \quad (|z| > a) \tag{4.90}$$

$$= \left(\frac{2\gamma}{a}\right)^{1/2} \left(e^{2\gamma} + 2\gamma + 1\right)^{-1/2} \cosh\left(\frac{\gamma z}{a}\right) \quad (|z| < a). \tag{4.91}$$

Momentum probability distributions: The momentum probability amplitudes associated with the bound state solutions described above are found by
performing a Fourier transform on the Schrödinger Equation (4.71), following the
derivation in Problem 4.2.

For the odd and even bound states, this leads to the following equations:

$$\left[\frac{p^2}{2m^*} - E\right]\phi(p) = \frac{2i\Gamma\phi(a)\sin\left(\frac{pa}{\hbar}\right)}{\sqrt{2\pi\hbar}} \quad \text{(odd)} \tag{4.92}$$

and

$$\left[\frac{p^2}{2m^*} - E\right]\phi(p) = \frac{2\Gamma\phi(a)\cos\left(\frac{pa}{\hbar}\right)}{\sqrt{2\pi\hbar}} \quad \text{(even)}. \tag{4.93}$$

The momentum probability $\phi(p)$ associated with the bound states is therefore
given by

$$\phi(p) = \frac{2i\Gamma\left(\frac{2\gamma}{a}\right)^{\frac{1}{2}}\sinh(\gamma)\sin\left(\frac{pa}{\hbar}\right)}{\left(\frac{p^2}{2m^*} - E\right)\sqrt{2\pi\hbar}\sqrt{e^{2\gamma} - 2\gamma - 1}} \quad \text{(odd)} \tag{4.94}$$

and

$$\phi(p) = \frac{2\Gamma\left(\frac{2\gamma}{a}\right)^{\frac{1}{2}}\cosh(\gamma)\cos\left(\frac{pa}{\hbar}\right)}{\left(\frac{p^2}{2m^*} - E\right)\sqrt{2\pi\hbar}\sqrt{e^{2\gamma} + 2\gamma + 1}} \quad \text{(even)}, \tag{4.95}$$

where

$$E = \frac{-m^*\Gamma^2}{2\hbar^2}\left(\frac{2\gamma}{\beta}\right)^2 \tag{4.96}$$

and β, γ satisfy Equations (4.82) and (4.83) for the odd and even bound states,
respectively.

Calculation of Δz and Δp: For the even bound state (4.93), using the shorthand notation

$$\alpha = \left(\frac{2\gamma}{a}\right)^{1/2}\left(e^{2\gamma} + 2\gamma + 1\right)^{-1/2}, \tag{4.97}$$

we obtain the following result:

$$(\Delta z)^2 = \langle z^2\rangle - \langle z\rangle^2 = \langle z^2\rangle = \alpha^2(I_1 + 2e^{2\gamma}\cosh^2\gamma I_2), \tag{4.98}$$

where

$$I_1 = \int_{-a}^{+a} dz\, z^2 \cosh^2\left(\frac{\gamma z}{a}\right) \tag{4.99}$$

and

$$I_2 = 2\int_0^{+a} dz\, z^2 e^{-\frac{2\gamma z}{a}}. \tag{4.100}$$

Performing the integration leads to

$$I_1 = 0.5\left(\frac{a}{\gamma}\right)^3\left[(\gamma^2 + 0.5)\sinh(2\gamma) - \gamma\cosh(2\gamma) + \frac{2}{3}\gamma^2\right] \tag{4.101}$$

and

$$I_2 = \frac{2\gamma}{a} \frac{e^{2\gamma} \cosh^2 \gamma (\gamma^2 + \gamma + 1)}{\gamma^3 [e^{2\gamma} + 2\gamma + 1]}. \tag{4.102}$$

Similarly,

$$(\Delta p)^2 = \langle p^2 \rangle - \langle p \rangle^2 = \langle p^2 \rangle = 2m^* \left[\frac{-\hbar^2 \gamma^2}{2m^* a^2} + 4\frac{\gamma}{a} \frac{\Gamma \cosh^2 \gamma}{(e^{2\gamma} + 2\gamma + 1)} \right]. \tag{4.103}$$

Using the last result, one easily gets the product $\Delta z \Delta p$, which can be shown to satisfy the Heisenberg uncertainty relation. A similar derivation gives the value of this product for the odd bound state (see suggested problems).

** Problem 4.7: Estimate of the ground state energy near the local minimum of a one-dimensional potential

Consider a particle in a one-dimensional continuous symmetric potential, i.e., $V(z) = V(-z)$, with a global minimum at $z = 0$. Starting with the generalized Heisenberg uncertainty relations (Problem 4.1), show that a lower bound of the energy of the ground state is given by $V_0 + \frac{\hbar}{2}\sqrt{\frac{\alpha}{m^*}}$, where $\alpha = \frac{d^2 V}{dz^2}\Big|_{z=0}$.

Solution: If $V(z)$ is continuous and symmetric with a local minimum at $z = 0$, a Taylor series expansion near the minimum gives

$$\langle V \rangle = V_0 + \frac{1}{2}\alpha \langle z^2 \rangle, \tag{4.104}$$

where $\langle \cdot \rangle$ stands for the average (or expectation) value calculated with the ground state wave function.

Futhermore, we have

$$\frac{1}{2m^*}\langle p^2 \rangle + \langle V \rangle = E, \tag{4.105}$$

where E is the energy of the ground state.

Hence,

$$\langle p^2 \rangle = 2m^*(E - \langle V \rangle). \tag{4.106}$$

Plugging the last relation in the Heisenberg uncertainty relation for position and momentum, and using the fact that $\langle p \rangle = 0$ in the ground state, we get

$$\Delta p \Delta z = \sqrt{\langle p^2 \rangle}\sqrt{\langle z^2 \rangle} \geq \frac{\hbar}{2}. \tag{4.107}$$

Using Equation (4.106), this leads to

$$\sqrt{2m^*}\sqrt{E - \langle V \rangle} \geq \frac{\hbar}{2\sqrt{\langle z^2 \rangle}}. \tag{4.108}$$

Hence,

$$E \geq V_0 + \frac{1}{2}\alpha\langle z^2\rangle + \frac{\hbar^2}{8m^*\langle z^2\rangle}. \tag{4.109}$$

The right-hand side of this inequality is minimum when

$$\langle z^2\rangle = \frac{\hbar}{2}\frac{1}{\sqrt{m^*\alpha}}. \tag{4.110}$$

Therefore, a lower bound E_b of the ground state energy is given by

$$E_b = V_0 + \frac{\hbar}{2}\sqrt{\frac{\alpha}{m^*}}. \tag{4.111}$$

In the case of the one-dimensional harmonic oscillator described by the potential $V(z) = V_0 + \frac{1}{2}m^*\omega^2 z^2$, the lower bound for the energy of the ground state is $V_0 + \frac{\hbar\omega}{2}$, which is the exact value.

**** Problem 4.8: A simple treatment of potential barrier penetration

Show that the penetration of a particle through a potential barrier can be interpreted as a climb over the barrier as a result of energy fluctuations expected from the Heisenberg uncertainty principle, rather than due to tunneling through the barrier.

Solution: The following argument was first given by Cohen [6]. Potential barrier penetration is an extremely difficult phenomenon to explain or even understand. The widely accepted view of this phenomenon is tunneling, which goes against our intuitive understanding of physics since it requires us to admit the notion of negative kinetic energy, which seems to be entirely unphysical.

There is an alternate view of barrier penetration which does not require us to admit the notion of negative kinetic energy and therefore is more palatable. This involves the uncertainty principle

$$\Delta E \Delta t \simeq \hbar. \tag{4.112}$$

One of the benefits of using the uncertainty principle is that it does not immediately assail our intuitive understanding of physics. Equation (4.112) tells us that the energy of a particle is subjected to short periods of fluctuations. Therefore, a particle confined by a potential barrier may momentarily gain enough energy to climb over the barrier. It is possible that the fluctuation may last long enough for the particle to pass over the entire length of the barrier, thereby completely penetrating the barrier.

The quantity ΔE in Equation (4.112) stands for the total fluctuation on either side of the energy E. The particle can gain some excess energy ϵ over the mean energy E over a timescale τ, which can be estimated as follows:

$$\epsilon \simeq \frac{\hbar}{\tau}. \tag{4.113}$$

It can be expected that the particle energy can fall below E as well, but those situations are not relevant to the discussion.

To refine the argument, we can define a "probability" P for the particle to have excess energy ϵ over a time interval τ as

$$P \sim e^{-\frac{2\epsilon\tau}{\hbar}}. \tag{4.114}$$

Equation (4.114) is completely *heuristic* and *ad hoc*, but it is nonetheless able to provide a simple physical understanding of what follows.

To study barrier penetration, let us consider a potential profile $V(z)$ that contains a region demarcated by the end points z_0 (the starting point) and z_n (the ending point), where the value of $V(z)$ is greater than E. Taking into account the fluctuation in energy, the kinetic energy of the particle in this region will be $E + \epsilon(z) - V(z)$, and the particle velocity $\nu(z)$ will therefore be given by

$$v(z) = \sqrt{\frac{2}{m}(E + \epsilon(z) - V(z))}. \tag{4.115}$$

The particle travels an infinitesimal distance dz, from z_0 to z_1, during the time interval $dz/v(z)$, where $dz = z_1 - z_0$. Hence, the probability of this event (see Equation (4.114)) is given by

$$P_1 \simeq e^{\left(-\frac{2\epsilon}{\hbar}\frac{dz}{v}\right)}. \tag{4.116}$$

We can easily find the probability of travel from z_1 to z_n by multiplying the probabilities of travel from z_0 to z_1, z_1 to z_2, ..., z_{n-1} to z_n. Each sojourn is treated as an independent event. This gives us the overall probability to "cross" the barrier as

$$P \simeq \prod_{i=1}^{n} e^{\left(-\frac{2\epsilon_i}{\hbar}\frac{dz_i}{v_i}\right)} = e^{\left(-\sum_i \frac{2\epsilon_i}{\hbar}\frac{dz_i}{v_i}\right)}. \tag{4.117}$$

In the limit $dz \to 0$, the last equation reduces to

$$P \simeq e^{-\int_{z_0}^{z_n} \frac{2\epsilon(z)}{\hbar v(z)} dz}. \tag{4.118}$$

If we use Equation (4.115) in Equation (4.118), we obtain

$$P \simeq e^{\left(-\sqrt{2m}\frac{I}{\hbar}\right)}, \tag{4.119}$$

where

$$I = \int_{z_0}^{z_n} \epsilon(z)(E + \epsilon(z) - V(z))^{-\frac{1}{2}} dz. \tag{4.120}$$

The function $\epsilon(z)$ needs to be on the order of the barrier height so that we are able to minimize I and maximize the probability. If we set the partial derivative of the integrand in Equation (4.120) with respect to ϵ equal to zero in order to minimize I, this leads to

$$[E + \epsilon(z) - V(z)]^{-\frac{1}{2}} - \frac{\epsilon(z)}{2}[E + \epsilon(z) - V(z)]^{-\frac{3}{2}} = 0, \tag{4.121}$$

or

$$E + \epsilon(z) - V(z) - \frac{\epsilon(z)}{2} = 0, \qquad (4.122)$$

yielding

$$\epsilon(z) = 2\left[V(z) - E\right]. \qquad (4.123)$$

If we subsitute Equation (4.123) in Equation (4.120), we finally get

$$I_{\min} = 2 \int_{z_0}^{z_n} \left(V(z) - E\right)^{\frac{1}{2}} dz. \qquad (4.124)$$

If we substitute Equation (4.124) into Equation (4.119), we obtain

$$P \simeq e^{-2\int_{z_0}^{z_n} \left(\frac{2m}{\hbar}(V(z)-E)\right)^{\frac{1}{2}} dz}. \qquad (4.125)$$

The minimization can be omitted and any simple function $\epsilon(z)$ can be used, leading to less complicated mathematics. The minimum of I is actually very broad, because if $V(z)$ is linear in z, and ϵ is a constant of any reasonable value different from the one in Equation (4.119), the integral I will differ in value by only about 20%.

This approach yields the same probability for traversing a barrier as the Wentzel–Kramers–Brillouin approach based on quantum tunneling (see Equation (4.125)) [7]. A particle climbing over a barrier during an energy fluctuation is intuitively more appealing than the strange notion of tunneling through the barrier.

Suggested problems

- Starting with the wave function of the bound state derived in Problem 3.4, calculate the value of $\Delta z \Delta p_z$ as a function of the height of the step potential ΔE_c. How does this product vary as ΔE_c is varied from 0 to four times $E_0 = \frac{m^* \Gamma^2}{2\hbar^2}$ (the magnitude of the bound state energy of the one-dimensional delta scatterer) when $\Delta E_c = 0$?

- Estimate the value of the product $\Delta z \Delta p_z$ for the zero-energy bound state of a particle in a box in the presence of an attractive delta scatterer as discussed in Problem 3.11. Show that it is in agreement with the Heisenberg uncertainty principle. For what value of the position z_0 of the scatterer in the well is this product minimum?

- Following the derivation in Problem 4.6, prove that the normalized wave function for the even bound state of the one-dimensional ionized hydrogen molecule is given by Equation (4.91).

- Starting with the results of Problem 4.3, calculate the standard deviations Δz, Δp for the odd bound state of the one-dimensional ionized hydrogen molecule. Using these results, derive the analytical expression for the product $\Delta z \Delta p$.

- Using the results of the previous problem, plot $\Delta z \Delta p$ versus the distance $2a$ between the two delta scatterers of the one-dimensional ionized hydrogen molecule. Use $\Gamma = 5\,\text{eV-Å}$ for the strength of the two delta scatterers. Show that the product $\Delta z \Delta p$ for both the even and odd bound states reaches a minimum as a function of the distance $2a$ between the two delta scatterers. Determine numerically the values of $2a$ at which the minimum of $\Delta z \Delta p$ is reached, the value of that minimum, and show that it is in agreement with the Heisenberg uncertainty principle.

- Starting with the generalized virial theorem (Problem 2.9), show that for a system described by the Hamiltonian $H = \frac{p^2}{2m^*} + V(z)$, the following inequality holds:

$$\left\langle z \frac{dV}{dz} \right\rangle \geq \frac{\hbar^2}{4m^* \langle z^2 \rangle}, \qquad (4.126)$$

where the average is taken over an eigenstate of the Hamiltonian H.

- Consider a particle in the ground state of the half-harmonic potential, i.e., $V(z) = \frac{1}{2} m^* \omega^2$ for $z > 0$ and $V(z) = \infty$ for $z < 0$. What is the normalized wave function associated with the ground state in this potential? Use this normalized wave function to calculate the standard deviations Δz and Δp and show that their product obeys the Heisenberg uncertainty principle.

- The wave function of a free particle of mass m^* moving in one dimension is given by

$$\phi(z, t = 0) = N \int_{-\infty}^{+\infty} dk\, e^{ikz - \frac{|k|}{k_0}}, \qquad (4.127)$$

where N and k_0 are positive constants.

(1) What is the probability $W(p, 0)$ that a measurement of the momentum at time $t = 0$ will give a value in the range $-P \leq p \leq +P$?

(2) Find an analytical expression for $W(p, t)$.

(3) Calculate the standard deviations $\Delta z(t)$, $\Delta p(t)$ and the wavepacket $\phi(z, t)$. Comment on the time evolution of the wavepacket.

(4) Calculate the uncertainty product $\Delta z\, \Delta p$ and show that it satisfies the Heisenberg uncertainty principle.

- Repeat Problem 4.4 using the following approximate expression for the wave function of the first excited state of a particle in a box:

$$\psi(z) = Nz \left(z - \frac{w}{2} \right)(z - w). \qquad (4.128)$$

This trial wave function is a good approximation for the first excited state because is has three nodes, one at each end of the box and one in center. It is antisymmetric (odd) with respect to the center of the box.

References

[1] Levi, A. F. J. (2006) *Applied Quantum Mechanics*, 2nd edition, Section 5.5.4, Cambridge University Press, Cambridge.

[2] Nielsen, M. A. and Chuang, I. L. (2000) *Quantum Computation and Quantum Information*, Box 2.4, p. 89, Cambridge University Press, New York.

[3] Goldenberg, L. and Vaidman, L. (1996) Applications of a simple quantum mechanical formula. *American Journal of Physics* 64, pp. 1059–1060.

[4] Lapidus, R. (1970) One-dimensional model of a diatomic ion. *American Journal of Physics* 51, pp. 905–908.

[5] Lapidus, R. (1983) One-dimensional hydrogen atom and hydrogen molecule ion in momentum space. *American Journal of Physics* 38, pp. 663–665.

[6] Cohen B. L. (1965) A simple treatment of potential barrier penetration. *American Journal of Physics* 33, pp. 97–98.

[7] Kroemer, H. (1994) *Quantum Mechanics for Engineering, Materials Science, and Applied Physics*, Chapter 6, Prentice Hall, Englewood Cliffs, NJ.

Suggested Reading

• Tyagi, N. K. (1963) New derivation of the Heisenberg uncertainty principle. *American Journal of Physics* 31, p. 624.

• Gersch, H. A. and Braden, C. H. (1982) Approximate energy levels and sizes of bound quantum systems. *American Journal of Physics* 50, pp. 53–59.

• Mayants, L. (1987) Note on "Correlation of quantum properties and the generalized Heisenberg inequality." *American Journal of Physics* 55, p. 1063.

• DeYoung, P. A., Jolivette, P. L., and Rouze, N. (1993) Experimental verification of the Heisenberg uncertainty principle: an advanced undergraduate laboratory. *American Journal of Physics* 61, pp. 560–563.

• Moshinsky, M. (1976) Diffraction in time and the time–energy uncertainty relation. *American Journal of Physics* 44, pp. 1037–1042.

• Sánchez-Velasco, E. S. (1994) A note on the representation of the commutation relations. *American Journal of Physics* 62, pp. 374–375.

• Moloney, M. J. (1982) Simple F-center argument from the uncertainty principle. *American Journal of Physics* 50, p. 557.

• Bligh, P. H. (1974) Note on the uncertainty principle and barrier penetration. *American Journal of Physics* 42, pp. 337–338.

- Weichel, H. (1976) The uncertainty principle and the spectral width of a laser beam. *American Journal of Physics* 44, pp. 839–840.

- Chisolm, E. D. (2001) Generalizing the Heisenberg uncertainty principle. *American Journal of Physics* 69, pp. 368–371.

- Wolsky, A. M. (1974) Kinetic energy, size, and the uncertainty principle. *American Journal of Physics* 42, pp. 760–763.

- Matteucci, G. (2010) The Heisenberg uncertainty principle demonstrated with an electron diffraction experiment. *European Journal of Physics* 31, pp. 1287–1293.

- Blado, G., Owens, C., and Meyers, V. (2014) Quantum wells and the generalized uncertainty principle. *European Journal of Physics* 35, p. 065011.

- Palenik, M. C. (2014) Quantum mechanics from Newton's second law and the canonical commutation relation $[X, P] = i\hbar$. *European Journal of Physics* 35, p. 045014.

- Ahmed, Z. (2014) Position–momentum uncertainty products. *European Journal of Physics* 35, p. 045015.

- Denur, J. (2010) The energy–time uncertainty principle and quantum phenomena. *American Journal of Physics* 78, pp. 1132–1145.

- Loeb, A. L. (1963) The Heisenberg uncertainty principle. *American Journal of Physics* 31, p. 945.

- Hilgevoord, J. (1996) The uncertainty principle for energy and time. *American Journal of Physics* 64, pp. 1451–1456.

- Raymer, M. G. (1994) Uncertainty principle for joint measurement of noncommuting variables. *American Journal of Physics* 62, pp. 986–993.

- de la Peña, L. (1980) Conceptually interesting generalized Heisenberg inequality. *American Journal of Physics* 48, pp. 775–776.

- Jauch, W. (1993) Heisenberg's uncertainty relation and thermal vibrations in crystals. *American Journal of Physics* 61, pp. 929–932.

Chapter 5: Current and Energy Flux Densities

This set of problems introduces the current density operator [1–5], which is applied to the study of various tunneling problems, including the case of a general one-dimensional heterostructure under bias (i.e., subjected to an electric field), the tunneling of an electron through an absorbing one-dimensional delta scatterer and potential well, and the calculation of the dwell time above a quantum well (QW). The dwell time is the time that an electron traversing a QW potential, with energy above the well's barrier, lingers within the well region. This chapter also includes an introduction to a quantum mechanical version of the energy conservation law based on the concept of quantum mechanical energy flux derived from the Schrödinger equation. Some basic tunneling problems are revisited using the conservation of energy flux principle.

---

* Problem 5.1: Current continuity equation in an open quantum system

Starting with the one-dimensional time-dependent Schrödinger equation and assuming that the potential energy has an imaginary part which makes the Hamiltonian non-Hermitian, i.e.,

$$V(z) + i\xi W(z), \tag{5.1}$$

where V(z), W(z), and ξ are real, show that the probability current density $J_z(z,t)$ satisfies the continuity equation

$$\frac{\partial}{\partial z}J_z(z,t) + \frac{\partial}{\partial t}\rho(z,t) = \frac{2}{\hbar}W(z)\xi\rho(z,t), \tag{5.2}$$

where ρ is the probability charge density. Assume a constant effective mass.

The fact that the right-hand side of the above equation is non-zero suggests violation of charge conservation. This is a consequence of using a non-Hermitian Hamiltonian, which allows for exchange of charge with the environment and is symptomatic of an open quantum system. Non-Hermitian Hamiltonians are sometimes used to model dissipative and irreversible processes.

Solution: Start with the time-dependent one-dimensional Schrödinger equation

$$H\psi = i\hbar\frac{\partial \psi}{\partial t}, \tag{5.3}$$

where

$$H = -\frac{\hbar^2}{2m^*}\frac{\partial^2}{\partial z^2} + V(z) + i\xi W(z), \tag{5.4}$$

Problem Solving in Quantum Mechanics: From Basics to Real-World Applications for Materials Scientists, Applied Physicists, and Devices Engineers, First Edition.
Marc Cahay and Supriyo Bandyopadhyay.
© 2017 John Wiley & Sons Ltd. Published 2017 by John Wiley & Sons Ltd.

and its complex conjugate

$$H^*\psi^* = -i\hbar\frac{\partial\psi^*}{\partial t}. \tag{5.5}$$

We multiply Equation (5.3) on the left by ψ^* to yield

$$\psi^* H\psi = i\hbar\psi^*\frac{\partial\psi}{\partial t}, \tag{5.6}$$

and Equation (5.5) on the left by ψ to yield

$$\psi H^*\psi^* = -i\hbar\psi\frac{\partial\psi^*}{\partial t}. \tag{5.7}$$

Using the explicit expression for the Hamiltonian H given in Equation (5.4), we get from Equation (5.3) that

$$-\frac{\hbar^2}{2m^*}\psi^*\frac{\partial^2\psi}{\partial z^2} + [V(z) + i\xi W(z)]\,\psi^*\psi = +i\hbar\psi^*\frac{\partial\psi}{\partial t}, \tag{5.8}$$

and from Equation (5.5) we get

$$-\frac{\hbar^2}{2m^*}\psi\frac{\partial^2\psi^*}{\partial z^2} + [V(z) - i\xi W(z)]\,\psi^*\psi = -i\hbar\psi\frac{\partial\psi^*}{\partial t}. \tag{5.9}$$

Substracting Equation (5.9) from Equation (5.8) leads to

$$-i\hbar\left(\frac{\hbar}{2m^*i}\right)\left[\psi^*\frac{\partial^2\psi}{\partial z^2} - \psi\frac{\partial^2\psi^*}{\partial z^2}\right] + 2i\xi W(z)\psi^*\psi = i\hbar\frac{\partial}{\partial t}(\psi^*\psi). \tag{5.10}$$

Using the definitions of the probability current density,

$$J_z(z,t) = \frac{\hbar}{2m^*i}\left[\psi^*(z,t)\frac{\partial\psi(z,t)}{dz} - \psi(z)\frac{\partial\psi^*(z,t)}{dz}\right], \tag{5.11}$$

and the probability charge density,

$$\rho(z,t) = \psi^*(z,t)\psi(z,t), \tag{5.12}$$

Equation (5.10) can be rewritten as

$$-i\hbar\frac{\partial}{\partial z}J_z - i\hbar\frac{\partial}{\partial t}(\psi^*\psi) = -2i\xi W(z)\psi^*\psi. \tag{5.13}$$

So, finally,

$$\frac{\partial J_z(z,t)}{\partial z} + \frac{\partial\rho(z,t)}{\partial t} = \frac{2\xi}{\hbar}W(z)\rho(z,t). \tag{5.14}$$

Clearly, a non-Hermitian Hamiltonian violates current continuity since the right-hand side of the above equation is non-zero. Equations (5.11)–(5.12) must be multiplied by q, the electronic charge, to get the current and charge densities, respectively.

Physical significance of this result: It is common practice to model dissipation in a quantum mechanical system by invoking a non-Hermitian Hamiltonian. This is unacceptable *unless* there is an underlying model for the system to exchange charge with the surroundings (such a system is called an open quantum system) which would conserve charge in the "universe" consisting of the system and the surroundings, but not in the system alone. It is always imperative to physically justify a non-Hermitian Hamiltonian before invoking it.

---

* Problem 5.2: Properties of the reflected wave in a scattering problem

Consider the steady-state tunneling problem shown in Figure 1.1 for the case of zero bias. Assume the wave function in the left contact is given by the coherent superposition of an incident and a reflected wave,

$$\phi(z) = e^{\nu z} + re^{-\nu z}, \tag{5.15}$$

where r is the complex reflection coefficient and ν is the complex propagation constant given by

$$\nu = \kappa + ik. \tag{5.16}$$

(a) Find the expressions for the expectation values of the charge and current density in this region of space (i.e., the left contact).

(b) Assume that $\kappa = 0$, i.e. the wave is not evanescent. In this case, show that in the absence of the reflected wave ($r = 0$), the charge density in the left contact region is spatially uniform. Show that in the presence of the reflected wave, it is non-uniform. Explain the significance of this result.

(c) Show that if there is no evanescent component to the wave function, then the current density in the left contact is the difference between the incident and reflected current densities.

(d) If the wave is purely evanescent (as in the case of tunneling through a potential barrier, i.e., $k = 0$), show that the current density will be zero if the reflection coefficient is purely real. Therefore, a non-zero tunneling current requires an imaginary or complex reflection coefficient.

Solution:
(a) The charge density in the left contact region is given by

$$\rho(z) = q|\phi(z)|^2 = q\left(e^{\nu z} + re^{-\nu z}\right)^* \left(e^{\nu z} + re^{-\nu z}\right), \tag{5.17}$$

where q is the electronic charge.

Expanding, we get

$$\rho(z) = q\left[e^{2\kappa z} + |r|^2 e^{-2\kappa z} + 2\mathrm{Re}(re^{-2ikz})\right]. \tag{5.18}$$

Note that the expectation value of the charge density is purely real, as expectation value of any physical quantity ought to be.

The current density associated with the wave function (5.15) is

$$
\begin{aligned}
J_z(z) &= \frac{q\hbar}{2m^*i}\left[\phi^*(z)\frac{d\phi(z)}{dz} - \phi(z)\frac{d\phi^*(z)}{dz}\right] \\
&= \frac{iq\hbar}{2m^*}\left[(e^{\nu z} + re^{-\nu z})(\nu e^{\nu z} - \nu re^{-\nu z})^* - \text{c.c.})\right],
\end{aligned}
\tag{5.19}
$$

where c.c. stands for complex conjugate. Expanding the last expression,

$$
J_z(z) = \frac{q\hbar k}{m^*}\left[e^{2\kappa z} - |r|^2 e^{-2\kappa z}\right] - \frac{2q\hbar\kappa}{m^*}\text{Im}\left(re^{-2ikz}\right),
\tag{5.20}
$$

where Im stands for the imaginary part.

(b) With $\kappa = 0$, we see from Equation (5.18) that

$$
\rho(z) = q\left[1 + |r|^2 + 2\text{Re}(re^{-2ikz})\right].
\tag{5.21}
$$

The spatially varying term (i.e., the term which depends on the coordinate z) vanishes if $r = 0$, i.e., when there is no reflected wave. In that case, the charge density is spatially uniform. However, if there is a reflected wave, then the incident and reflected waves interfere to cause spatial modulation of the charge density in the region, as seen from Equation (5.20). Without the reflected wave, there is no interference and hence no spatial modulation. The spatial modulation is therefore a consequence of interference between two waves.

(c) If $\kappa = 0$, then from Equation (5.20) we get

$$
J_z = \frac{q\hbar k}{m^*}(1 - |r|^2).
\tag{5.22}
$$

Therefore the total current is the difference of the forward-traveling component and the backward-traveling (reflected) component. It is also spatially invariant, as it must be in steady-state transport.

(d) If the wave is purely evanescent and $k = 0$, then

$$
J_z = \frac{2q\hbar\kappa}{m^*}\text{Im}(r) = 0
\tag{5.23}
$$

if r is purely real.

Physical significance of this result: No tunneling can occur if the reflection coefficient is purely real.

* Problem 5.3: Conservation of current density and the general scattering problem

Start with the definition of the quantum mechanical steady-state current density derived in Problem 5.1:

$$J_z(z) = \frac{q\hbar}{2m^*i} \left[\phi^*(z)\frac{d\phi}{dz} - \phi(z)\frac{d\phi^*}{dz} \right], \tag{5.24}$$

and the time-independent Schrödinger equation derived in Problem 1.1. Assume a constant effective mass and a conduction band energy profile depending only on the z-coordinate.

Show that $\frac{dJ_z(z)}{dz} = 0$, i.e., J_z is independent of position.

For the tunneling problem with a left incident electron, as shown in Figure 1.1, use $E_c(0) = 0$ to express k_0 and k_L in terms of the total energy E of the incident electron in the left contact. The quantity k_0 is the component of the incident wave's wavevector in the z-direction (in the left contact) and the quantity k_L is the component of the transmitted wave's wavevector in the z-direction (in the right contact).

The tunneling probability is defined as the ratio of the outgoing (i.e., in the right contact) current density to the incoming current density. Express the tunneling probability in terms of k_0, k_L, and t, the transmission amplitude.

The reflection probability is defined as the ratio of the reflected current density to the incoming current density. What is the expression for the reflection probability in terms of k_0, k_L, and r, the reflection amplitude?

Solution: Starting with the results of Problem 1.1, $\phi(z)$ satisfies

$$\frac{d}{dz}\left[\frac{1}{\gamma(z)}\frac{d}{dz} \right]\phi(z) + \frac{2m_c^*}{\hbar^2}\left\{ E_p + E_t[1 - \gamma^{-1}(z)] - E_c(z) \right\}\phi(z) = 0. \tag{5.25}$$

Since the effective mass is spatially invariant, $m^*(z) = m^*$, and hence $\gamma(z) = 1$. As a result, the Schrödinger equation becomes

$$\frac{d^2}{dz^2}\phi(z) + \beta^2(z)\phi(z) = 0, \tag{5.26}$$

with

$$\beta^2(z) = \frac{2m_c^*}{\hbar^2}(E_p - E_c(z)). \tag{5.27}$$

Therefore,

$$\begin{aligned}
\frac{dJ_z(z)}{dz} &= \frac{q\hbar}{2m^*i}\left[\frac{d\phi^*}{dz}\frac{d\phi}{dz} + \phi^*\frac{d^2\phi}{dz^2} - \frac{d\phi}{dz}\frac{d\phi^*}{dz} - \phi\frac{d^2\phi^*}{dz^2} \right] \\
&= \frac{q\hbar}{2m^*i}\left[\phi^*(z)(-\beta^2\phi(z)) - \phi(z)(-\beta^{*2}\phi^*(z)) \right]. \tag{5.28}
\end{aligned}$$

Since $\beta^2 = \beta^{*2}$, $\frac{dJ_z}{dz} = 0$ and J_z is a constant. In the left contact, the total energy E is the sum of the transverse plus longitudinal energy E_p,

$$E = E_p + \frac{\hbar^2 k_t^2}{2m^*}, \tag{5.29}$$

where k_t is the transverse component of the electron's wavevector. Therefore,

$$\frac{\hbar^2 k_0^2}{2m^*} = E - \frac{\hbar^2 k_t^2}{2m^*} = E - E_t = E_p \tag{5.30}$$

and

$$k_0 = \frac{1}{\hbar}\sqrt{2m^* E_p}. \tag{5.31}$$

In the right contact,

$$E = \frac{\hbar^2 k_L^2}{2m^*} + \frac{\hbar^2 k_t^2}{2m^*} + E_c(L), \tag{5.32}$$

i.e.,

$$\frac{\hbar^2 k_L^2}{2m^*} + E_t - eV_{\text{bias}} = E = E_t + E_p, \tag{5.33}$$

or

$$\frac{\hbar^2 k_L^2}{2m^*} = E_p + eV_{\text{bias}} = \frac{\hbar^2 k_0^2}{2m^*} + eV_{\text{bias}}. \tag{5.34}$$

Therefore,

$$k_L = \frac{1}{\hbar}\sqrt{2m^*(E_p + eV_{\text{bias}})}. \tag{5.35}$$

So, for $z > L$, where the wave function is $te^{ik_L(z-L)}$,

$$J_z(z) = \frac{q\hbar}{2m^* i}\left[ik_L te^{+ik_L(z-L)}t^* e^{-ik_L(z-L)} \right.$$
$$\left. - te^{ik_L(z-L)}t^*(-ik_L)e^{-ik_L(z-L)}\right], \tag{5.36}$$

and therefore, at $z = L$,

$$J_z(z = L) = \frac{q\hbar}{2m^* i}\left[2ik_L|t|^2\right] = \frac{q\hbar k_L}{m^*}|t|^2. \tag{5.37}$$

Similarly, at $z = 0$,

$$J_z^{\text{inc}} = \frac{q\hbar k_0}{m^*}, \tag{5.38}$$

and the reflected current density at $z = 0$ is given by

$$J_z^{\text{refl}} = \frac{q\hbar k_0}{m^*}|r|^2. \tag{5.39}$$

The transmission probability is, by definition,

$$T = \frac{J_z^{\text{trans}}}{J_z^{\text{inc}}} = \frac{k_L}{k_0}|t|^2 = \frac{k_L}{k_0}|\phi(z = L)|^2, \tag{5.40}$$

and the reflection probability is given by

$$R = \frac{J_z^{\text{refl}}}{J_z^{\text{inc}}} = |r|^2. \tag{5.41}$$

In the case of coherent ballistic transport, the incident current density must be either reflected or transmitted and therefore the following equality must hold:

$$1 = T + R = \frac{k_L}{k_0}|t|^2 + |r|^2. \tag{5.42}$$

---

* Problem 5.4: Definition of current amplitude

In a region of constant potential energy E_c, the solution to the one-dimensional Schrödinger equation can be written as

$$\phi(z) = A^+ e^{ikz} + A^- e^{-ikz}, \tag{5.43}$$

where $k = \frac{1}{\hbar}\sqrt{2m^*(E - E_c)}$. The coefficients A^+ and A^- are the amplitudes of the right- and left-propagating solutions, respectively.

We have seen that under steady-state conditions, the current density is spatially invariant and is given by

$$J_z = \frac{q\hbar^2}{2m^*i}\left[\phi^*(z)\frac{d\phi}{dz} - \phi(z)\frac{d\phi^*}{dz}\right]. \tag{5.44}$$

The goal of this problem is to calculate the current density amplitudes associated with the left- and right-propagating plane waves in Equation (5.43). Defining the quantities

$$\phi^+(z) = \frac{1}{2}\sqrt{\frac{q\hbar k}{m^*}}\left[\phi(z) + \frac{1}{ik}\frac{d\phi(z)}{dz}\right] \tag{5.45}$$

and

$$\phi^-(z) = \frac{1}{2}\sqrt{\frac{q\hbar k}{m^*}}\left[\phi(z) - \frac{1}{ik}\frac{d\phi(z)}{dz}\right], \tag{5.46}$$

show that the current density associated with $\phi(z)$ can be calculated as follows:

$$J_z(z) = (\phi^+(z))^*\phi^+(z) - (\phi^-(z))^*\phi^-(z), \tag{5.47}$$

where ϕ^+, ϕ^- are referred to as the left-propagating and right-propagating current density amplitudes.

For the wave function given in Equation (5.43), calculate the explicit expressions for

$$J^+ = (\phi^+)^*\phi^+, \tag{5.48}$$

the right-propagating current density amplitude, and

$$J^- = (\phi^-)^*\phi^-, \tag{5.49}$$

the left-propagating current density amplitude.

Solution: Starting with Equations (5.45) and (5.46), we get

$$\phi(z) = \sqrt{\frac{m^*}{q\hbar k}}\left[\phi^+(z) + \phi^-(z)\right], \tag{5.50}$$

and

$$\frac{d\phi(z)}{dz} = i\sqrt{\frac{m^*k}{q\hbar}}\left[\phi^+(z) - \phi^-(z)\right]. \tag{5.51}$$

Hence,

$$
\begin{aligned}
J_z(z) &= \frac{q\hbar}{2m^*i}\left[\phi^*(z)\frac{d\phi(z)}{dz} - \phi(z)\frac{d\phi^*(z)}{dz}\right]\\[2mm]
&= \frac{q\hbar}{2m^*i}\left[\sqrt{\frac{m^*}{q\hbar k}}\left[(\phi^+)^* + (\phi^-)^*\right]i\sqrt{\frac{m^*}{q\hbar k}}\left[\phi^+ - \phi^-\right]\right.\\[2mm]
&\quad\left. - \sqrt{\frac{m^*}{q\hbar k}}(\phi^+ + \phi^-)(-i)\sqrt{\frac{m^*}{q\hbar k}}\left[(\phi^+)^* - (\phi^-)^*\right]\right]\\[2mm]
&= \frac{1}{2}\left[(\phi^+)^*\phi^+ - (\phi^+)^*\phi^+ + (\phi^-)^*\phi^+ - (\phi^-)^*\phi^-\right.\\[2mm]
&\quad\left. + \phi^+(\phi^+)^* - \phi^+(\phi^-)^* + \phi^-(\phi^+)^* - \phi^-(\phi^-)^*\right]\\[2mm]
&= (\phi^+)^*\phi^+ - (\phi^-)^*\phi^-. \tag{5.52}
\end{aligned}
$$

Plugging in the wave function (5.43) into Equations (5.45) and (5.46), we obtain

$$\phi^+(z) = \sqrt{\frac{q\hbar k}{m^*}}A^+e^{ikz} \tag{5.53}$$

and

$$\phi^-(z) = \sqrt{\frac{q\hbar k}{m^*}}A^-e^{-ikz}. \tag{5.54}$$

Hence,

$$J^+ = \frac{q\hbar k}{m^*}|A^+|^2 \tag{5.55}$$

and

$$J^- = \frac{q\hbar k}{m^*}|A^-|^2, \tag{5.56}$$

which are the values of the current densities associated with the right- and left-moving portions of the wave functions in Equation (5.43), respectively.

The concept of current density amplitude is used in a set of problems on the scattering matrix in Chapter 8.

---

* Problem 5.5: Reflection and transmission probabilities across a potential step

Consider a potential step as shown in Figure 5.1. The effective mass on the left and right side of the step is equal to m_1^* and m_2^*, respectively. The step height is ΔE_c.

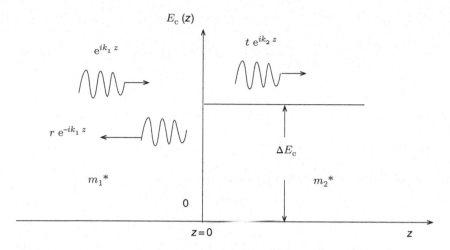

Figure 5.1: Illustration of electron impinging from the left on a potential step with height ΔE_c. The effective mass is assumed to be m_1^* and m_2^* on the left and right side of the step, respectively.

Start with the general time-independent Schrödinger equation for an electron moving in an arbitrary potential energy profile $E_\mathrm{c}(z)$ and with a spatially varying effective mass $m^*(z)$ derived in Problem 1.1.

Write down the Schrödinger equation for the z-component of the wave function $\phi(z)$ on the left and right sides of the potential step assuming that the electron is incident from the left contact with a transverse kinetic energy $E_\mathrm{t} = \frac{\hbar^2 k_\mathrm{t}^{\,2}}{2m_1^*}$.

Assume a plane wave is incident from the left and that the total energy of the incident electron is large enough so that it is transmitted on the other side. Write down the analytical form of the solution to the Schrödinger equation on either side of the junction.

By matching the wave function $\phi(z)$ at $z = 0$ and also $\frac{1}{m^*(z)}\frac{\mathrm{d}\phi(z)}{\mathrm{d}z}$ at $z = 0$, calculate the reflection and transmission amplitudes of the incident wave.

Calculate the reflection and transmission probabilities across the step starting with the the quantum mechanical expression for the current densities of the incident, reflected, and transmitted beams.

Prove that the sum of the reflection and transmission probabilities is equal to unity.

Solution: Starting with the results of Problem 1.1 and using m_1^* as the effective mass in the contact, the Schrödinger equation for $z < 0$ is

$$\frac{\mathrm{d}^2\phi(z)}{\mathrm{d}z^2} + \frac{2m_1^*}{\hbar^2}\left(E - E_\mathrm{t}\right)\phi(z) = 0, \tag{5.57}$$

where E, E_t are the total and transverse energy of the electron, respectively.

For $z > 0$, we have

$$\frac{d^2\phi(z)}{dz^2} + \frac{2m_2^*}{\hbar^2}\left(E - \frac{E_t}{\gamma} - \Delta E_c\right)\phi(z) = 0, \tag{5.58}$$

where $\gamma = m_2^*/m_1^*$.

For a plane wave incident from the left to be transmitted across the step, we must have

$$E > \frac{E_t}{\gamma} + \Delta E_c. \tag{5.59}$$

For $z < 0$, the solution to the Schrödinger equation is

$$\phi_I = e^{ik_1 z} + re^{-ik_1 z}, \tag{5.60}$$

with

$$k_1 = \frac{1}{\hbar}\sqrt{2m_1^*(E - E_t)}. \tag{5.61}$$

For $z > 0$, the solution to the Schrödinger equation is

$$\phi_{II} = te^{ik_2 z}, \tag{5.62}$$

with

$$k_2 = \frac{1}{\hbar}\sqrt{2m_2^*\left(E - \frac{E_t}{\gamma} - \Delta E_c\right)}. \tag{5.63}$$

Continuity of the wave function at $z = 0$ gives

$$1 + r = t. \tag{5.64}$$

Continuity of $\frac{1}{m^*(z)}\frac{d\phi}{dz}$ at $z = 0$ requires

$$\frac{ik_1}{m_1^*}(1 - r) = \frac{ik_2 t}{m_2^*}, \tag{5.65}$$

which can be rewritten as

$$1 - r = \left(\frac{k_2}{k_1}\frac{m_1}{m_2}\right)t. \tag{5.66}$$

Using Equations (5.64) and (5.66), we get

$$t = \frac{2}{\left[1 + \left(\frac{k_2}{k_1}\right)\left(\frac{m_1^*}{m_2^*}\right)\right]} \tag{5.67}$$

and

$$r = \frac{1 - \frac{k_2}{k_1}\frac{m_1^*}{m_2^*}}{1 + \frac{k_2}{k_1}\frac{m_1^*}{m_2^*}}. \tag{5.68}$$

The proof that $|r|^2 + \frac{k_2}{k_1}\frac{m_1^*}{m_2^*}|t|^2 = 1$, obtained by equating the incident current density to the sum of the reflected and transmitted current densities, is left as an exercise. The quantities $|r|^2$ and $\frac{k_2}{k_1}\frac{m_1^*}{m_2^*}|t|^2$ are the reflection and transmission probabilities across the potential step, respectively.

** Problem 5.6: Tunneling across an absorbing delta scatterer

The Schrödinger equation describing propagation of an electron through a region containing an absorbing one-dimensional δ-scatterer is given by

$$-\frac{\hbar^2}{2m^*}\phi'' + [V_0\delta(z) - iW_0\delta(z)]\phi = E\phi, \qquad (5.69)$$

where, once again, we adopt the convention that single prime represents a first derivative with respect to position and double prime represents the second derivative. Here, W_0 is the strength of the imaginary portion (absorbing) of the δ-scatterer.

Consider an electron incident from the left on the absorbing potential and show that the reflection (r) and transmission (t) amplitudes satisfy the relation

$$|r|^2 + |t|^2 + A = 1, \qquad (5.70)$$

where the absorbing probability is given by

$$A = \frac{2m^*W_0}{\hbar^2 k_0} \frac{1}{\left[\frac{m^{*2}V_0^2}{\hbar^4 k_0^2} + \left(1 + \frac{m^*W_0}{k_0\hbar^2}\right)^2\right]}, \qquad (5.71)$$

with $k_0 = \frac{1}{\hbar}\sqrt{2m^*E}$.

Solution: As was done in several previous problems, we integrate the Schrödinger Equation (5.69) on a small interval around $z = 0$. This leads to

$$\phi'(0^+) - \phi'(0_-) = \frac{2m^*}{\hbar^2}[V_0 - iW_0]\phi(0). \qquad (5.72)$$

For an electron incident from the left, continuity of the wave function across the δ-scatterer leads to

$$1 + r = t. \qquad (5.73)$$

Multiplying the last equation on both sides by ik_0 gives

$$ik_0t - ik_0r = ik_0. \qquad (5.74)$$

Furthermore, a second relation between the r and t amplitudes is obtained from Equation (5.72),

$$ik_0t - ik_0(1 - r) = \frac{2m^*}{\hbar^2}(V_0 - iW_0)t, \qquad (5.75)$$

which can be rewritten as

$$\left[ik_0 - \frac{2m^*}{\hbar^2}(V_0 - iW_0)\right]t + ik_0r = ik_0. \qquad (5.76)$$

Solving Equations (5.74) and (5.76) for r and t, we get the transmission probability

$$T = |t|^2 = \frac{k_0^2}{\frac{m^{*2}V_0^2}{\hbar^4} + \left(k_0 + \frac{m^*W_0}{\hbar^2}\right)^2}, \qquad (5.77)$$

and the reflection probability

$$R = |r|^2 = \frac{\frac{m^{*2}V_0^2}{\hbar^4} + \frac{m^{*2}W_0^2}{\hbar^4}}{\frac{m^{*2}V_0^2}{\hbar^4} + \left(k_0 + \frac{m^*W_0}{\hbar^2}\right)^2}. \qquad (5.78)$$

Substituting Equations (5.77) and (5.78) into Equation (5.70) leads to the following result for the absorption probability:

$$A = \frac{2m^*W_0k_0/\hbar^2}{\frac{m^{*2}V_0^2}{\hbar^4} + \left(k_0 + \frac{m^*W_0}{\hbar^2}\right)^2} = \frac{2m^*W_0}{\hbar^2 k_0} \frac{1}{\left[\frac{m^{*2}V_0^2}{\hbar^4 k_0^2} + \left(1 + \frac{m^*W_0}{k_0\hbar^2}\right)^2\right]}. \qquad (5.79)$$

Plots of T, R, and A versus incident electron energy for a repulsive delta scatterer with strength V_0 equal to 0.1 eV-Å for different values of W_0 are shown in Figures 5.2 and 5.3. The Matlab code to generate these figures is given in Appendix G.

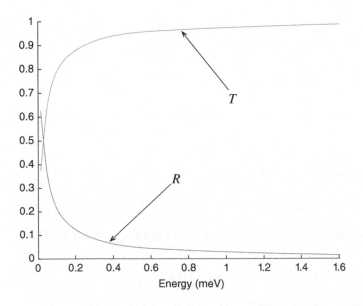

Figure 5.2: Plot of the transmission probability T, reflection probability R, and absorption probability A as a function of electron incident kinetic energy for a repulsive delta scatterer with strength V_0 equal to 0.1 eV-Å and W_0 equal to 0 (no absorption). The effective mass is assumed to be $m^* = 0.067m_0$.

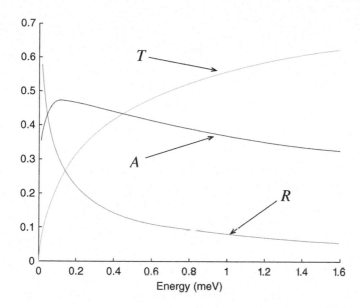

Figure 5.3: Plot of the transmission probability T, reflection probability R, and absorption probability A as a function of electron incident kinetic energy for a delta scatterer with strength V_0 equal to 0.1 eV-Å and W_0 equal to 0.2 eV-Å. The effective mass is assumed to be $m^* = 0.067m_0$.

*** Problem 5.7: Tunneling across an absorbing barrier

Consider the problem of an electron incident from the left on the absorbing well shown in Figure 5.4. In the barrier, the potential energy has a real and an imaginary part, i.e., $V(z) = V_0 - iW_0$, where W_0 is real and positive. The quantity V_0 is real and negative for a potential well.

Starting with the one-dimensional time-independent Schrödinger equation and assuming a constant effective mass, show that the reflection (r) and transmission (t) amplitudes satisfy the relation

$$|r|^2 + |t|^2 + A = 1, \qquad (5.80)$$

where A is the absorbing coefficient in the well and is given by

$$A = \frac{2m^*W_0}{\hbar^2 k} \int_0^W \phi_{\mathrm{II}}(z)\phi_{\mathrm{II}}{}^*(z)\mathrm{d}z, \qquad (5.81)$$

where $\phi_{\mathrm{II}}(z)$ is the solution to the Schrödinger equation in the well region.

Solution: The solution to the one-dimensional time-independent Schrödinger equation

$$-\frac{\hbar^2}{2m^*}\frac{\mathrm{d}^2\phi}{\mathrm{d}z^2} + V(z)\phi = E\phi \qquad (5.82)$$

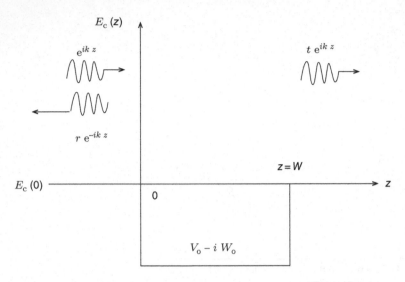

Figure 5.4: Illustration of an electron impinging from the left on an absorbing well. The effective mass is assumed to be the same throughout.

corresponding to an electron incident from the left is given by

$$\phi_{\mathrm{I}} = e^{ikz} + re^{-ikz} \tag{5.83}$$

for $z < 0$ and

$$\phi_{\mathrm{III}}(z) = te^{ikz} \tag{5.84}$$

for $z > W$, where

$$k = \frac{1}{\hbar}\sqrt{2m^*E}, \tag{5.85}$$

E being the kinetic energy of the incident electron. In the well region, the solution to the Schrödinger equation is given by

$$\phi_{\mathrm{II}} = Ce^{i\alpha z} + De^{-i\alpha z}, \tag{5.86}$$

where α is a complex number such that

$$\alpha^2 = \frac{2m}{\hbar^2}(E - V_0 + iW_0). \tag{5.87}$$

The selection of the appropriate signs for the real and imaginary parts of α will be discussed later.

Imposing the continuity of the wave function and its derivative at $z = 0$, we obtain the following two equations:

$$1 + r = C + D, \tag{5.88}$$

$$\frac{k}{\alpha}(1 - r) = C - D. \tag{5.89}$$

Similarly, the continuity of the wave function and its derivative at $z = W$ results in the additional relations:

$$Ce^{i\alpha W} + De^{-i\alpha W} = te^{i\alpha W}, \tag{5.90}$$

$$\frac{\alpha}{k}\left(Ce^{i\alpha W} - De^{-i\alpha W}\right) = te^{i\alpha W}. \tag{5.91}$$

Subtraction of Equation (5.89) from Equation (5.88) gives

$$(1+r) - \frac{k}{\alpha}(1-r) = 2D. \tag{5.92}$$

Subtraction of Equation (5.91) from Equation (5.90) and then multiplication of both sides by $e^{-i\alpha W}$ gives

$$C\left(1 - \frac{\alpha}{k}\right) + De^{-2i\alpha W}\left(1 + \frac{\alpha}{k}\right) = 0. \tag{5.93}$$

Multiplication of Equation (5.88) by $(1 - \frac{\alpha}{k})$ yields

$$(1+r)\left(1 - \frac{\alpha}{k}\right) = (C+D)\left(1 - \frac{\alpha}{k}\right). \tag{5.94}$$

Substitution of the value of $C\left(1 - \frac{\alpha}{k}\right)$ from Equation (5.92) in Equation (5.93) leads to

$$(1+r)\left(1 - \frac{\alpha}{k}\right) = D\left[\left(1 - e^{-2i\alpha W}\right) - \frac{\alpha}{k}\left(1 + e^{-2i\alpha W}\right)\right]. \tag{5.95}$$

Substituting the value of D from this last equation into Equation (5.92) and solving for r leads to

$$r = \frac{\left[\left(1 - \left(\frac{k}{\alpha}\right)^2\right)\left(1 - e^{-2i\alpha W}\right)\right]}{\left[\left(1 + \frac{k}{\alpha}\right)^2 e^{-2i\alpha W} - \left(1 - \frac{k}{\alpha}\right)^2\right]}. \tag{5.96}$$

Equation (5.91) can be rewritten as

$$\left(Ce^{i\alpha W} - De^{-i\alpha W}\right) = \frac{k}{\alpha}te^{ikW}. \tag{5.97}$$

Subtraction of Equation (5.97) from Equation (5.90) yields

$$2De^{-i\alpha W} = te^{ikW}\left(1 - \frac{k}{\alpha}\right). \tag{5.98}$$

Substituting the value of $2D$ from Equation (5.92) in this last equation, we obtain

$$\left[(1+r) - \frac{k}{\alpha}(1-r)\right]e^{-i\alpha W} = te^{ikW}\left(1 - \frac{k}{\alpha}\right). \tag{5.99}$$

Finally, substitution of the value of r from Equation (5.96) in this last equation gives the transmission amplitude t as

$$t = \frac{4\frac{k}{\alpha}}{\left[\left(1+\frac{k}{\alpha}\right)^2 e^{i(k-\alpha)W} - \left(1-\frac{k}{\alpha}\right)^2 e^{i(k+\alpha)W}\right]}. \tag{5.100}$$

In the well region, the Schrödinger equation is

$$\frac{d^2\phi_{II}}{dz^2} + \frac{2m}{\hbar^2}(E - V_0 + iW_0)\phi_{II} = 0. \tag{5.101}$$

Taking the complex conjugate of this last equation, we get

$$\frac{d^2\phi_{II}^*}{dz^2} + \frac{2m^*}{\hbar^2}(E - V_0 - iW_0)\phi_{II}^* = 0. \tag{5.102}$$

Now,

$$\left[\phi_{II}\frac{d\phi_{II}^*}{dz} - \phi_{II}^*\frac{d\phi_{II}}{dz}\right]\bigg|_0^W = \int_0^W \left(\phi_{II}\frac{d^2\phi_{II}^*}{dz^2} - \phi_{II}^*\frac{d^2\phi_{II}}{dz^2}\right)dz. \tag{5.103}$$

Substituting for $\frac{d^2\phi_{II}}{dz^2}$ and $\frac{d^2\phi_{II}^*}{dz^2}$ from Equations (5.101) and (5.102), respectively, in the last equation, we obtain

$$\left[\frac{d\phi_{II}^*}{dz}\phi_{II} - \frac{d\phi_{II}}{dz}\phi_{II}^*\right]\bigg|_0^W = \frac{4m^*iW_0}{\hbar^2}\int_0^W \phi_{II}^*\phi_{II}dz. \tag{5.104}$$

Using the expressions for the wave function on both sides of the interface, the left-hand side of Equation (5.104) becomes

$$\left[\phi_{II}\frac{d\phi_{II}^*}{dz} - \phi_{II}^*\frac{d\phi_{II}}{dz}\right]\bigg|_0^W = 2ik\left(1 - |r|^2 - |t|^2\right). \tag{5.105}$$

Substitution of this result in Equation (5.104) leads to

$$|r|^2 + |t|^2 + A = 1, \tag{5.106}$$

with

$$A = \frac{2m^*W_0}{\hbar^2 k}\int_0^W \phi_{II}(z)\phi_{II}^*(z)dz. \tag{5.107}$$

The explicit form of the absorption coefficient can be found by performing the integral in the last equation using Equation (5.86). This leads to

$$\int_0^W \phi_{II}^*\phi_{II}dz = CC^*\left[\frac{e^{-2\alpha_i W}-1}{-2\alpha_i}\right] + DD^*\left[\frac{e^{2\alpha_i W}-1}{2\alpha_i}\right]$$
$$+ CD^*\left[\frac{e^{2\alpha_r W}-1}{2i\alpha_r}\right] + C^*D\left[\frac{e^{-2i\alpha_r W}-1}{-2i\alpha_r}\right], \tag{5.108}$$

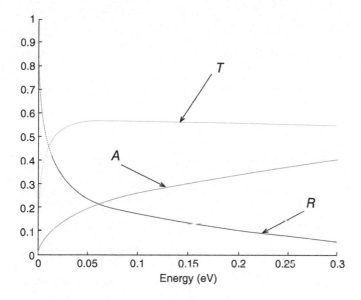

Figure 5.5: Plot of the transmission probability T, reflection probability R, and absorption probability A as a function of the electron incident kinetic energy for a potential well of width of $50\,\text{Å}$, depth V_0 equal to $-0.3\,\text{eV}$, and absorbing potential W_0 equal to $0.1\,\text{eV}$. The effective mass is assumed to be $m^* = 0.067 m_0$.

where α_r and α_i are the real and imaginary parts of α, respectively, and the following shorthand notations were used:

$$D = \frac{1}{2}\left(1 - \frac{k}{\alpha}\right) t e^{ikL} e^{i\alpha W}, \tag{5.109}$$

$$C = \frac{1}{2}\left(1 + \frac{k}{\alpha}\right) t e^{ikL} e^{-i\alpha W}. \tag{5.110}$$

Figure 5.5 is a plot of the transmission probability T, reflection probability R, and absorption probability A as functions of the electron incident kinetic energy for a potential well of width of $50\,\text{Å}$, depth V_0 equal to $-0.3\,\text{eV}$, and absorbing potential W_0 equal to $0.1\,\text{eV}$. The effective mass is assumed to be $m^* = 0.067 m_0$. The sum $T + R + A = 1$ was checked numerically.

* Problem 5.8: Energy conservation law

Preliminary: *We start with the time-independent Schrödinger equation*

$$-\frac{\hbar^2}{2m^*}\nabla^2 \psi + V(\vec{r})\psi = E\psi, \tag{5.111}$$

where ψ is subject to the condition

$$\int d\vec{r}^3 \psi^*\psi = 1. \tag{5.112}$$

Multiplying the Schrödinger equation on both sides by ψ^* and integrating over all space, we get

$$E = \int d^3\vec{r}\,\psi^*\left[-\frac{\hbar^2}{2m^*}\nabla^2\psi + V(\vec{r})\psi \right]. \tag{5.113}$$

We can integrate the first term by parts and use Green's theorem to get

$$\int d^3\vec{r}\,\psi^*\nabla^2\psi = \int_S d\vec{s}\,\psi^*\vec{\nabla}\psi - \int d\vec{r}^3\vec{\nabla}\psi^*\cdot\vec{\nabla}\psi. \tag{5.114}$$

The normalization integral exists if and only if, at large r,

$$\psi \sim r^{-\frac{3}{2}-\epsilon}, \tag{5.115}$$

where $\epsilon > 0$. The surface integral then vanishes if $S \to \infty$ and the energy becomes

$$E = \int d^3\vec{r}\left\{ \frac{\hbar^2}{2m^*}\vec{\nabla}\psi^*\cdot\vec{\nabla}\psi + \psi^*V(r)\psi \right\}. \tag{5.116}$$

Starting with the concept of energy density defined above and the time-dependent Schrödinger equation, prove the following law of conservation of energy:

$$\frac{\partial w}{\partial t} + \vec{\nabla}\cdot\vec{S} = 0, \tag{5.117}$$

where the energy flux density w is

$$w = \frac{\hbar^2}{2m^*}\vec{\nabla}\psi^*\cdot\vec{\nabla}\psi + \psi^*V(\vec{r})\psi, \tag{5.118}$$

and $\vec{S}$ is the energy flux vector

$$\vec{S} = -\frac{\hbar^2}{2m^*}\left(\frac{\partial\psi^*}{\partial t}\vec{\nabla}\psi + \frac{\partial\psi}{\partial t}\vec{\nabla}\psi^* \right). \tag{5.119}$$

Solution: Taking the time derivative of w in Equation (5.118), we get

$$\frac{\partial w}{\partial t} = \frac{\hbar^2}{2m^*}\left(\vec{\nabla}\frac{\partial\psi^*}{\partial t}\cdot\vec{\nabla}\psi + \vec{\nabla}\psi^*\cdot\vec{\nabla}\frac{\partial\psi}{\partial t} \right) + V(\vec{r})\left[\left(\frac{\partial\psi^*}{\partial t}\right)\psi + \psi^*\left(\frac{\partial\psi}{\partial t}\right) \right]. \tag{5.120}$$

Since

$$\vec{\nabla}\frac{\partial\psi^*}{\partial t}\cdot\vec{\nabla}\psi = \vec{\nabla}\cdot\left(\frac{\partial\psi^*}{\partial t}\vec{\nabla}\psi \right) - \frac{\partial\psi^*}{\partial t}\nabla^2\psi \tag{5.121}$$

and

$$\vec{\nabla}\psi^*\cdot\vec{\nabla}\frac{\partial\psi}{\partial t} = \vec{\nabla}\cdot\left(\frac{\partial\psi}{\partial t}\vec{\nabla}\psi^* \right) - \frac{\partial\psi^*}{\partial t}\psi\nabla^2, \tag{5.122}$$

we can rewrite $\frac{\partial w}{\partial t}$ as

$$
\begin{aligned}
\frac{\partial w}{\partial t} = \;& \vec{\nabla} \cdot \frac{\hbar^2}{2m^*}\left(\frac{\partial \psi^*}{\partial t}\vec{\nabla} + \frac{\partial \psi}{\partial t}\vec{\nabla}\psi^*\right) - \frac{\hbar^2}{2m^*}\left(\frac{\partial \psi^*}{\partial t}\right)\nabla^2\psi \\
& -\frac{\hbar^2}{2m^*}\left(\frac{\partial \psi}{\partial t}\right)\nabla^2\psi^* + \left(\frac{\partial \psi^*}{\partial t}\right)V(\vec{r})\psi + \left(\frac{\partial \psi}{\partial t}\right)V(\vec{r})\psi^*. \quad (5.123)
\end{aligned}
$$

The last four terms add up to zero if we use the time-dependent Schrödinger equation and its complex conjugate. Hence,

$$
\frac{\partial w}{\partial t} + \vec{\nabla}\cdot\vec{S} = 0, \quad (5.124)
$$

with w and $\vec{S}$ defined above. This is the quantum mechanical expression of the energy conservation law.

For the time-independent problem, this last equation leads to $\vec{\nabla}\cdot\vec{S} = 0$. For a one-dimensional scattering problem, that means that the energy flux S_z is independent of position.

---

* Problem 5.9: Energy flux of a plane wave

For a free particle solution of the one-dimensional Schrödinger equation, show that the energy flux

$$
S_z = -\frac{\hbar^2}{2m^*}\left(\frac{\partial \psi^*}{\partial t}\frac{d\psi}{dz} + \frac{\partial \psi}{\partial t}\frac{d\psi^*}{dz}\right) \quad (5.125)
$$

can be rewritten as

$$
S_z = \frac{\hbar^3}{2m^{*2}}\mathrm{Im}\left[\left(\frac{d^2\psi}{dz^2}\right)\left(\frac{d\psi^*}{dz}\right)\right], \quad (5.126)
$$

where Im stands for imaginary part.

Show that for a plane wave e^{ikz}, S_z is given by $\left(\frac{\hbar^2 k^2}{2m^*}\right)\hbar k$, i.e., the product of the kinetic energy of the particle and its momentum.

Solution: Using the time-dependent Schrödinger equation $H\psi = -\frac{\hbar}{i}\frac{\partial \psi}{\partial t}$ and its complex conjugate $H\psi^* = \frac{\hbar}{i}\frac{\partial \psi^*}{\partial t}$, the energy flux density becomes

$$
S(z,t) = -\frac{\hbar^2}{2m^*}\left[\frac{i}{\hbar}H\psi^*\left(\frac{d\psi}{dz}\right) - \frac{i}{\hbar}H\psi\left(\frac{d\psi^*}{dz}\right)\right]. \quad (5.127)
$$

For a free particle, the Hamiltonian is given by $H = -\frac{\hbar^2}{2m^*}\frac{d^2}{dz^2}$. Therefore, we obtain

$$S_z = \frac{\hbar^3}{4m^{*2}i}\left[\left(\frac{d^2\psi}{dz^2}\right)\left(\frac{d\psi^*}{dz}\right) - \left(\frac{d^2\psi^*}{dz^2}\right)\left(\frac{d\psi}{dz}\right)\right], \tag{5.128}$$

which can be rewritten as

$$S = \frac{\hbar^3}{2m^{*2}}\text{Im}\left[\left(\frac{d^2\psi}{dz^2}\right)\left(\frac{d\psi^*}{dz}\right)\right]. \tag{5.129}$$

For a plane wave, $\psi = e^{ikz}$,

$$\frac{d^2\psi}{dz^2} = -k^2 e^{ikz} \tag{5.130}$$

and

$$\frac{d^2\psi^*}{dz} = -ike^{-ikz}. \tag{5.131}$$

Hence,

$$S_z = \left(\frac{\hbar^2 k^2}{2m^*}\right)\frac{\hbar k}{m^*}. \tag{5.132}$$

---

* Problem 5.10: Relation between energy flux vector and probability current density for the time-independent Schrödinger equation

If ψ is a solution of the time-independent three-dimensional Schrödinger equation with energy E, prove that the energy flux vector is E times the probability current density vector $\vec{J}$. Assume a constant effective mass throughout.

Solution: As shown in Problem 5.8, the energy flux vector is given by

$$\vec{S} = -\frac{\hbar^2}{2m^*}\left[\frac{\partial\psi^*}{\partial t}\vec{\nabla}\psi + i\frac{\partial\psi}{\partial t}\vec{\nabla}\psi^*\right]. \tag{5.133}$$

Using the time-dependent Schrödinger equation and its complex conjugate $H\psi = -\frac{\hbar}{i}\frac{\partial\psi}{\partial t}$ and $H\psi^* = \frac{\hbar}{i}\frac{\partial\psi^*}{\partial t}$, we get

$$\vec{S} = -\frac{\hbar^2}{2m^*}\left[\frac{i}{\hbar}H\psi^*\vec{\nabla}\psi + \left(\frac{-i}{\hbar}\right)H\psi\vec{\nabla}\psi^*\right], \tag{5.134}$$

or

$$\vec{S} = \frac{\hbar}{2m^*i}\left[H\psi^*\vec{\nabla}\psi - H\psi\vec{\nabla}\psi^*\right]. \tag{5.135}$$

For stationary states $\frac{\partial E}{\partial t} = 0$, $H\psi = E\psi$ and therefore

$$\vec{S} = E\frac{\hbar}{2m^*i}\left[\psi^*\vec{\nabla}\psi - \psi\vec{\nabla}\psi^*\right] \tag{5.136}$$

and $\vec{S} = E\vec{J}$, i.e., the energy flux vector $\vec{S}$ is E times the probability current density $\vec{J}$.

* **Problem 5.11: Relation between the energy flux density and quantum mechanical wave impedance**

Rewrite the expression of the energy flux vector defined in Problem 5.8 in terms of the quantum mechanical wave impedance $Z_{QM}(z)$ for solutions of the time-independent Schrödinger equation with a potential energy profile $E_c(z)$ varying along the z-axis only.

Solution: For the time-independent Schrödinger equation with a spatially varying $E_c(z)$, it was shown in the previous problem that the energy flux density can be written as

$$S_z = EJ_z, \tag{5.137}$$

where J_z is the probability current density,

$$J_z = \frac{\hbar}{2m^*i}\left[\phi^*\frac{\mathrm{d}\phi}{\mathrm{d}z} - \phi\frac{\mathrm{d}\phi^*}{\mathrm{d}z}\right]. \tag{5.138}$$

Since the quantum mechanical wave impedance is defined as (see Problem 1.5)

$$Z_{QM}(z) = \frac{2\hbar}{m^*i}\frac{1}{\phi}\frac{\mathrm{d}\phi}{\mathrm{d}z}, \tag{5.139}$$

we get

$$S_z = E\frac{1}{2}\left[Z_{QM}(z) + Z_{QM}^*(z)\right]\phi^*(z)\phi(z) = E\,\mathrm{Re}\left(Z_{QM}(z)\right)\rho(z), \tag{5.140}$$

where $\mathrm{Re}\left(Z_{QM}(z)\right)$ is the real part of the quantum mechanical wave impedance and $\rho(z)$ is the probability density, $\phi^*(z)\phi(z)$.

---

* **Problem 5.12: Continuity of the energy flux across a potential step**

Consider an electron incident from the left on a potential step of height ΔE_c as shown in Figure 5.3. Assume the electron's effective mass is constant throughout and that the kinetic energy component associated with motion in the z-direction is larger than ΔE_c. Express the continuity of the energy flux at $z = 0$. Discuss the resulting equation.

Solution: Since the energy flux S_z is independent of z (as shown in Problem 5.8), we calculate its value on either side of $z = 0$ using the wave functions associated with the scattering problem shown in Figure 5.3.

For $z > 0$, the energy flux density is given by

$$S_z = \frac{\hbar}{2m^*i}\left[\left(-\frac{\hbar^2}{2m}\frac{\mathrm{d}^2}{\mathrm{d}z^2} + \Delta\right)\psi^*\frac{\mathrm{d}\psi}{\mathrm{d}z} - \left(-\frac{\hbar^2}{2m^*}\frac{\mathrm{d}^2}{\mathrm{d}z^2} + \Delta\right)\psi\frac{\mathrm{d}\psi^*}{\mathrm{d}z}\right]. \tag{5.141}$$

Since $\psi = t e^{ik_2 z}$ for $z > 0$, where t is the transmission amplitude and $k_2 = \frac{1}{\hbar}\sqrt{2m^*(E_p - \Delta E_c)}$, where E_p is the kinetic energy of the incident electron, we get

$$S_z = \frac{\hbar k_2}{m^*}|t|^2 \left[\frac{\hbar^2}{2m^*}k_2{}^2 + \Delta \right]. \tag{5.142}$$

For $z < 0$, we find

$$S_z = \frac{\hbar k + 1}{m^*}(1 - |r|^2) \frac{\hbar^2 k_1{}^2}{2m^*}. \tag{5.143}$$

Equating the two expressions for S_z for $z < 0$ and $z > 0$ at $z = 0$, we find

$$k_2|t|^2 \left(\frac{\hbar k_2{}^2}{2m^*} + \Delta \right) = k_1 (1 - |t|^2) \frac{\hbar^2 k^2}{2m^*}. \tag{5.144}$$

Conservation of the total energy (kinetic and potential) across the step requires

$$\frac{\hbar^2 k_2{}^2}{2m^*} + \Delta = \frac{\hbar^2 k_1{}^2}{2m^*}. \tag{5.145}$$

Hence, Equation (5.144) is equivalent to

$$|r|^2 + \frac{k_2}{k_1}|t|^2 = 1, \tag{5.146}$$

which is the same as the equation expressing the conservation of the current density across a step, i.e., the fraction of reflected particles $|r|^2$ and transmitted particles $\frac{k_2}{k_1}|t|^2$ must add up to unity in the case of coherent transport (see Problem 5.2).

** Problem 5.13: General tunneling problem using the concept of energy flux

For the tunneling problem in the general one-dimensional conduction band energy profile $E_c(z)$ (see Figure 5.2) with an applied bias V_{bias} across the structure, use the concept of energy flux conservation discussed in Problem 5.8 and show the general result

$$|r|^2 + \frac{k_F}{k_0}|t|^2 = 1 \tag{5.147}$$

between the reflection r and transmission t amplitudes for an electron with incident energy E, where

$$k_0 = \frac{1}{\hbar}\sqrt{2m^* E} \tag{5.148}$$

and

$$k_F = \frac{1}{\hbar}\sqrt{2m^*(E + eV_{\text{bias}})}, \tag{5.149}$$

E being the total energy of the incident electron.

Solution: We use the fact that for a steady-state tunneling problem, the energy flux density S_z is spatially invariant (see Problem 5.8) and write

$$S_z(0) = S_z(L). \tag{5.150}$$

Furthermore, we use the results of Problem 5.11 in which it was shown that

$$S_z = E\,\mathrm{Re}\left(Z_{\mathrm{QM}}(z)\right)\rho(z), \tag{5.151}$$

where $Z_{\mathrm{QM}}(z)$ is the quantum mechanical wave impedance $\frac{\hbar}{m^*i}\frac{\frac{d\phi}{dz}}{\phi(z)}$ (see Problem 1.5) and $\rho(z)$ is the probability density.

Therefore,

$$S_z(L) = E\frac{\hbar k_F}{m^*}|t|^2, \tag{5.152}$$

and

$$S_z(0) = E\left(\frac{\hbar k_0}{m^*}\right)\left\{\frac{1}{2}\frac{(2-2|t|^2)}{1+2\mathrm{Re}(r)+|r|^2}\right\}\left\{|1+r|^2\right\}, \tag{5.153}$$

where the first curly bracket is $\mathrm{Re}\,(Z(0))$ and the second is $\rho(0)$.

Using the above results, Equation (5.150) leads to

$$|r|^2 + \frac{k_F}{k_0}|t|^2 = 1, \tag{5.154}$$

an equality typically derived by utilizing the fact that the magnitude of the incident current density must equal the sum of the current densities of the reflected and transmitted beams (see Problem 5.3).

** Problem 5.14: Dwell time above a potential well

The dwell time of a particle across a region of width W is given by

$$\tau_{\mathrm{d}} = \frac{\int_0^W |\phi|^2 \mathrm{d}z}{J_{\mathrm{inc}}}, \tag{5.155}$$

where $J_{\mathrm{inc}} = \frac{\hbar k_0}{m_1^*}$ is the probability current density associated with a beam of electrons incident from the left contact, where the effective mass is assumed to be m_1^*, as shown in Figure 5.6.

Calculate the dwell time as a function of the energy of the incident electron for a particle impinging on a square well of depth $-V_0$ if the effective mass in the barrier region m_2^* is different from the one in the contacts m_1^*.

Solution: As shown in Problem 1.2, in the presence of a varying effective mass we must enforce the continuity of

$$\phi(z) \tag{5.156}$$

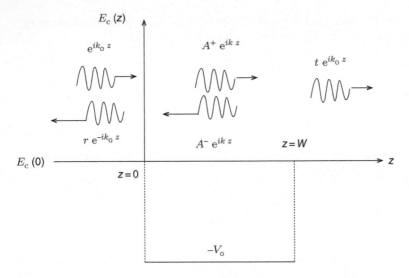

Figure 5.6: Illustration of an electron impinging from the left on a potential well. The effective mass is assumed to be the same throughout.

and

$$\frac{1}{\gamma(z)} \frac{\mathrm{d}\phi}{\mathrm{d}z}. \tag{5.157}$$

For the scattering problem depicted in Figure 5.6, continuity of (5.156) and (5.157) at $z = 0$ and $z = W$ leads to four equations for the parameters A^+, A^-, r, and t:

$$1 + r = A^+ + A^-, \tag{5.158}$$

$$1 - r = \frac{k}{k_0} \frac{m_1^*}{m_2^*} \left(A^+ - A^- \right), \tag{5.159}$$

$$A^+ \mathrm{e}^{ikW} + A^- \mathrm{e}^{-ikW} = t \mathrm{e}^{ikW}, \tag{5.160}$$

and

$$A^+ \mathrm{e}^{ikW} - A^- \mathrm{e}^{-ikW} = \frac{k_0}{k} \frac{m_2^*}{m_1^*} t \mathrm{e}^{ik_0 W}. \tag{5.161}$$

Using the last two equations we can express A^+ and A^- in terms of the transmission amplitude t as

$$A^+ = \left(1 + \frac{k_0}{k} \frac{m_2^*}{m_1^*} \right) \frac{t}{2} \mathrm{e}^{i(k_0 - k)W}, \tag{5.162}$$

$$A^- = \left(1 - \frac{k_0}{k} \frac{m_2^*}{m_1^*} \right) \frac{t}{2} \mathrm{e}^{i(k_0 + k)W}. \tag{5.163}$$

The square of the magnitude of the wave function inside the well region is given by

$$|\Phi|^2 = \left[\frac{\beta_+^2}{4} + \frac{\beta_-^2}{4} + \frac{\beta_- \beta_+}{2} \cos 2k(W - z) \right] |t|^2, \tag{5.164}$$

where the following shorthand notations were used:

$$\beta_+ = 1 + \frac{m_2^* \, k_0}{m_1^* \, k},\tag{5.165}$$

$$\beta_- = 1 - \frac{m_2^* \, k_0}{m_1^* \, k}.\tag{5.166}$$

Evaluating the integral in Equation (5.155), we get

$$\tau_{\mathrm{d}} = \frac{m_1^*}{\hbar k_0} \left[A_0 W + A_1 W \frac{\sin(2kW)}{2kW} \right],\tag{5.167}$$

where

$$A_0 = \left(\frac{\beta_+{}^2 + \beta_-{}^2}{4} \right) |t|^2,\tag{5.168}$$

$$A_1 = \left(\frac{\beta_+ \beta_-}{2} \right) |t|^2.\tag{5.169}$$

Plugging back Equations (5.161)–(5.162) into Equations (5.158)–(5.159) and solving for t and r, we obtain

$$t = \frac{\left(\beta_+{}^2 - \beta_-{}^2 \right) e^{i(k-k_0)W}}{\beta_+{}^2 - \beta_-{}^2 e^{2ikW}},\tag{5.170}$$

$$r = \frac{2i\beta_+ \beta_- \sin(kW)}{\beta_+{}^2 e^{-ikW} - \beta_-{}^2 e^{ikW}}.\tag{5.171}$$

Suggested problems

- If $\vec{J}$ is the probability current density for a particle of mass m^* in a potential field, show that the expectation value of its angular momentum $\vec{L} = \vec{r} \times \vec{p}$ is given by

$$\vec{L} = \langle \vec{r} \times \vec{p} \rangle = m^* \int \mathrm{d}^3 \vec{r} \left(\vec{r} \times \vec{J} \right).\tag{5.172}$$

- For a particle of spin $1/2$, show that the expectation value of the spin operator,

$$\langle \vec{S} \rangle = \frac{\hbar}{2} \int \mathrm{d}^3 \vec{r} \, \Psi^\dagger \vec{\sigma} \Psi,\tag{5.173}$$

can be written as

$$\langle \vec{S} \rangle = m^* \int \mathrm{d}^3 \vec{r} \left(\vec{r} \times \vec{J}_{\mathrm{S}} \right),\tag{5.174}$$

where the spin probability current density is given by

$$\vec{J}_{\mathrm{S}} = \vec{\nabla} \times \vec{V}_{\mathrm{S}}\tag{5.175}$$

with

$$\vec{V}_{\mathrm{S}} = \frac{\hbar}{4m^*} \Psi^\dagger \vec{\sigma} \Psi.\tag{5.176}$$

$\vec{V}_{\mathrm{S}}$ is referred to as the vector potential of the spin probability current density $\vec{J}_{\mathrm{S}}$.

- Write Matlab code to compute the reflection, transmission, and absorption coefficients for a quantum well of width 50 Å and depth $V_0 = -0.3\,\text{eV}$. Assume $W_0 = 0.1\,\text{eV}$ and the effective mass is the same everywhere $(m^* = 0.067m_0)$. Compute $|r|^2$, $|t|^2$, and A as a function of the incident energy and show that Equation (5.80) is satisfied.

- Starting with the results of Problem 5.5, prove that $|r|^2 + \frac{k_2}{k_1}\frac{m_1^*}{m_2^*}|t|^2 = 1$ by equating the incident current density to the sum of the reflected and transmitted current densities.

- Derive an analytical expression for the quantum mechanical impedance associated with a plane wave moving from left to right in a region where $E_c(z)$ is constant. Assume a constant effective mass throughout.

- Write Matlab code to compute the dwell time above a quantum well as a function of the kinetic energy of the electron incident from the left contact (see Figure 5.4). Allow for the effective masses in the contact and well region and the depth and width of the well to be adjustable parameters. For some parameters of your choice, compare the dwell time above the quantum well to its classical counterpart, i.e., $\tau_{\text{cl}} = W / \left(\frac{\hbar k_{\text{w}}}{m_2^*}\right)$, where $\hbar k_{\text{w}}/m_2^*$ is the velocity of the electron in the well region. Comment on the difference between the quantum mechanical expression for τ and its classical counterpart.

- Repeat Problem 5.14 with the addition of repulsive delta scatterers at $z = 0$ and $z = W$. By varying the strength of the repulsive delta scatterers, study how the dwell time through the rectangular barrier changes.

- Suppose an electron is incident on an infinite potential barrier defined as follows:

$$E_c(z) = 0 \text{ for } z < 0, \tag{5.177}$$

$$E_c(z) = \infty \text{ for } z \geq 0. \tag{5.178}$$

What is the value of the quantum mechanical wave impedance $Z(0)$ (see Problem 5.1)? Interpret the result physically.

References

[1] Levi, A. F. J. (2006) *Applied Quantum Mechanics*, 2nd edition, Section 3.5, Cambridge University Press, Cambridge.

[2] Cohen-Tannoudji, C., Diu, B., and Laloe, F. (2000) *Quantum Mechanics*, Chapter 3, Section D, Hermann, Paris.

[3] Landau, L. D. and Lifshitz, E. M. (1965) *Quantum Mechanics*, 2nd edition, pp. 435–437, Pergamon, Oxford.

[4] Greiner, W. (1989) *Quantum Mechanics*, pp. 241–242, Springer-Verlag, Berlin.

[5] Lévy-Leblond, J. M. (1987) The total probability current and the quantum period. *American Journal of Physics* 55, pp. 146–149.

Suggested Reading

- Draper, J. E. (1979) Use of $|\psi|^2$ and flux to simplify analysis of transmission past rectangular barriers and wells. *American Journal of Physics* 47, pp. 525–530.

- Uma Maheswari, A., Mahadevan, S., Prema, P., Shastry, C. S., and Agarwalla, S. K. (2007) Transmission and scattering by an absorptive potential. *American Journal of Physics* 75, pp. 245–253.

- Kaiser, H., Neidhardt, H., and Rehberg, J. (2002) Density and current of a dissipative Schrodinger operator. *J. Math. Phys.* 43, p. 5325.

- Manolopoulos, D. E. (2002) Derivation and reflection properties of a transmission-free absorbing potential. *J. Chem. Phys.* 117, p. 9552.

- Mita, K. (2000) Virtual probability current associated with the spin. *American Journal of Physics* 68, pp. 259–264.

Chapter 6: Density of States

The concept of density of states (DOS), or the number of energy states available for a particle to occupy within the energy range E and $E + dE$, is one of the most important concepts in statistical mechanics [1]. The DOS is needed to compute the average value of various physical quantities at equilibrium, and it also appears in the calculation of scattering rates associated with an electron scattering due to microscopic interactions with different entities in a medium. Among many other things, it is also needed to compute the current density due to electrons flowing through a device under bias.

The next set of problems illustrates the calculation of the DOS for systems of fermions and bosons either under equilibrium or under steady-state (non-equilibrium) conditions arising from the presence of a finite bias across a device. These problems include a study of the dependence of the DOS on the spatial dimensions of confined systems containing either electron or photon gases.

*** Problem 6.1: Density of states in quantum confined structures (from three to zero dimensions)

Consider the quantum confined geometries shown in Figure 6.1; 2D: two-dimensional electron gas (2DEG) or "quantum well" (QW); 1D: one-dimensional electron gas (1DEG) or "quantum wire"; 0D: zero-dimensional electron gas ("quantum dot" or "quantum box"). Calculate the energy dependence of the DOS in these structures at equilibrium and compare them to the three-dimensional bulk sample shown in the upper left corner of Figure 6.1. Assume a parabolic energy dispersion relation in the lowest conduction band, i.e., $E(\vec{k}) = \frac{\hbar^2 k^2}{2m^*}$, where k is the magnitude of the electron wave vector and m^* is the electron's effective mass.

Solution: (a) **3D:** Consider a uniform homogeneous semiconductor at equilibrium with the bottom of the conduction band denoted by E_{c0} and with a parabolic $E(\vec{k})$ dispersion relation (see Figure 6.2), given by

$$E_k = E_{c0} + \frac{\hbar^2 k^2}{2m^*}. \tag{6.1}$$

The solutions of the three-dimensional effective mass Schrödinger equation (i.e., the electron wave functions in the semiconductor) are plane waves expressed as

$$\phi_k(\vec{r}) = \frac{1}{\sqrt{\Omega}} e^{i\vec{k}\cdot\vec{r}}, \tag{6.2}$$

normalized over a volume $\Omega = L^3$, where L is the side of a cube large compared to the lattice unit cell of the semiconductor.

Problem Solving in Quantum Mechanics: From Basics to Real-World Applications for Materials Scientists, Applied Physicists, and Devices Engineers, First Edition.
Marc Cahay and Supriyo Bandyopadhyay.

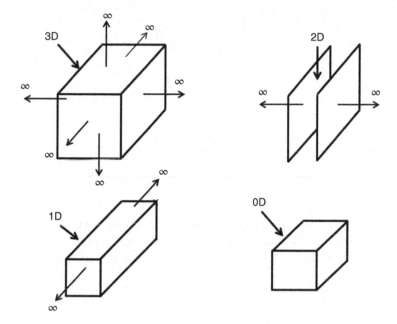

Figure 6.1: Illustration of the formation of a quantum dot (bottom right) through the gradual squeezing of a bulk piece of semiconductor (upper left). When the dimension of the bulk structure is reduced in one direction to a size comparable to the de Broglie wavelength, the resulting electron gas is referred to as a two-dimensional electron gas (2DEG) because the carriers are free to move in two directions only. If quantum confinement occurs in two directions, as illustrated in the bottom left figure, the resulting electron gas is referred to as a one-dimensional electron gas (1DEG) since an electron in this structure is free to move in one direction only. If confinement is imposed in all three directions (bottom right frame), we get a quantum dot (0DEG).

(b) **2D:** In a QW, quantum confinement in one direction (x-direction) leads to the formation of subbands whose energy dispersion relations are given by

$$E_{m,k_y,k_z} = E_{m,k_t} = \epsilon_m + \frac{\hbar^2}{2m^*}\left(k_y^2 + k_z^2\right) = \epsilon_m + \frac{\hbar^2 k_t^2}{2m^*} \qquad (6.3)$$

for the mth subband whose subband bottom energy is ϵ_m. Here, $\vec{k}_t = (k_y, k_z)$.

The solutions of the two-dimensional effective mass Schrödinger equation are

$$\phi_{m,k_t}(\vec{\rho}, x) = \phi_{m,k_y,k_z}(x,y,z) = \frac{1}{\sqrt{A}}e^{ik_y y}e^{ik_z z}\xi_m(x) = \frac{1}{\sqrt{A}}e^{i\vec{k}_t \cdot \vec{\rho}}\xi_m(x), \qquad (6.4)$$

normalized over an area $A = L^2$.

(c) **1D:** In a quantum wire, quantum confinement in two transverse (y, z) directions leads to the formation of subbands, each of which is labeled by two

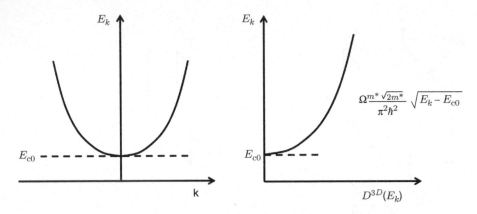

Figure 6.2: (Left) Parabolic energy dispersion relation close to the bottom of the conduction band (E_{c0}) of a typical semiconductor. (Right) Corresponding energy dependence of the three-dimensional DOS in a bulk semiconductor.

indices p, q because of the confinement in two transverse directions. Their energy dispersion relations are given by

$$E_{p,q,k_x} = \epsilon_{p,q} + \frac{\hbar^2 k_x^2}{2m^*}, \qquad (6.5)$$

where $\epsilon_{p,q}$ is the energy at the bottom of the corresponding subband.

The solutions of the one-dimensional effective mass Schrödinger equation are

$$\phi^{p,q,k_x}(x,y,z) = \frac{1}{\sqrt{L}} e^{ik_x x} \chi_p(y) \zeta_q(z), \qquad (6.6)$$

normalized over a length L. Here, $\chi_p(y)$ is the envelope wave function in the (p, q)th subband in the y-direction and $\zeta_q(z)$ is that in the z-direction.

In 3D, assuming periodic boundary conditions for $\phi_k(\vec{r})$, i.e.,

$$\phi_k(x + L, y + L, z + L) = \phi_k(x, y, z), \qquad (6.7)$$

the allowed values of $\vec{k} = (k_x, k_y, k_z)$ are given by

$$k_x = n_x \frac{2\pi}{L}, \qquad (6.8)$$

$$k_y = n_y \frac{2\pi}{L}, \qquad (6.9)$$

and

$$k_z = n_z \frac{2\pi}{L}, \qquad (6.10)$$

where n_x, n_y, n_z are positive or negative integers.

Therefore adjacent k states are separated by $2\pi/L$ in any of the three coordinate directions. In the wavevector magnitude range k to $k + dk$, the number of available k-states will therefore be $\frac{dk}{2\pi/L} = L dk/(2\pi)$.

In 3D, this will be $d^3\vec{k}L^3/(2\pi)^3 = d^3\vec{k}\Omega/(2\pi)^3$, where Ω is the volume in real space and $d^3\vec{k}$ is the volume in k space (spherical coordinates), i.e., $d^3\vec{k} = 4\pi k^2 dk$. In 2D, it will be $d^2\vec{k}_t L^2/(2\pi)^2 = d^2\vec{k}_t A/(2\pi)^2$, where A is the area in real space and $d^2\vec{k}_t$ is the area in k_t space (circular coordinates), i.e., $d^2\vec{k}_t = 2\pi k_t dk_t$. By the same token, the number of available states in 1D will be $dk_x L/(2\pi)$, but this would be wrong because of a subtlety. In 1D we have to account for the direction of k_x since there can be two directions, corresponding to $\pm k_x$. Therefore, we should multiply the number of states by a factor of two and hence the number of available states is $dk_x L/(\pi)$.

In each k state, one can accommodate a maximum of two electrons of opposite spins owing to the Pauli exclusion principle. Hence the number of *electron* states available in the k-space volume $d^3\vec{k}$, or the k-space area $d^2\vec{k}_t$, or the k-space length dk_x, is

$$\frac{d^3\vec{k}\Omega}{4\pi^3}\,(\text{3D}),$$

$$\frac{d^2\vec{k}_t A}{2\pi^2}\,(\text{2D}),$$

$$\frac{dk_x 2L}{\pi}\,(\text{1D}). \tag{6.11}$$

By definition, these numbers are, respectively, $D^{\text{3D}}(k)d^3\vec{k}$, $D^{\text{2D}}(k_t)d^2\vec{k}_t$, and $D^{\text{1D}}(k_x)dk_x$, where $D^{\text{3D}}(k)$, $D^{\text{2D}}(k_t)$, and $D^{\text{1D}}(k_x)$ are the *density of states in k space* for three, two, and one dimensions, respectively. Therefore,

$$D^{\text{3D}}(k) = \frac{\Omega}{4\pi^3},$$

$$D^{\text{2D}}(k_t) = \frac{A}{2\pi^2},$$

$$D^{\text{1D}}(k_x) = \frac{2L}{\pi}. \tag{6.12}$$

Note that these quantities are all independent of wavevector.

In equilibrium, the probability of a k state being occupied by an electron is given by the Fermi–Dirac factor $f(E_k) = [\exp((E_k - E_\text{F})/kT) + 1]^{-1}$. Therefore, if we wish to find the number of electrons $N(k)dk$ in the wavevector range k to $k+dk$, that number will be (in the 2D and 1D cases, we assume only any one subband)

$$f(E_k)D^{\text{3D}}d^3\vec{k} = f(E_k)\frac{d^3\vec{k}\Omega}{4\pi^3} \qquad (\text{3D}),$$

$$f(E_{k_t})D^{\text{2D}}(k_t)d^2\vec{k}_t = f(E_k)\frac{d^2\vec{k}_t A}{2\pi^2} \qquad (\text{2D}),$$

$$f(E_{k_x})D^{\text{1D}}(k_x)dk_x = f(E_k)\frac{dk_x 2L}{\pi} \qquad (\text{1D}). \tag{6.13}$$

Now, let the wavevector range k to $k + dk$ correspond to the energy range E_k to $E_k + dE_k$. We will define the energy-dependent density of states $D(E_k)$ such that the number of electrons $N(E_k)dE_k$ occupying the energy range E_k to $E_k + dE_k$ is $D(E_k)f(E_k)dE_k$. Since electron number is *conserved*, we must have $N(k)dk = N(E_k)dE_k$, and hence

$$f(E_k)\frac{d^3\vec{k}\Omega}{4\pi^3} = D^{3D}(E_k)f(E_k)dE_k \quad (3D),$$

$$f(E_k)\frac{d^2\vec{k}_t A}{2\pi^2} = D^{2D}(E_{k_t})f(E_k)dE_{k_t} \quad (2D),$$

$$f(E_k)\frac{dk_x 2L}{\pi} = D^{1D}(E_{k_x})f(E_k)dE_{k_x} \quad (1D), \qquad (6.14)$$

which translates to

$$D^{3D}(E_k)dE_k = \frac{4\pi k^2 dk\Omega}{4\pi^3},$$

$$D^{2D}(E_{k_t})dE_{k_t} = \frac{2\pi k_t dk_t A}{2\pi^2},$$

$$D^{1D}(E_{k_x})dE_{k_x} = \frac{2dk_x L}{\pi}. \qquad (6.15)$$

This allows us to write (for any one subband in the 2D and 1D cases)

$$D^{3D}(E_k)\frac{dE_k}{dk} = \frac{4\pi k^2\Omega}{4\pi^3},$$

$$D^{2D}(E_{k_t})\frac{dE_{k_t}}{dk_t} = \frac{2\pi k_t A}{2\pi^2},$$

$$D^{1D}(E_{k_x})\frac{dE_{k_x}}{dk_x} = \frac{2L}{\pi}. \qquad (6.16)$$

We now invoke the dispersion relation in Equations (6.1), (6.3), and (6.5) to calculate the derivatives and express k, k_t, or k_x in terms of E_k, E_{k_t}, or E_{k_x} (again assuming a specific subband for the 2D and 1D cases). The derivative is $\frac{dE_k}{dk} = \frac{\hbar^2 k}{m^*} = \frac{\hbar^2\sqrt{2m^* E_k}}{m^*}$ for the 3D case, and the reader can easily repeat for the 2D and 1D cases. Thus, we get

$$D^{3D}(E_k) = \frac{m^*\sqrt{2m^*}\Omega}{\pi^2\hbar^2}\sqrt{E_k - E_{c0}},$$

$$D^{2D}(E_{k_t}) = \frac{m^* A}{\pi^2\hbar^2},$$

$$D^{1D}(E_{k_x}) = \frac{2m^* L}{\pi\hbar^2\sqrt{2m^*}}\frac{1}{\sqrt{E_{p,q,k_x} - \epsilon_{p,q}}}. \qquad (6.17)$$

Note that the energy-dependent DOS is proportional to $\sqrt{E_k - E_{c0}}$ in 3D, independent of energy in 2D, and inversely proportional to $\sqrt{E_{p,q,k_x} - \epsilon_{p,q}}$ in 1D.

The energy-dependent three-dimensional DOS is plotted as a function of energy in Figure 6.2.

If we wish to calculate the electron density in 3D, 2D, or 1D, then we will find that since the total number of electrons $N = \int N(E)dE$,

$$n^{3D} = N/\Omega = \frac{\int_{E_{c0}}^{\infty} D^{3D}(E_k)f(E_k)dE_k}{\Omega},$$

$$n^{2D} = N/A = \frac{\sum_m \int_{\epsilon_m}^{\infty} D^{2D}(E_{k_t})f(E_k)dE_{k_t}}{A},$$

$$n^{1D} = N/L = \frac{\sum_{p,q} \int_{\epsilon_{p,q}}^{\infty} D^{1D}(E_{k_x})f(E_{k_x})dE_{k_x}}{L}. \quad (6.18)$$

3D: Using Equation (6.17) in Equation (6.18) and using the Fermi–Dirac function for $f(E_k)$, we get that, for the 3D case,

$$n^{3D} = \frac{2}{\sqrt{\pi}}N_c F_{\frac{1}{2}}(\xi), \quad (6.19)$$

where

$$N_c = \frac{1}{4\hbar^3}\left(\frac{2m^*k_BT}{\pi}\right)^{\frac{3}{2}} \quad (6.20)$$

and

$$\xi = \frac{(E_F - E_{c0})}{k_BT}, \quad (6.21)$$

where k_B is the Boltzmann constant and T is the absolute temperature.

In Equation (6.19), $F_{\frac{1}{2}}(\xi)$ is given by

$$F_{\frac{1}{2}}(\xi) = \int_{E_{c0}}^{+\infty} dE \frac{\sqrt{E - E_{c0}}}{1 + e^{\frac{E - E_F}{k_BT}}}. \quad (6.22)$$

2D: For the 2D case, we find the sheet carrier concentration as

$$n^{2D} = \frac{\sum_m \int_{\epsilon_m}^{\infty} D^{2D}(E_{k_t})f(E_{k_t})dE_{k_t}}{A}$$

$$= \sum_m \frac{m^*}{\pi^2\hbar^2}\int_{\epsilon_m}^{\infty} f(E_{k_t})dE_{k_t}$$

$$= \frac{m^*}{\pi^2\hbar^2}\sum_m \int_0^{\infty}\Theta(E - \epsilon_m)f(E)dE, \quad (6.23)$$

where Θ is the Heaviside (or unit step) function. The last integral can be evaluated analytically since $f(E)$ is the Fermi–Dirac function. This yields that the sheet carrier concentration due to M occupied subbands is

$$n^{2D} = \frac{m^*}{\pi\hbar^2}k_BT\ln\left[\prod_{m=1}^{m=M}\left(1 + e^{-\frac{\epsilon_m - E_F}{k_BT}}\right)\right], \quad (6.24)$$

where Π denotes product. This analytical expression for n^{2D} (sheet carrier concentration) is valid for any shape of the confining potential in the x-direction. Only the numerical values of ϵ_m must be determined through a solution of the Schrödinger equation

$$-\frac{\hbar^2}{2m^*}\frac{d^2\xi_m(x)}{dx^2} + E_c(x)\xi_m(x) = E_m\xi_m(x). \tag{6.25}$$

1D: In the 1D case, we find the linear carrier concentration as

$$
\begin{aligned}
n^{1D} &= \frac{\sum_{p,q}\int_{\epsilon_{p,q}}^{\infty} D^{1D}(E_{k_x})f(E_{k_x})dE_{k_x}}{L}\\
&= \sum_{p,q}\int_{\epsilon_{p,q}}^{\infty}\frac{2m^*}{\pi\hbar^2\sqrt{2m^*}}\frac{1}{\sqrt{E_{p,q,k_x}-\epsilon_{p,q}}}f(E_{p,q,k_x})d\left(E_{p,q,k_x}-\epsilon_{p,q}\right)\\
&= \sum_{p,q}\int_{0}^{\infty}\frac{2m^*}{\pi\hbar^2\sqrt{2m^*}}\frac{1}{\sqrt{E}}f(E+\epsilon_{p,q})dE.
\end{aligned}
\tag{6.26}
$$

The integral here is an improper integral since the integrand diverges for $E = 0$. These points are called van Hove singularites, and the 1D DOS diverges at the singularities that occur when $E_{p,q,k_x} = \epsilon_{p,q}$.

0D: In this case, we are dealing with a quantum box with quantum confinement in all three directions (see lower right frame in Figure 6.1). The electron energy is completely discretized. Each discrete level is labeled by three indices l, m, and n, corresponding to confinement in the x-, y-, and z-directions.

The electron density in any subband is given by

$$n_{l,m,n}^{0D} = \int_{0}^{\infty} dE D^{0D}(E)f(E), \tag{6.27}$$

where the zero-dimensional DOS is simply

$$D^{0D}(E) = 2\sum_{l,m,n}\delta(E-\epsilon_{l,m,n}), \tag{6.28}$$

where δ is the Dirac delta function and the factor 2 has been included since each $\epsilon_{n,m,l}$ energy level can be occupied by two electrons with opposite spin.

** Problem 6.2: Onset of degeneracy in confined systems [2]

In a semiconductor bulk sample at equilibrium, the statistics of electrons is well described by the Maxwell–Boltzmann distribution if the Fermi level is at least $3\,k_BT$ below the bottom of the conduction band. Such a semiconductor is said to be "non-degenerate." On the other hand, if the Fermi level is less than $3\,k_BT$ below the conduction band, the semiconductor is "degenerate" and the Fermi–Dirac distribution must be used to describe the statistics of electrons. In a quantum-confined system (2D, 1D, or 0D), which is different from bulk, we will use as a criterion for the onset of degeneracy (transition from a Maxwell–Boltzmann to a Fermi–Dirac distribution of carriers) the condition that the Fermi level coincides with the lowest energy level for the appearance of free propagating states in a sample. We will then extend this criterion to bulk systems and modify the criterion for the onset of degeneracy to the condition that the Fermi level coincides with the conduction band edge (instead of being less than $3k_BT$ below the conduction band edge. The Fermi level placement in the energy band diagram is determined by the carrier concentration; so, there is a critical carrier concentration for the onset of degeneracy.

The goal of this problem is to show that the critical carrier concentration for the onset of degeneracy increases with stronger confinement. More specifically, the followingstatements will be proved:

(a) The ratio of the critical electron concentration in a quantum well to that in the bulk for the onset of degeneracy is proportional to λ_D/W, where λ_D is the thermal de Broglie wavelength $\left(\lambda_D = \hbar/\sqrt{2m^*k_BT}\right)$ and W is the thickness of the QW.

(b) In a quasi one-dimensional structure (quantum wire) of cross-sectional area A, the critical concentration for the onset of degeneracy is proportional to $\lambda_D{}^2/A$.

Solution: (a) In a 3D sample, using the results of the previous problem, the critical concentration for the onset of degeneracy is found by using $E_c = 0$ and setting $E_F = 0$ in Equation (6.19):

$$n_{\text{crit}}^{3D} = \frac{(2m^*k_BT)^{3/2}}{2\pi^2\hbar^3}F_{1/2},\tag{6.29}$$

where

$$F_{1/2}(0) = \int_0^\infty \frac{x\sqrt{x}}{1+e^x}dx.\tag{6.30}$$

The quantity n_{crit}^{3D} can be expressed in terms of the thermal de Broglie wavelength:

$$n_{\text{crit}}^{3D} = F_{1/2}(0)/(2\pi^2\lambda_D{}^3).\tag{6.31}$$

In a QW whose lowest energy level is located at an energy E_1 above the bottom of the conduction band, if we assume that only one subband is occupied, the electron sheet concentration in the well is given by (see previous problem)

$$n_s = \frac{m^*k_BT}{\pi\hbar^2}\ln\left(1+e^{\frac{E_F-E_1}{k_BT}}\right).\tag{6.32}$$

Neglecting the upper subbands is a good approximation since the criterion for the onset of degeneracy in the QW is found by setting $E_F = E_1$, the lowest energy for the existence of free propagating states in the QW. Hence, in a QW the critical sheet carrier concentration for the onset of degeneracy is given by

$$n_s^{\text{crit}} = \frac{m^*k_BT}{\pi\hbar^2}\ln 2.\tag{6.33}$$

The critical electron density per unit volume in a QW of width W is related to $n_{\mathrm{s}}^{\mathrm{crit}}$ as follows:

$$n_{\mathrm{crit}}(W,T) = n_{\mathrm{s}}^{\mathrm{crit}}/W = \frac{m^* k_{\mathrm{B}} T}{\pi \hbar^2} \frac{\ln 2}{W}. \tag{6.34}$$

Hence, the ratio of the critical concentration for the onset of degeneracy in a bulk and QW is given as

$$n_{\mathrm{crit}}(W,T)/n_{\mathrm{crit}}^{3\mathrm{D}} = \frac{\pi \ln 2}{F_{1/2}(0)} \left(\frac{\lambda_{\mathrm{D}}}{W}\right) = 3.21 \frac{\lambda_{\mathrm{D}}}{W}. \tag{6.35}$$

(b) In a quantum wire, the 1D DOS in the lowest energy subband with energy bottom located at E_{11} is given by

$$D^{1\mathrm{D}}(E) = \frac{\sqrt{2m^*}}{\pi \hbar} \frac{1}{\sqrt{E - E_{11}}}. \tag{6.36}$$

The contribution to the electron concentration per unit length of the quantum wire is therefore given by (see previous problem)

$$n_{\mathrm{l}} = \frac{\sqrt{2m^*}}{\pi \hbar} \int_{E_{11}}^{\infty} \mathrm{d}E \frac{1}{\sqrt{E - E_{11}}} \frac{1}{1 + e^{\frac{(E - E_{11}) - E_{\mathrm{F}}}{k_{\mathrm{B}} T}}}. \tag{6.37}$$

Setting $E_{\mathrm{F}} = 0$ and making a variable substitution with $x = (E - E_{11})/k_{\mathrm{B}} T$, the onset of degeneracy in a quantum wire occurs at a critical density per unit length equal to

$$n_{\mathrm{l}}^{\mathrm{crit}} = \frac{\sqrt{2m^*}}{\pi \hbar} F_{-1/2}(0), \tag{6.38}$$

where

$$F_{-1/2}(0) = \int_{0}^{+\infty} \mathrm{d}x \frac{x^{-1/2}}{1 + e^x}. \tag{6.39}$$

Using Equations (6.31) and (6.38), we finally get

$$\frac{n_{\mathrm{l}}^{\mathrm{crit}}/A}{n_{\mathrm{crit}}^{3\mathrm{D}}} = 2\pi (F_{-1/2}(0)/F_{1/2}(0)) \lambda_{\mathrm{d}}^2/A \propto \frac{\lambda_{\mathrm{d}}^2}{A}. \tag{6.40}$$

---

* Problem 6.3: Sheet carrier concentration in a two-dimensional electron gas with a few subbands occupied [1]

(a) *Show that the sheet carrier concentration n_{s} in a 2DEG of a high electron mobility transistor (HEMT) is given by*

$$n_{\mathrm{s}} = \frac{m^*}{\pi \hbar^2} k_{\mathrm{B}} T \ln \left[\left(1 + e^{\frac{E_{\mathrm{F}} - E_1}{k_{\mathrm{B}} T}}\right) \left(1 + e^{\frac{E_{\mathrm{F}} - E_2}{k_{\mathrm{B}} T}}\right) \right], \tag{6.41}$$

when only two subbands are occupied in the 2DEG. Here, E_1 and E_2 are the energy bottoms of the two lowest subbands. First, read the brief introduction to the HEMT device in Appendix E.

(b) Starting with the result of part (a), show that at low temperature

$$n_s = \frac{m^*}{\pi\hbar^2}(E_F - E_1) \tag{6.42}$$

when the second subband is unoccupied, and

$$n_s = \frac{m^*}{\pi\hbar^2}(E_2 - E_1) + 2\frac{m^*}{\pi\hbar^2}(E_F - E_2) \tag{6.43}$$

when both subbands are occupied.

Solution: (a) From Equation (6.24), we immediately get the stated result.

(b) If only one subband in the 2DEG is occupied,

$$n_s = k_B T \ln\left(1 + e^{\frac{E_F - E_1}{k_B T}}\right). \tag{6.44}$$

If $k_B T \ll E_F - E_1$, then using the result $\ln(1 + x) \approx x$ when $x \ll 1$, we get

$$n_s = \frac{m^*}{\pi\hbar^2}(E_F - E_1). \tag{6.45}$$

When the second subband is occupied (but the third one is unoccupied), we get, from Equation (6.24),

$$n_s = \frac{m^*}{\pi\hbar^2}k_B T \left[\ln\left(1 + e^{\frac{E_F - E_1}{k_B T}}\right) + \ln\left(1 + e^{\frac{E_F - E_1}{k_B T}}\right)\right]. \tag{6.46}$$

If $k_B T \ll E_F - E_1$, $E_F - E_2$ (i.e., at low enough temperature), then

$$\ln\left(1 + e^{\frac{E_F - E_1}{k_B T}}\right) = \frac{E_F - E_1}{k_B T} \tag{6.47}$$

and

$$\ln\left(1 + e^{\frac{E_F - E_2}{k_B T}}\right) = \frac{E_F - E_2}{k_B T}. \tag{6.48}$$

Hence,

$$n_s = \frac{m^*}{\pi\hbar^2}(E_F - E_1) + \frac{m^*}{\pi\hbar^2}(E_F - E_2) = \frac{m^*}{\pi\hbar^2}(2E_F - E_1 - E_2)$$
$$= \frac{m^*}{\pi\hbar^2}(E_2 - E_1) + 2\frac{m^*}{\pi\hbar^2}(E_F - E_2). \tag{6.49}$$

---

* Problem 6.4: Fraction of ionized impurities in a QW

The electrons contributing to the sheet carrier concentration in a HEMT come from ionized donor impurities (dopant atoms) in the gate insulator (see Appendix E). Not all the impurities may ionize at low temperatures so that the electrons generated by the dopants may be fewer in number than the impurities. This is sometimes called carrier freeze-out. We study this phenomenon in this problem.

Assuming that only one subband is occupied in a 100 Å-wide GaAs QW uniformly doped with donors at a concentration $N_D = 10^{17}$ cm^{-3}, what is the fraction α of ionized impurities if the Fermi level is $3 k_B T$ below the ground state energy level? Assume $T = 4.2$ K. Model the QW as a box with infinite walls (see Problem 3.5). The effective mass of electrons in GaAs is 0.067 times the free electron mass of 9.1×10^{-31} kg.

Solution: As shown in Problem 6.3, with only one subband occupied the electron sheet concentration is given by

$$n_s = \frac{m^*}{\pi \hbar^2} k_B T \ln \left[\left(1 + e^{\frac{E_F - E_1}{k_B T}} \right) \right]. \tag{6.50}$$

If a fraction α of the impurities is ionized and all the resulting electrons transfer to the two-dimensional electron gas in the HEMT, then the sheet concentration n_s is given by

$$n_s = N_D{}^+ W = \alpha N_D W. \tag{6.51}$$

Using the last two equations and solving for α,

$$\alpha = \frac{m^*}{N_D W \pi \hbar^2} k_B T \ln \left[1 + e^{\frac{E_F - E_1}{k_B T}} \right]. \tag{6.52}$$

For $E_1 - E_F = 3 k_B T$, $\alpha = 0.5\%$.

** Problem 6.5: Intrinsic carrier concentration in a two-dimensional electron gas [1]

Consider Figure 6.3, which shows the energy dependence of the two-dimensional DOS of electrons and holes in a 2D semiconductor single layer, where only the first subband for both electrons and holes is assumed to be occupied. The quantities $a = m_e/(\pi \hbar^2)$ and $b = m_h/(\pi \hbar^2)$, where m_e and m_h are the effective masses of electrons and holes, respectively.

(a) Assuming the well is undoped, obtain an expression for the Fermi level E_F at room temperature in terms of a, b, and the temperature T. Assume Boltzmann statistics to be valid. When is E_F exactly equal to the midgap energy, $\frac{(E_c + \Delta_e + E_v + \Delta_h)}{2}$?

(b) Obtain the expression for n_i, the intrinsic carrier concentration, in terms of a, b, $k_B T$, and the effective bandgap energy $E_g = E_c + \Delta_e - E_v - \Delta_h$.

Hint: Start with the approximate expressions for the electron (n) and hole (p) concentrations in terms of $g_c(E)$ and $g_v(E)$, where $g_c(E)$ is the energy-dependent two-dimensional density of states for electrons and $g_v(E)$ is that for holes (subscripts

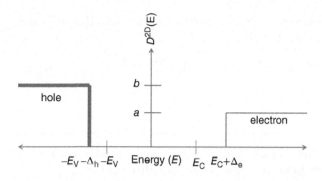

Figure 6.3: Density of states of electrons and holes as a function of energy in a two-dimensional electron and hole gas. Here, Δ_e is the first subband energy for electrons and Δ_h is that for holes.

c and v denote conduction and valence bands). Also, assume Boltzmann statistics of carriers, i.e.,

$$f(E) = e^{\frac{E_F - E}{k_B T}}. \tag{6.53}$$

Solution: (a) The electron and hole concentrations are given by

$$n = \int_{E_c}^{\infty} g_c(E) f(E) dE, \tag{6.54}$$

$$p = \int_{-\infty}^{E_v} g_v(E)[1 - f(E)] dE. \tag{6.55}$$

Using Boltzmann's approximation, $f(E) = e^{\frac{(E_F - E)}{k_B T}}$, $g_c(E) = a\Theta(E - E_c - \Delta_e)$, and $g_v(E) = b\Theta(E_v + \Delta_h - E)$, where $\Theta(\eta)$ is the unit step (Heaviside) function, leads to

$$n = \int_{E_c + \Delta_e}^{\infty} a e^{\frac{(E_F - E)}{k_B T}} dE = a k_B T e^{\frac{E_F - E_c - \Delta_e}{k_B T}}. \tag{6.56}$$

Similarly,

$$p = \int_{-\infty}^{E_v + \Delta_h} b e^{\frac{(E - E_F)}{k_B T}} dE = b k_B T e^{\frac{E_v + \Delta_h - E_F}{k_B T}}. \tag{6.57}$$

If the sample is intrinsic, then $n = p = n_i$. Therefore,

$$a k_B T e^{\left(\frac{E_F - E_c}{k_B T}\right)} = b k_B T e^{\left(\frac{E_v - E_F}{k_B T}\right)}, \tag{6.58}$$

from which we derive

$$E_F = \frac{E_c + \Delta_e + E_v + \Delta_h}{2} + \frac{k_B T}{2} \ln\left(\frac{b}{a}\right). \tag{6.59}$$

Hence, $E_F = \frac{E_c + \Delta_e + E_v + \Delta_h}{2}$ whenever $a = b$, i.e., the Fermi level is exactly half way through the effective bandgap when the electron and hole effective masses are the same.

(b) The intrinsic carrier concentration is given by $n_i = \sqrt{np}$. Hence, using Equations (6.56) and (6.57), we obtain

$$n_i = k_B T \sqrt{abe}^{\frac{(E_c + \Delta_e - E_v - \Delta_h)}{2k_B T}} = k_B T \sqrt{abe}^{-\frac{E_g + \Delta_e + \Delta_h}{2k_B T}}, \tag{6.60}$$

where E_g is the bulk bandgap $E_c - E_v$.

** Problem 6.6: Charge density, electric field, and electrostatic potential energy profile in a QW with infinite barriers at equilibrium

Consider a QW of width W. Assume that the QW is uniformly doped with donors whose concentration (i.e., volume density) is N_D. Also assume it is at room temperature and all impurities are ionized.

Consider the situation when only one subband is occupied in the QW. Derive analytical expressions for the total charge concentration $\rho(z)$, the electric field $E(z)$, and the electrostatic potential $V(z)$ across the well. Assume the boundary conditions $E(0) = E(W) = 0$, and $V(0) = V(W) = 0$. According to Gauss's law, this is equivalent to assuming that the QW is electrically neutral.

Plot $\rho(z)$, $E(z)$, and $V(z)$ as a function of z for $T = 300\,K$, $m^* = 0.067m_0$, $W = 100\,Å$, and $N_D = 10^{17}\,cm^{-3}$.

Solution: Because the wave function varies across the width of the QW, the volume carrier density is not constant, but varies with the coordinate z (the sheet carrier concentration is still constant). Consequently, the electron volume concentration will be given by (see Equation (6.24))

$$n(z) = \sum_m \sigma_m |\xi_m(z)|^2, \tag{6.61}$$

where $\xi_m(z)$ is the wave function in the mth subband and

$$\sigma_m = \frac{m^*}{\pi \hbar^2} k_B T \ln \left[1 + e^{-\frac{\epsilon_m - E_F}{k_B T}} \right]. \tag{6.62}$$

If only one subband is occupied, the spatial variation of the electron volume concentration will be given by

$$n(z) = \sigma_1 |\xi_1(z)|^2, \tag{6.63}$$

where, because of the infinite square well potential,

$$\xi_1(z) = \sqrt{\frac{2}{W}} \sin \left(\frac{\pi z}{W} \right). \tag{6.64}$$

In writing down the last equation, we assumed (as a first approximation) that the potential inside the QW is spatially invariant.

Therefore, the electron volume concentration in the first subband is

$$n(z) = \rho(z)/q = \frac{2\sigma_1}{W} \sin^2\left(\frac{\pi z}{W}\right).$$ (6.65)

The total charge concentration in the QW is

$$\rho(z) = q\left[N_\mathrm{D} - n(z)\right].$$ (6.66)

Since the QW is electrically neutral and all the impurities are ionized, we must have

$$n_\mathrm{s} = \frac{1}{q}\int_{-\infty}^{+\infty} \rho(z)\mathrm{d}z = \int_{-\infty}^{+\infty} n(z)\mathrm{d}z = \sigma_1 \int_{-\infty}^{+\infty} |\xi_1(z)|^2 \mathrm{d}z = N_\mathrm{D}W.$$ (6.67)

Since the wave function is normalized, the integral in the above equation is equal to unity. Hence, using the last equation, we find $\sigma_1 = N_\mathrm{D}W$. As a result, the spatial dependence of the total charge concentration in the QW is given, from Equations (6.65) and (6.66), by

$$\rho(z) = q\left[10^{17} - 2\times 10^{17} \sin^2\left(\frac{\pi z}{W}\right)\right].$$ (6.68)

Hence,

$$\rho(z) = 10^{17} q \cos\left(\frac{2\pi z}{W}\right) \mathrm{cm}^{-3}.$$ (6.69)

From Poisson's equation,

$$\frac{\mathrm{d}E(z)}{\mathrm{d}z} = \frac{\rho(z)}{\epsilon},$$ (6.70)

where ϵ is the dielectric constant in the QW. Therefore, $E(z) = \int_0^z \frac{\rho(z)}{\epsilon}\mathrm{d}z + E(0)$ and, since $E(0) = 0$, the spatial dependence of the electric field inside the QW is given by

$$E(z) = \frac{qN_\mathrm{D}W}{2\pi\epsilon} \sin\left(\frac{2\pi z}{W}\right).$$ (6.71)

It is easy to verify from the above expression that $E(W) = 0$, which means that the QW is electrically neutral.

The electrostatic potential energy is found from the relation

$$\frac{\mathrm{d}V(z)}{\mathrm{d}z} = -E(z).$$ (6.72)

A simple integration (taking into account that $V(0) = 0$) leads to

$$V(z) = \frac{eN_\mathrm{D}W^2}{(2\pi)^2\epsilon}\left[\cos\left(\frac{2\pi z}{W}\right) - 1\right],$$ (6.73)

which satisfies the boundary condition $V(W) = 0$.

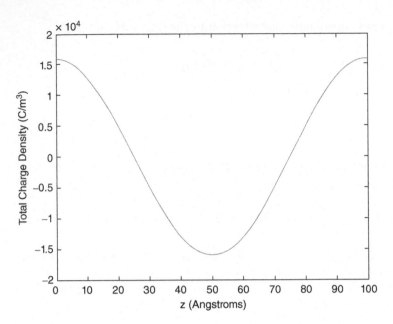

Figure 6.4: Spatial dependence of the total charge concentration, $\rho(z)$ (in C/m^3), in the QW (see Equation (6.69)).

Using the above results, Figures 6.4–6.6 show plots of $\rho(z)$, $E(z)$, and $V(z)$ across a QW for the following set of parameters: $T = 300\,\mathrm{K}$, $m^* = 0.067m_0$, $W = 100\,\text{Å}$, and $N_{\mathrm{D}} = 10^{17}\,\mathrm{cm}^{-3}$.

In this problem, we started with an initially flat conduction band energy profile. The solution above shows that there is in fact a spatial variation of the electrostatic potential energy across the QW. This leads to a modification of the conduction band energy profile $E_{\mathrm{c}}(z)$ inside the QW which is given by $E_{\mathrm{c}}(z) = \chi - qV(z)$, where χ is the electron affinity in the QW. The new $E_{\mathrm{c}}(z)$ must be used to re-solve the Schrödinger equation to recalculate its new eigenvalues and associated eigenfunctions. The electric field and the electrostatic potential energy must then be calculated again and the iterative procedure continued until a self-consistent solution is found. The set of equations to be solved self-consistently is given in detail in Problem 6.8.

** Problem 6.7: Gate capacitance of a HEMT

For a HEMT device, the gate capacitance per unit area is defined as

$$\frac{C}{ZL_{\mathrm{g}}} = \frac{\mathrm{d}(qn_{\mathrm{s}})}{\mathrm{d}V_{\mathrm{g}}},\tag{6.74}$$

where L_{g} and Z are the length and width of the gate, respectively (see Appendix E).

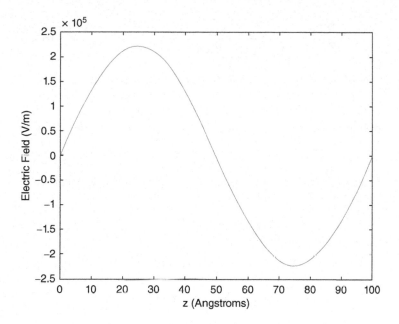

Figure 6.5: Spatial dependence of the electric field $E(z)$ (in V/m) across the QW (see Equation (6.71)). The dielectric constant of GaAs was set equal to $12.9\epsilon_0$.

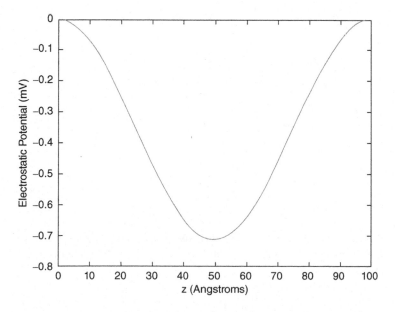

Figure 6.6: Spatial dependence of the electrostatic potential $V(z)$ (in mV) across the QW (see Equation (6.73)). The dielectric constant of GaAs was set equal to $12.9\epsilon_0$.

Show that

$$\frac{C}{ZL_{\mathrm{g}}} = \frac{\frac{\epsilon_{\mathrm{s}}}{d}}{\left[1 + \frac{\epsilon_{\mathrm{s}}}{d}\frac{1}{q^2}\left(\frac{\mathrm{d}E_{\mathrm{F}}}{\mathrm{d}n_{\mathrm{s}}}\right)\right]}, \tag{6.75}$$

where q is the magnitude of the charge of the electron, ϵ_{s} is the dielectric constant of the AlGaAs layer, and E_{F} is the Fermi energy in the 2DEG at the AlGaAs/GaAs interface. The latter is the Fermi level far into the substrate if we assume no leakage current, i.e., no current flow in the z-direction; d is the thickness of the AlGaAs layer (see Appendix E).

Solution: We start with the results of Appendix E where the electron sheet concentration in the 2DEG was found to be given by

$$qn_{\mathrm{s}} = \frac{\epsilon_{\mathrm{s}}}{d}\left(V_{\mathrm{G}} - V_{\mathrm{T}}\right), \tag{6.76}$$

where V_{G} is the gate voltage, and the threshold voltage V_{T} is given explicitly by

$$V_{\mathrm{T}} = \phi_{\mathrm{m}} + \frac{1}{q}\left(E_{\mathrm{F}} - \Delta E_{\mathrm{c}}\right) - V_{\mathrm{d}}, \tag{6.77}$$

where the various quantities on the right-hand side are defined in Appendix E.

Using Equations (6.74) and (6.76), we get

$$\frac{C}{ZL_{\mathrm{g}}} = \frac{\mathrm{d}(qn_{\mathrm{s}})}{\mathrm{d}V_{\mathrm{G}}} = \frac{\epsilon_{\mathrm{s}}}{d}\frac{\mathrm{d}}{\mathrm{d}V_{\mathrm{g}}}(V_{\mathrm{G}} - V_{\mathrm{T}}), \tag{6.78}$$

or

$$\frac{C}{ZL_{\mathrm{g}}} = \frac{\epsilon_{\mathrm{s}}}{d}\left(1 - \frac{\mathrm{d}V_{\mathrm{T}}}{\mathrm{d}V_{\mathrm{G}}}\right). \tag{6.79}$$

But,

$$\frac{\mathrm{d}V_{\mathrm{T}}}{\mathrm{d}V_{\mathrm{G}}} = \frac{1}{q}\frac{\mathrm{d}E_{\mathrm{F}}}{\mathrm{d}V_{\mathrm{G}}} - \frac{\mathrm{d}V_{\mathrm{d}}}{\mathrm{d}V_{\mathrm{G}}} = \frac{1}{q}\frac{\mathrm{d}E_{\mathrm{F}}}{\mathrm{d}n_{\mathrm{s}}}\frac{\mathrm{d}n_{\mathrm{s}}}{\mathrm{d}V_{\mathrm{G}}}. \tag{6.80}$$

Rearranging, we finally get

$$\frac{C}{ZL_{\mathrm{g}}} = \frac{\frac{\epsilon_{\mathrm{s}}}{d}}{\left[1 + \frac{\epsilon_{\mathrm{s}}}{d}\frac{1}{q^2}\left(\frac{\mathrm{d}E_{\mathrm{F}}}{\mathrm{d}n_{\mathrm{s}}}\right)\right]}. \tag{6.81}$$

Therefore, the capacitance per unit square area is not just ϵ_{s}/d, i.e., the parallel plate capacitance per unit area associated with the AlGaAs layer. There is a correction in the denominator proportional to

$$\frac{\epsilon_{\mathrm{s}}}{d}\frac{1}{q^2}\left(\frac{\mathrm{d}E_{\mathrm{F}}}{\mathrm{d}n_{\mathrm{s}}}\right). \tag{6.82}$$

This correction is purely quantum mechanical in origin. It is due to the finite extent of the different wave functions associated with the lowest energy subbands participating in the population of the 2DEG. The average location of the electron

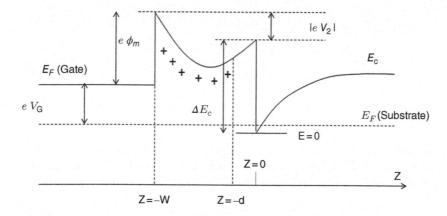

Figure 6.7: Energy band profile under the gate of a HEMT device consisting of an AlGaAs/GaAs heterostructure. The substrate is held at ground and a voltage $V_{\rm G}$ is applied to the gate.

charge density in the 2DEG is located away from the AlGaAs/GaAs interface. The capacitance per unit area can be rewritten as

$$\frac{C}{ZL_{\rm g}} = \frac{\epsilon_{\rm s}}{d'}, \tag{6.83}$$

with $d' = d\left[1 + \frac{\epsilon_{\rm s}}{d}\frac{1}{q^2}\left(\frac{{\rm d}E_{\rm F}}{{\rm d}n_{\rm s}}\right)\right]$, which is larger than d. An estimate of the correction requires a calculation of the quantity $\frac{{\rm d}E_{\rm F}}{{\rm d}n_{\rm s}}$, which must be obtained numerically using the self-consistent procedure outlined in the next problem.

*** Problem 6.8: Self-consistent calculations for a HEMT device

Write down the set of equations which must be solved simultaneously and self-consistently to calculate the carrier sheet concentration $n_{\rm s}$ versus the applied gate voltage for the HEMT structure whose energy band diagram (in the direction perpendicular to the hetero-interfaces) is shown in Figure 6.7. The AlGaAs layer contains some ionized donors of (spatially varying) concentration $N_{\rm D}^{+}(z)$ due to modulation doping of the device, and the substrate is assumed to be doped with acceptor concentration $N_{\rm A}(z)$ (typically uniform).

Solution: If we take into account the difference between the dielectric constants of the AlGaAs and the GaAs layers, the Poisson equation describing the electrostatic potential variation between the gate and back of the substrate is given by

$$\frac{{\rm d}}{{\rm d}z}\left[\epsilon(z)\frac{{\rm d}V}{{\rm d}z}\right] = \frac{q}{\epsilon}\left[N_{\rm D}^{+}(z) - N_{\rm A}^{-}(z) - n_{\rm el}(z)\right], \tag{6.84}$$

where q is the magnitude of the charge of the electron and $n_{\rm el}(z)$ is the electron concentration in the channel given by

$$n_{\text{el}}(z) = \sum_n \sigma_n |\xi_n(z)|^2, \qquad (6.85)$$

where

$$\sigma_n = \frac{m^* k_B T}{\pi \hbar^2} \ln\left[1 + e^{\frac{E_F{}^{\text{sub}} - E_m}{k_B T}}\right], \qquad (6.86)$$

where $E_F{}^{\text{sub}}$ is the Fermi level in the substrate, which is spatially invariant because of the lack of current flow through the substrate.

The wave functions $\xi_n(z)$ are the solutions to the Schrödinger equation in the channel of the HEMT. Taking into account the variation of the effective mass across the AlGaAs/GaAs interface, the following Schrödinger equation must be solved:

$$-\hbar^2 \frac{d}{dz}\left[\frac{1}{2m^*(z)} \frac{d\xi_n}{dz}\right] + E_c(z)\xi_n(z) = E_n \xi_n(z), \qquad (6.87)$$

where the conduction band edge is given by

$$E_c(z) = E_0 - \chi(z) - eV(z). \qquad (6.88)$$

Here, $\chi(z)$ is the spatially varying electron affinity (different in the AlGaAs and GaAs layers) and E_0 is a reference level in the energy band diagram. At the AlGaAs/GaAs interface, the conduction band discontinuity must be taken into account:

$$E_c(d_-) = E_c(d_+) + \Delta E_c. \qquad (6.89)$$

The set of equations above must be solved simultaneously. The Poisson equation must be solved taking into account the fact that at the gate contact, $V = V_G$. Typically, V is set equal to zero far into the bulk.

A good approximation for the shape of the energy potential $E_c(z)$ near the AlGaAs/GaAs interface is a triangular well. The wave functions corresponding to the first few bound states in the triangular well can then be calculated using Airy functions if the AlGaAs barrier is considered as infinite (see Problem 3.12) or using a variational procedure (see Chapter 10). A more accurate treatment requires a numerical solution of the Schrödinger equation.

The set of equations given above must be solved simultaneously to take into account the effects of space-charge in the device. Typically, the set of equations is solved iteratively starting with an initial guess for the electrostatic potential, using the analytical treatment presented in Appendix E. Examples of self-consistent calculations of the conduction band energy profile and electron sheet concentration as a function of the applied gate bias can be found in Refs. [3, 4].

*** Problem 6.9: Electron charge density profile in a current-carrying nanoscale device

Derive a general expression for the electron charge density profile in a current-carrying nanoscale device taking into account the spatial variation of the effective mass along the direction of current flow (see Figure 5.1).

Solution: Assuming ballistic transport and that the conduction band and effective mass vary along the z-direction only (direction of current flow), the Schrödinger equation inside the device is given by (see Problem 1.1)

$$\frac{d}{dz}\left[\frac{1}{\gamma(z)}\frac{d}{dz}\phi(z)\right] + \frac{2m_c^*}{\hbar^2}\left[E - \frac{E_t}{\gamma(z)} - E_c(z)\right]\phi(z) = 0, \tag{6.90}$$

where $\gamma(z) = \frac{m^*(z)}{m_c^*}$, m_c^* being the effective mass in the contacts, $E = E_t + E_p$, $E_p = \frac{\hbar^2 k_0^2}{2m_c^*}$, k_0 is the electron wave vector in the direction of current flow (z-axis) in the contacts, and $E_t = \frac{\hbar^2(k_x^2+k_y^2)}{2m_c^*}$ is the transverse kinetic energy in the contacts.

To calculate the total charge density in a nanoscale device due to electrons incident from the left contact, we perform the following integration assuming that the conduction band energy profile and effective mass vary along the z-direction only:

$$n^{1-r}(z) = \frac{1}{4\pi^3}\int d^3\vec{k}\,|\psi_k^{1-r}(z)|^2 f(E_k), \tag{6.91}$$

where $f(E_k) = \left[1 + e^{\frac{(E_k-E_F)}{k_BT}}\right]^{-1}$, $E_k = E_c(0) + \frac{\hbar^2 k^2}{2m_c^*}$, and $k^2 = (k_x^2 + k_y^2) + k_z^2 = k_t^2 + k_z^2$. The label 1–r is used to indicate that the electrons are incident from the left contact and are traveling to the right (left-to-right current component).

Using cylindrical coordinates in k space, we obtain

$$n^{1-r}(z) = \int_0^{+\infty}\frac{dk_z}{2\pi}\int_0^{+\infty}\frac{dk_t k_t}{\pi}F(k_z, k_t), \tag{6.92}$$

where

$$F(k_z, k_t) = \left[\exp\left(\frac{E_c(0) - E_F + \frac{\hbar^2}{2m_c^*}(k_z^2 + k_t^2)}{k_BT}\right) + 1\right]^{-1}|\psi_k^{1-r}(z)|^2. \tag{6.93}$$

The quantity $|\psi_k^{1-r}|^2$ depends on both k_z and k_t since it is the solution of Equation (6.90). As an approximation, if we replace E_t by k_BT, the average kinetic energy in the transverse direction, the charge density associated with electrons incident from the left contact is given by

$$n^{1-r}(z) = \int_0^{+\infty}\frac{dk_z}{2\pi}|\psi_{k_z,k_BT}^{1-r}(z)|^2\sigma^{1-r}(k_z), \tag{6.94}$$

where $\psi_{k_z,k_BT}^{1-r}(z)$ is the solution to the Schrödinger equation with E_t replaced by k_BT, and

$$\sigma^{1-r}(k_z) = \int_0^{+\infty}\frac{dk_t k_t}{\pi}\left[\exp\left(\frac{E_c(0) - E_F + \frac{\hbar^2}{2m_c^*}(k_x^2 + k_t^2)}{k_BT}\right) + 1\right]^{-1}. \tag{6.95}$$

This integral can be evaluated exactly by making the following change of variables:

$$U = \left(E_{\mathrm{c}}(0) - E_{\mathrm{F}} + \frac{\hbar^2}{2m_{\mathrm{c}}^*} k_z^2 \right) / k_{\mathrm{B}}T, \tag{6.96}$$

$$V = \hbar^2 k_t^2 / 2m^* k_{\mathrm{B}}T. \tag{6.97}$$

This leads to

$$dU = \frac{\hbar^2}{m^* k_{\mathrm{B}}T} k_t dk_t \tag{6.98}$$

and

$$\sigma^{\mathrm{l\text{-}r}}(k_z) = \frac{m^* k_{\mathrm{B}}T}{\pi \hbar^2} \int_0^{+\infty} dU \left[e^{U+V} + 1 \right]^{-1}. \tag{6.99}$$

Next, we use the fact that

$$\frac{\mathrm{d}}{\mathrm{d}U} \ln(1 + e^{U+V}) = \frac{e^{U+V}}{1 + e^{U+V}}. \tag{6.100}$$

Hence,

$$\sigma^{\mathrm{l\text{-}r}}(k_z) = \frac{m^* k_{\mathrm{B}}T}{\pi \hbar^2} \ln(1 + e^{-U}), \tag{6.101}$$

i.e.,

$$\sigma^{\mathrm{l\text{-}r}}(k_z) = \frac{m^* k_{\mathrm{B}}T}{\pi \hbar^2} \ln \left[1 + \exp \left[\left(E_{\mathrm{F}} - E_{\mathrm{c}}(0) - \frac{\hbar^2 k_z^2}{2m^*} \right) / k_{\mathrm{B}}T \right] \right]. \tag{6.102}$$

The contribution to the total charge density in the device coming from electrons incident from the right contact is given by $n^{\mathrm{r\text{-}l}}(z)$, which is obtained from $n^{\mathrm{l\text{-}r}}(x)$ by making the following substitutions in Equation (6.91):

$$|\psi^{\mathrm{l\text{-}r}}_{k_z, k_{\mathrm{B}}T}(z)|^2 \rightarrow |\psi^{\mathrm{r\text{-}l}}_{k_z, k_{\mathrm{B}}T}(z)|^2, \tag{6.103}$$

$$\sigma^{\mathrm{l\text{-}r}}(k_z) \rightarrow \sigma^{\mathrm{r\text{-}l}}(k_z). \tag{6.104}$$

$\sigma^{\mathrm{r\text{-}l}}(k_z)$ is identical to $\sigma^{\mathrm{l\text{-}r}}(k_z)$ with $E_{\mathrm{c}}(0)$ replaced by $E_{\mathrm{c}}(L)$, the bottom of the conduction band in the right contact.

The current density associated with electrons incident from the left contact is obtained by making the following substitution in Equation (6.94):

$$|\psi^{\mathrm{r\text{-}l}}_{k_z, k_{\mathrm{B}}T}(z)|^2 \rightarrow \frac{-q\hbar k_z}{m_{\mathrm{c}}^*} T^{\mathrm{l\text{-}r}}(k_z, k_t), \tag{6.105}$$

where

$$T^{\mathrm{l\text{-}r}}(k_z, k_t) = \frac{k_z(L)}{k_z(0)} \left| \psi^{\mathrm{l\text{-}r}}_{k_z(0), k_t}(L) \right|^2, \tag{6.106}$$

as shown in Problem 5.3.

In this last equation, $k_z(0)$ [$k_z(L)$] is the z component of the electron wave vector in the left (right) contact, respectively. The quantity $\psi^{\mathrm{l\text{-}r}}_{k_z(0), k_t}(L)$ is the wave function amplitude at $z = L$ for an electron incident from the left contact with wave vector $k_z(0)$ and transverse momentum k_t. This amplitude is obtained by solving the Schrödinger Equation (6.90) for an electron incident from the left contact.

Therefore, the current density J^{l-r} associated with the flux of electrons incident from the left contact is given by

$$J^{l-r} = \frac{-q\hbar}{m^*} \int_0^{+\infty} \frac{dk_z}{2\pi} k_z \int_0^{+\infty} \frac{dk_t k_t}{2\pi} T^{l-r}(k_z, k_t) f(E_k), \qquad (6.107)$$

where the second integral must be performed numerically since $T^{l-r}(k_z, k_t)$ is a function of both the k_z and k_t components of the electrons incident from the contact. Once again, if we replace the transverse energy by its average value $k_B T$, $T^{l-r}(k_z, E_t)$ can be pulled out of the second integral in Equation (6.107) and the current density associated with the electron incident from the left contact becomes

$$J^{l-r} = \frac{-q\hbar}{m_c^*} \int_0^{+\infty} \frac{dk_z}{2\pi} k_z T^{l-r}(k_z, E_t = k_B T) \int_0^{+\infty} \frac{dk_t k_t}{2\pi} f(E_k). \qquad (6.108)$$

The integral over the transverse momentum can be performed exactly as we have shown earlier, leading to

$$J^{l-r} = \frac{-q\hbar}{m_c^*} \int_0^{+\infty} \frac{dk_z}{2\pi} k_z T^{l-r}(k_z, E_t = k_B T) \sigma^{l-r}(k_z), \qquad (6.109)$$

where $\sigma^{l-r}(k_z)$ is given by Equation (6.104).

By analogy, for the electrons incident from the right contact,

$$J^{r-l} = \frac{-q\hbar}{m_c^*} \int_0^{+\infty} \frac{dk_z}{2\pi} k_z T^{r-l}(k_z) \sigma^{r-l}(k_x), \qquad (6.110)$$

where the transmission probability for electrons incident from the right contact is given by

$$T^{r-l}(k_x) = \frac{k_z(0)}{k_z(L)} \left| \psi^{r-l}_{k_z(L), k_B T}(0) \right|^2. \qquad (6.111)$$

Since the wave functions in the left and right contacts are mutually incoherent, the total current density flowing through the device is then the *difference* of the two oppositely flowing current densities,

$$J_{tot} = J^{l-r} - J^{r-l}. \qquad (6.112)$$

** **Problem 6.10: Self-consistent calculation of the current–voltage characteristics of a nansocale device under the assumption of ballistic transport**

Write down the set of equations to be solved to calculate the current–voltage characteristics of a nanoscale device under the assumption of ballistic transport and assuming that the contacts are heavily doped with donors (neglect the hole carrier concentration).

Solution:

Calculation of the total electron carrier concentration and current density: In order to calculate the spatially dependent electron concentration throughout the device, we must add the electron concentrations associated with electrons incident from the left and right contacts since the wave functions in the two contacts are mutually incoherent. Hence,

$$n(z) = n^{\text{l-r}}(z) + n^{\text{r-l}}(z), \tag{6.113}$$

where either term on the right-hand side can be evaluated using the expression derived in the previous problem. Similarly, the total current density is obtained by subtracting the current densities associated with the fluxes incident from the two contacts, using the results of the previous problem.

Solution of the Poisson equation: The total electron concentration $n(z)$ given above modifies the electrostatic potential $V(z)$ between the two contacts. Neglecting the contribution from holes, we must solve the Poisson equation

$$\frac{\mathrm{d}}{\mathrm{d}z}\left[\epsilon(z)\frac{\mathrm{d}}{\mathrm{d}z}V(z)\right] = -q[N_{\mathrm{D}}^{+}(z) - n(z)], \tag{6.114}$$

while imposing continuity of $V(z)$ and $\epsilon(z)\frac{\mathrm{d}}{\mathrm{d}z}$, and using the values of the potential in the contacts as boundary conditions. Typically, one contact is assumed to be at ground and the bias across the device is applied to the other contact.

To find the self-consistent solution to the set of equations given above, the iterative scheme shown in Figure 6.8 can be used.

The total electron density is first calculated for a specific conduction band energy profile. A good initial guess is to use the conduction band energy profile neglecting the effects of space charge and assuming a linear drop of the electrostatic potential between the two contacts.

The Poisson equation is then solved numerically and the resulting electrostatic potential energy profile is added to the electron affinity (which can be spatially varying) to get the new conduction band energy profile. The total electron density in the device is then recalculated from a solution of the Schrödinger equation and the procedure is repeated until a self-consistent solution is obtained. The number of iterations depends on the requested accuracy for any quantity of interest. Since the main focus is on the current–voltage characteristics, the iterative procedure is typically carried out until the current is obtained with a predetermined accuracy. This requires a calculation of the current density at each step of the iteration process, as indicated in Figure 6.8.

---

* Problem 6.11: Electron density near a perfectly reflecting potential wall

Consider a one-dimensional electron gas obeying Maxwell–Boltzmann statistics and impinging from the left on a perfectly reflecting wall located at z = 0, as shown in

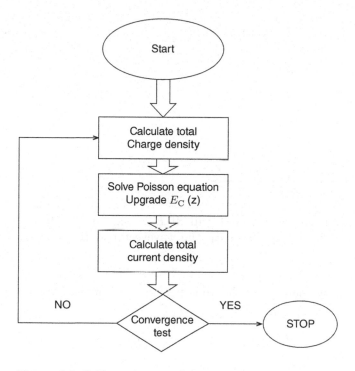

Figure 6.8: Self-consistent scheme to calculate the current–voltage characteristics in a nanoscale device under the approximation of ballistic transport.

Figure 6.9. Calculate the spatial dependence of the electron density as a function of z, for z < 0. Assume a constant effective mass m^*.

Solution: The spatial dependence of the (linear) electron density is given by

$$n_1(z) = \sum_{k_z} |\phi_{k_z}(z)|^2 f(E_{k_z}),$$ (6.115)

where

$$f(E_{k_z}) = e^{-\frac{E(k_z) - E_F}{k_B T}}$$ (6.116)

is the Maxwell–Boltzmann factor and $E(k_z) = \hbar^2 k_z^2 / 2m^*$ is the kinetic energy of an incident electron. Furthermore, the wave function of an incident electron is given by

$$\phi_{k_z}(z) = e^{ik_z z} + re^{-ik_z z}.$$ (6.117)

Since the electron is incident on a perfectly reflecting wall, $\phi_{k_z}(0) = 0$ for all k_z and therefore $r = -1$ for all k_z.

Converting the sum into an integral in Equation (6.115) using the one-dimensional density of states, we get

$$n_1(z) = 2\frac{L}{2\pi} \int_0^\infty dk_z |\phi_{k_z}(z)|^2 f(E_{k_z}).$$ (6.118)

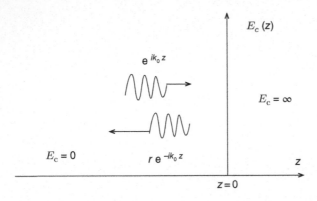

Figure 6.9: Reflection from a infinite potential wall. If $E_c = \infty$ for $z > 0$, the reflection coefficient $r = -1$ for all values of k_0, the wavevector of the electron incident from the left.

The linear electron density (number of electrons per unit length) is therefore given by

$$n_l(z) = \frac{n(z)}{L} = \frac{1}{\pi} \int_0^\infty \mathrm{d}k_z |e^{ik_z z} - e^{-ik_z z}|^2 f(E_{k_z})$$
$$= \frac{2}{\pi} \int_0^\infty \mathrm{d}k_z \left[1 - \cos(2k_z z)\right] f(E_{k_z}). \tag{6.119}$$

This integral can be performed exactly using the following results:

$$\int_0^\infty e^{-r^2 z^2} \mathrm{d}z = \frac{\sqrt{\pi}}{2r}, \tag{6.120}$$

$$\int_0^\infty \cos(mz) e^{-a^2 z^2} \mathrm{d}z = \frac{\sqrt{\pi}}{2a} e^{-\frac{m^2}{4a^2}}. \tag{6.121}$$

This leads to the final result

$$n_l(z) = \frac{1}{\sqrt{\pi}\lambda}(1 - e^{-\frac{z^2}{\lambda^2}}) e^{\frac{E_F}{k_B T}}, \tag{6.122}$$

where $\lambda = \frac{\hbar}{\sqrt{2m^* k_B T}}$ is the thermal de Broglie wavelength. This quantity characterizes the length scale over which the electron density changes from its value far from the interface, $\frac{1}{\sqrt{\pi}\lambda} e^{\frac{E_F}{k_B T}}$, as a result of the quantum mechanical reflection at the interface.

A plot of the normalized electron density $n_l(z)/(\frac{1}{\lambda\sqrt{\pi}} e^{\frac{E_F}{k_B T}})$ for different temperatures is shown in Figure 6.10, assuming that the electron's effective mass is $m^* = 0.067 m_0$. Because the thermal de Broglie wavelength is inversely proportional to the square root of the temperature, the effects of the quantum mechanical reflection from the potential wall are felt farther into the bulk of the one-dimensional electron gas at lower temperatures.

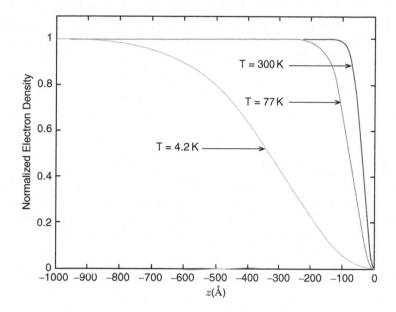

Figure 6.10: Plot of the normalized electron density $n_l(z)/\left(\frac{1}{\lambda\sqrt{\pi}}e^{\frac{E_F}{k_B T}}\right)$ for three different temperatures for an electron with an effective mass $m^* = 0.067m_0$. From bottom to top, the temperature is set equal to 4.2 K, 77 K, and 300 K, respectively.

** Problem 6.12: Richardson–Dushman equation in 3D

Calculate the current density flowing across a metal/vacuum interface modeled using the conduction band diagram shown in Figure 6.11. Assume that the transmission probability of electrons impinging on the interface is unity if their electron kinetic energy is above $E_F + \Phi$, and zero otherwise. The quantities E_F and Φ are the Fermi energy and work function of the metal, respectively.

Solution: For electrons impinging from the contact, the current density flowing into vacuum is given by (see Problem 6.9)

$$J_z = -q\sum_{k_z} \frac{\hbar k_z}{m^*}|T(k_z)|^2 f(E),\qquad (6.123)$$

where $|T(k_z)|^2$ is the transmission probability of the electron impinging with wave vector component k_z in the direction perpendicular to the interface. Since the electrons have the same effective mass m_0 inside and outside the metal, the transmission probability is independent of the electron's transverse wave vector k_t.

Because the work function is typically a few eV, the Fermi–Dirac occupation probability $f(E)$ can be approximated by the Maxwell–Boltzmann distribution,

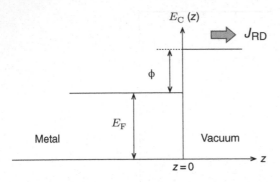

Figure 6.11: Richardson–Dushman thermionic current across a metal/vacuum interface modeled as a potential step of height $E_F + \Phi$. E_F and Φ are the Fermi energy and work function of the metal, respectively.

$$f(E) = e^{-\frac{E - E_F}{k_B T}}, \tag{6.124}$$

where E is the total kinetic energy of the electron, i.e.,

$$E = \frac{\hbar^2 k_z{}^2}{2m_0} + \frac{\hbar^2 k_t{}^2}{2m_0}. \tag{6.125}$$

Converting the sum into an integral in Equation (6.123) (see Problem 6.1), we get

$$J_z = -q\frac{2}{(2\pi)^3} \int d^3\vec{k} \frac{\hbar k_z}{m^*} |T(k_z)|^2 f(E). \tag{6.126}$$

Next, changing to cylindrical coordinates from spherical coordinates (cylinder axis along k_x direction), Equation (6.126) becomes

$$J_z = -q\frac{2}{(2\pi)^3} \int_0^{2\pi} d\phi \int_0^{+\infty} dk_t k_t \int_{k_{z,\min}}^{+\infty} dk_z \frac{\hbar k_z}{m^*} f(E), \tag{6.127}$$

where

$$k_{z,\min} = \frac{1}{\hbar}\sqrt{2m(E_F + \Phi)}. \tag{6.128}$$

Since $|T(k_z)|^2$ is independent of k_t and assumed to be unity for electrons with k_z above $k_{z,\min}$, we can separate the integrals over k_z and k_t in Equation (6.127) leading to

$$J_z = -q\frac{2}{(2\pi)^2} e^{\frac{E_F}{k_B T}} \int_0^{+\infty} dk_t k_t e^{-\frac{\hbar^2 k_t{}^2}{2m^* k_B T}} \int_{k_{z,\min}}^{+\infty} dk_z k_z e^{-\frac{\hbar^2 k_z{}^2}{2m^* k_B T}}. \tag{6.129}$$

The two integrations can be performed exactly, leading to the final result

$$J_z = J_{RD} = A^* T^2 e^{-\frac{\Phi}{k_B T}}, \tag{6.130}$$

where $A^* = \frac{4\pi q m^*}{h^3} k_B{}^2$. This last result is referred to as the Richardson–Dushman equation.

** Problem 6.13: Richardson–Dushman equation in 1D

Repeat the previous problem in 1D, i.e., assuming that the electrons have only a k_z component.

Solution: In this case, the thermionic current must be calculated using the expression

$$J_z = -q \sum_{k_z} \frac{\hbar k_z}{m^*} |T(k_z)|^2 f_{\mathrm{MB}}(E),$$ (6.131)

where $f_{\mathrm{MB}}(E_z) = \mathrm{e}^{-\frac{E_z - E_F}{k_B T}}$ is the Maxwell–Boltzmann factor and $E_z = \frac{\hbar^2 k_z^2}{2m^*}$ is the kinetic energy along the z-axis.

Converting the sum into an integral, since $|T(k_z)|^2 = 1$ for $k_z \geq k_{z,\min}$, with $k_{z,\min} = \frac{1}{\hbar}\sqrt{2m(E_F + \Phi)}$, we get

$$J_z = -\frac{q}{\pi} \mathrm{e}^{\frac{E_F}{k_B T}} \int_{k_{z,\min}}^{+\infty} \mathrm{d}k_z\, k_z \mathrm{e}^{-\frac{\hbar^2 k_z^2}{2m^* k_B T}}.$$ (6.132)

Performing the integration leads to the one-dimensional version of the Richardson–Dushman equation:

$$J_z^{\mathrm{1D}} = \frac{2qk_B T}{h} \mathrm{e}^{-\frac{\Phi}{k_B T}}.$$ (6.133)

Notice that the temperature dependence in front of the exponential has changed from T^2 to T going from 3D to 1D. This expression can be used to calculate the thermionic current emitted from metallic carbon nanotubes, which can be approximated as one-dimensional wires.

** Problem 6.14: Heat conduction across a metal/vacuum interface

In Chapter 3, the concept of energy flux was introduced and applied to the study of several tunneling problems. The goal of this problem is to calculate the heat flux across a metal/vacuum interface as shown in Figure 6.11. The latter is defined as follows:

$$J_Q = -2\frac{q}{(2\pi)^3} \int \mathrm{d}\vec{k} \frac{\hbar k_z}{m_0}(E - E_F)|T(k_z)|^2 f_{\mathrm{MB}}(E),$$ (6.134)

where E is the total kinetic energy of the incident electron on the metal/vacuum interface, E_F is the Fermi energy of the metal, $|T(k_z)|^2$ is the transmission probability across the interface, and f_{MB} is the Maxwell–Boltzmann factor.

Show that J_Q is given by

$$J_Q = J_{\mathrm{RD}}(\phi + 2k_B T),$$ (6.135)

where J_{RD} is the Richardson–Dushman result given in Equation (6.130), ϕ is the work function of the metal, k_B is Boltzmann's constant, and T is the temperature of the metal.

Solution: We first rewrite the heat flux in Equation (6.134) as

$$J_Q = J_Q' - E_F J_{RD}, \tag{6.136}$$

where

$$J_Q' = -2\frac{q}{(2\pi)^3} \int d\vec{k} \frac{\hbar k_z}{m_0} \left(E_z + \frac{\hbar^2 k_t^2}{2m_0} \right) |T(k_z)|^2 f_{MB}(E), \tag{6.137}$$

where E_z is the kinetic energy associated with longitudinal (or z-component) of motion and $\frac{\hbar^2 k_t^2}{2m_0}$ is the kinetic energy associated with the transverse component of motion. We then separate the above result into two integrals as follows:

$$J_{Q,1}' = -2\frac{q}{(2\pi)^3} \int d\vec{k} \frac{\hbar k_z}{m_0} E_z |T(k_z)|^2 f_{MB}(E), \tag{6.138}$$

$$J_{Q,2}' = -2\frac{q}{(2\pi)^3} \int d\vec{k} \frac{\hbar k_z}{m_0} \frac{\hbar^2 k_t^2}{2m_0} |T(k_z)|^2 f_{MB}(E). \tag{6.139}$$

Taking into account the fact that the transmission probability $|T(k_z)|^2$ is independent of k_t and assumed to be unity for electrons with k_z above $k_{z,min} = \frac{1}{\hbar}\sqrt{2m_0(E_F + \Phi)}$ (see Problem 6.12), we can separate the integrals over k_z and k_t in Equations (6.138) and (6.139) leading to

$$J_{Q,1}' = -q\frac{2}{(2\pi)^2} e^{\frac{E_F}{k_B T}} \int_0^{+\infty} dk_t k_t e^{-\frac{\hbar^2 k_t^2}{2m_0 k_B T}} \int_{k_{z,min}}^{+\infty} dk_z k_z e^{-\frac{\hbar^2 k_z^2}{2m_0 k_B T}}, \tag{6.140}$$

$$J_{Q,2}' = -q\frac{2}{(2\pi)^2} e^{\frac{E_F}{k_B T}} \int_0^{+\infty} dk_t k_t \frac{\hbar^2 k_t^2}{2m_0} e^{-\frac{\hbar^2 k_t^2}{2m_0 k_B T}}$$
$$\int_{k_{z,min}}^{+\infty} dk_z \frac{\hbar k_z}{m_0} e^{-\frac{\hbar^2 k_z^2}{2m_0 k_B T}}. \tag{6.141}$$

Changing variables from k_t to $E_t = \frac{\hbar^2 k_t^2}{2m_0}$ and k_z to $\frac{\hbar^2 k_z^2}{2m_0}$, the integrations in the expressions for $J_{Q,1}'$ and $J_{Q,2}'$ can be performed exactly, leading to

$$J_{Q,1}' = J_{RD}(E_F + \phi + k_B T), \tag{6.142}$$
$$J_{Q,2}' = J_{RD} k_B T. \tag{6.143}$$

Regrouping the results (6.136), (6.142), and (6.143), we finally get the expression for the heat flux across the metal/vacuum interface:

$$J_Q = J_{RD}(\phi + 2k_B T). \tag{6.144}$$

This equation has been used extensively in the design of thermionic converters and refrigerators [5].

*** Problem 6.15: Blackbody radiation in 3D

The birth of quantum mechanics occurred in 1900 with Max Planck's derivation of the correct expression for the experimentally measured energy per unit volume per frequency interval (or energy spectral distribution) of the blackbody radiation,

$$u(\nu) = \frac{8\pi h \nu^3}{c^3} \frac{1}{e^{\frac{h\nu}{k_B T}} - 1}, \tag{6.145}$$

where h is Planck's constant, ν is the frequency of the electromagnetic radiation in the cavity, c is the speed of light in vacuum, k_B is Boltzmann's constant, and T is the temperature inside the cavity in Kelvin. Planck was able to derive this important relation by assuming that the energy exchange between the electromagnetic waves inside a cavity and its walls occurs via emission and absorption of discrete quanta of energy. As a result, the energy of the electromagnetic radiation with frequency ν inside the blackbody cavity exists only in multiples of $h\nu$.

Using the concept of DOS for the photons trapped inside the cavity, derive the expression for the energy spectral density in Equation (6.145). Next, derive the analytical expression of the energy density per unit wavelength of the blackbody radiation and plot it for cavity temperatures of $T = 300\,K$, $1000\,K$, and $5000\,K$.

Show that the maximum of the energy spectral density u in Equation (6.145) occurs when

$$3 - 3\exp(-x) = x, \tag{6.146}$$

where $x = h\nu/k_B T$.

Solve this equation numerically and show that the maximum occurs for x around 2.82, i.e., at a frequency $\nu_{\max}$ given by

$$h\nu_{\max}/k_B T \sim 2.82. \tag{6.147}$$

This last relation is referred to as Wien's displacement law.

Solution: In the blackbody, the photons form standing waves in the cavity and are in thermal equilibrium with the walls, which continuously absorb and emit the photons. By piercing a small hole in one side of the blackbody outer shell, we can determine the spectral distribution of the blackbody radiation, i.e., the fraction of total radiated energy with frequency between ν and $\nu + d\nu$.

Assuming the photons form standing waves within the cavity, and that the latter are not much disturbed by the small hole created to observe the photon spectrum, one can show that the wavevector component of the photons in the rectangular cavity must obey the relations $k_x l = n_x \pi$, $k_y l = n_y \pi$, and $k_z l = n_z \pi$, where n_x, n_y, and n_z are positive integers and l is the linear dimension of the cavity. These values (or quantization of photon wavevectors) ensure the presence of a standing wave with integer multiples of half wavelength in all directions inside the cavity. Each triplet

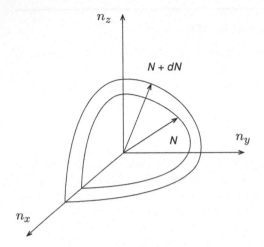

Figure 6.12: Only one-eighth of the spherical shell between the two spheres of radii N and $N + dN$ must be taken into account to calculate the density of distinct phonon modes present in the blackbody cavity within the corresponding energy range.

(n_x, n_y, and n_z) represents an electromagnetic mode of oscillation. These modes are photon states and are doubly degenerate since for each $\vec{k} = (k_x, k_y, k_z)$ there are two independent polarization directions possible for the standing waves.

For each node, the relation between the frequency and the photon wavevector is given by

$$\nu = \frac{c}{\lambda} = \frac{ck}{2\pi} = \frac{c|\vec{k}|}{2\pi}. \tag{6.148}$$

With the components of the photon wavevectors given above, we get

$$\nu = \frac{c}{2l}N, \text{ with } N = \sqrt{n_x^2 + n_y^2 + n_z^2}. \tag{6.149}$$

We must now count the number of photon modes within the two spheres of radii N and $N + dN$.

Since both positive and negative values of the photon $\vec{k}$ components belong to the same photon standing wave, we must count as distinct only the photon modes present in one-eighth of the spherical shell shown in Figure 6.12. The volume of that portion of the shell between the two spheres of radii N and $N + dN$ is given by

$$dM = \frac{1}{8}4\pi \, N^2 dN = \frac{\pi N^2}{2}dN. \tag{6.150}$$

From Equation (6.149), we get

$$\nu = \frac{cN}{2l} \rightarrow N = \frac{2l\nu}{c}. \tag{6.151}$$

Therefore,

$$dN = \frac{2l}{c}d\nu,$$

$$dM = \frac{\pi}{2}\left(\frac{2l\nu}{c}\right)^2\frac{2l}{c}d\nu = \frac{4\pi l^3}{c^3}\nu^2 d\nu. \tag{6.152}$$

Taking into account the two independent polarization directions for each node, we must multiply dM by 2. So the number of photon states in the shell is

$$dS = 2dM = \frac{8\pi V}{c^3}\nu^2 d\nu = g(\nu)d\nu, \tag{6.153}$$

where $V = l^3$ is the volume of the cavity and $g(\nu)$ is the photon DOS in the frequency interval dν:

$$g(\nu) = \frac{8\pi V}{c^3}\nu^2. \tag{6.154}$$

The occupation probability of photons must obey Bose–Einstein statistics because photons are bosons. Hence, this probability is

$$f_{\text{BE}}(E) = \frac{1}{e^{E/k_{\text{B}}T} - 1}. \tag{6.155}$$

Consequently, the number of photons, dn_ν, with frequencies between ν and $\nu + d\nu$ is

$$dn_\nu = f_{\text{BE}}(h\nu)g(\nu)d\nu = \frac{1}{e^{\frac{h\nu}{k_{\text{B}}T}} - 1}\frac{8\pi V}{c^3}\nu^2 d\nu. \tag{6.156}$$

The amount of energy carried by those dn_ν photons is found by multiplying the previous result with the photon energy $h\nu$, yielding

$$dE_\nu = h\nu dn_\nu = \frac{8\pi V h}{c^3}\frac{\nu^3}{e^{\frac{h\nu}{k_{\text{B}}T}} - 1}d\nu, \tag{6.157}$$

which we rewrite in the more condensed form

$$dE_\nu = F(\nu)d\nu, \tag{6.158}$$

where

$$F(\nu) = \frac{8\pi V h}{c^3}\frac{\nu^3}{e^{\frac{h\nu}{k_{\text{B}}T}} - 1} \tag{6.159}$$

is the spectral energy distribution we were looking for. This expression was first derived by Max Planck.

To derive the spectral energy distribution as a function of wavelength λ, we must find the function $G(\lambda)$ such that the total electromagnetic energy inside the black box is given by

$$E = \int_0^{+\infty} G(\lambda)d\lambda. \tag{6.160}$$

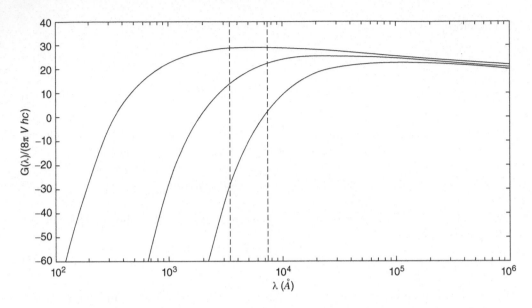

Figure 6.13: Plot of the normalized spectral energy distribution $G(\lambda)/(8\pi V hc)$ versus wavelength in Equation (6.161) for three different temperatures. From left to right, the curves correspond to $T = 5000$, 1000, and $300\,\mathrm{K}$.

Starting with Equations (6.158)–(6.159) and performing a change of variable using the relation $\nu = c/\lambda$, we get

$$G(\lambda) = \frac{8\pi V hc}{\lambda^5 \left[e^{\frac{hc}{k_\mathrm{B} T \lambda}} - 1 \right]}. \tag{6.161}$$

Figure 6.13 is a plot of the normalized spectral energy distribution, i.e., $G(\lambda)/(8\pi V hc)$ for $T = 300$, 1000, and $5000\,\mathrm{K}$.

Using the results above, the total electromagnetic energy inside the black box cavity whose walls are maintained at a temperature T is given by

$$E = \int_0^\infty \mathrm{d}E_\nu = \frac{8\pi V h}{c^3} \int_0^\infty \frac{\nu^3}{e^{\frac{h\nu}{k_\mathrm{B} T}} - 1}\,\mathrm{d}\nu = \frac{8\pi V k_\mathrm{B}{}^4 T^4}{c^3 h^3} \int_0^\infty \frac{q^3}{e^q - 1}\,\mathrm{d}q, \tag{6.162}$$

where $q = h\nu/(k_\mathrm{B} T)$. Using the result

$$\int_0^\infty \frac{q^p}{e^q - 1}\,\mathrm{d}q = \Gamma(p+1)\zeta(p+1), \tag{6.163}$$

where Γ is the Euler gamma function and ζ is the zeta function of Riemann, we get

$$E/V = \sigma T^4, \tag{6.164}$$

where $\sigma = 4.71\,\mathrm{keV/K^4\,m^3}$. This last equation is referred to as the Stefan–Boltzmann law.

Starting with the expression of the spectral energy distribution in Equation (6.159), we calculate the frequency at which it reaches a maximum:

$$\frac{\mathrm{d}F(\nu)}{\mathrm{d}\nu} = 0 \rightarrow \frac{\mathrm{d}}{\mathrm{d}\nu}\left[\frac{8\pi vh}{c^3}\frac{\nu^3}{\mathrm{e}^{\frac{h\nu}{kT}}-1}\right] = 0. \tag{6.165}$$

This leads to the following transcendental equation:

$$\frac{8\pi vh}{c^3}\frac{\nu^2[\mathrm{e}^{\frac{h\nu}{kT}}\left(3-\frac{h\nu}{kT}\right)-3]}{[\mathrm{e}^{\frac{h\nu}{kT}}-1]^2} = 0. \tag{6.166}$$

Hence, the maximum is reached when

$$\mathrm{e}^y(3-y)-3 = 0, \tag{6.167}$$

where

$$y = \frac{h\nu_{\mathrm{max}}}{kT}. \tag{6.168}$$

Matlab code to solve Equation (6.167) is given in Appendix G. The maximum in the spectral energy distribution occurs at a frequency given by

$$\lambda_{\mathrm{max}} \rightarrow h\nu_{\mathrm{max}} \approx 2.82\,k_{\mathrm{B}}T. \tag{6.169}$$

This last relation is referred to as Wien's displacement law.

** Problem 6.16: Blackbody radiation in 1D

In this problem, we repeat the study of the blackbody radiation assuming a one-dimensional world, i.e., a cavity extending along the z-axis only and photons traveling back and forth along that direction while being in equilibrium with the two walls at the end of the cavity held at a temperature T.

Use the one-dimensional DOS for photons trapped inside the cavity and derive the expression for the energy spectral density for this 1D case.

Show that the maximum of the energy spectral density occurs when

$$5 - 5\exp(-y) = y, \tag{6.170}$$

where $y = h\nu/k_{\mathrm{B}}T$.

Solve this equation and find for what value of y the maximum occurs. Use this result to write the one-dimensional version of Wien's displacement law.

Solution: Assuming that the photons form standing waves in the direction of the one-dimensional box, the wavevector component of the photons must obey the relation $k_x l = n_x \pi$, where n_x is a positive integer. These values ensure the presence of standing waves whose wavelengths λ_n are integral submultiples of twice the cavity length in the x-direction, i.e. $\lambda_{n_x} = 2l/n_x$. Each n_x represents an electromagnetic mode of oscillation. These modes are doubly degenerate since for each k_x there are two independent polarization directions possible for the standing waves.

For each node, the relation between its frequency and the photon wavevector is given by

$$\nu = \frac{c}{\lambda} = \frac{ck_x}{2\pi} = \frac{c}{2l}n_x. \tag{6.171}$$

Hence, the number of photon modes between ν and $\nu + d\nu$ is given by

$$dN_x = \frac{2l}{c}d\nu. \tag{6.172}$$

Multiplying by a factor of 2 to take into account the two independent polarizations, we finally obtain the number of modes within the frequencies ν and $\nu + d\nu$ as

$$dM = 2dN_x = g(\nu)d\nu, \tag{6.173}$$

where $g(\nu)$ is the one-dimensional photon DOS.

Each mode is occupied by a photon with probability given by Bose–Einstein statistics. Using Bose–Einstein statistics, the amount of energy per frequency interval is therefore given by

$$dE_\nu = h\nu dn_\nu = \frac{4l}{c}\frac{h\nu}{e^{\frac{h\nu}{k_B T}} - 1}d\nu. \tag{6.174}$$

The total electromagnetic energy per unit length inside the one-dimensional black box cavity whose walls are maintained at a temperature T is therefore given by

$$E/l = \frac{1}{l}\int_0^\infty dE_\nu = \frac{4}{c}\int_0^\infty \frac{h\mu}{e^{\frac{h\nu}{k_B T}} - 1}d\nu = \frac{4k_B^2 T^2}{ch}\int_0^\infty \frac{x}{e^x - 1}dx, \tag{6.175}$$

where $x = h\nu/(k_B T)$. Hence,

$$E/l \propto T^2, \tag{6.176}$$

which is the one-dimensional version of the Stefan–Boltzmann law.

Proceeding as in Problem 6.15, we next calculate the location of the maximum of the energy spectral density in the 1D case by setting the derivative of the integral in Equation (6.175) to zero. This leads to the following transcendental equation:

$$5 - 5\exp(-y) = y, \tag{6.177}$$

where $y = h\nu/k_B T$. Matlab code using an iterative solution of Equation (6.177) is given in Appendix G, leading to $y_{max} = 4.965$.

In this case, the maximum of the spectral energy distribution occurs at a frequency given by

$$h\nu_{max} \approx 4.965\, k_B T. \tag{6.178}$$

This is Wien's displacement law for the 1D case. This result can be used to study the importance of radiative loss at the tip of single-walled carbon nanotubes during field emission.

Suggested problems

- You will need the following results: $F_{1/2}(0) = 0.678$ and $F_{-1/2}(0) = 1.072$.

  (1) Using the results of Problem 6.2, compute the critical concentration for the onset of degeneracy in a bulk sample at $T = 300\,\text{K}$ and $T = 1\,\text{K}$ for $m^* = m_0$, the free electron mass.

  (2) For a metallic thin film of width $100\,\text{Å}$, use Equation (6.35) and compute the ratio $n_{\text{crit}}(W, T)/n_{\text{crit}}^{3D}$ for $T = 300\,\text{K}$. Use $m^* = m_0$, the free electron mass.

  (3) For a metallic wire with cross section $A = 10^{-12}\,\text{cm}^2$, use Equation (6.40) and compute the ratio $\frac{n_i^{\text{crit}}/A}{n_{\text{crit}}^{3D}}$ at $T = 1\,\text{K}$. Use $m^* = m_0$, the free electron mass.

- Repeat the problem above if $N_D = 10^{19}\,\text{cm}^{-3}$, and determine the fraction of ionized impurities in the QW if the Fermi level is $3\,k_B T$ above the bottom of the second subband in the QW and $T = 300\,\text{K}$.

- Starting with the results of Problem 6.5, assume that an electrostatic potential is applied to the 2D semiconductor single layer V_{ch} through the application of a gate potential. Assume the device is operated at high enough temperature that $n = p = n_i$. In the presence of V_{ch}, the expression for n and p derived in Problem 6.5 can be obtained via the substitution $E_0 \rightarrow E_0 - qV_{\text{ch}}$ and $E_0 \rightarrow E_0 + qV_{\text{ch}}$, for n and p, respectively (where q is the charge of the electron).

  Derive an analytical expression for the 2D channel quantum capacitance $C_q = \frac{\partial Q}{\partial V_{\text{ch}}}$.

- Using the results of Problem 6.1, the electron concentration in a heavily doped n-type bulk material is given by

$$n = \frac{2N_c}{\sqrt{\pi}} F_{1/2}(\xi), \qquad (6.179)$$

where $\xi = \frac{E_F - E_c}{k_B T}$.

If $\xi > 5$, $F_{1/2}(\xi)$ can be well approximated by $\frac{2}{3}\xi^{3/2}$.

If the n-type region of Si, Ge, and GaAs samples is doped heavily with donors with an ionization energy of $15\,\text{meV}$, find the value of N_D (doping concentration in cm^{-3}) such that the Fermi level will be exactly $5\,k_B T$ above the conduction band at room temperature.

Use the following values: $N_c(\text{Si}) = 3.22 \times 10^{19}\,\text{cm}^{-3}$, $N_c(\text{Ge}) = 1.03 \times 10^{19}\,\text{cm}^{-3}$, and $N_c(\text{GaAs}) = 4.21 \times 10^{17}\,\text{cm}^{-3}$.

- Repeat Problem 6.11 assuming electrons from a three-dimensional electron gas impinging from the left on the infinite potential wall at $z = 0$. The electrons obey Maxwell–Boltzmann statistics. Assume a constant effective mass m^*.

- Repeat Problem 6.4 assuming that the two lowest subbands are occupied. Give analytical expressions for the total charge concentration $\rho(z)$, the electric field $E(z)$, and the electrostatic potential $V(z)$ across the well. Assume the boundary conditions are $E(0) = 0$ and $V(0) = 0$.

- According to Problem 6.15, the peak in the energy spectral distribution curve shifts upward in frequency as the temperature increases. What is the equilibrium temperature inside a blackbody cavity when the maximum in the energy spectral distribution is to be located at wavelengths $1000\,\text{Å}$, $5000\,\text{Å}$, and $10\,\mu\text{m}$, which correspond to the ultraviolet, visible, and mid-infrared region of the electromagnetic spectrum, respectively?

- Following the solution of Problems 6.15 and 6.16, repeat the study of the blackbody radiation in a two-dimensional space, i.e., a square cavity extending along the x and y axes only (i.e., in the $z = 0$ plane) where the photons are in equilibrium with the two walls at the end of the cavity held at a temperature T.

  Use the two-dimensional DOS for photons trapped inside the cavity and derive the expression for the energy spectral density for this 2D case. Plot your results for cavity temperatures of $T = 1000$, 3000, and $5000\,\text{K}$.

  Find the frequency associated with the maximum of the energy spectral density and write the two-dimensional version of Wien's displacement law.

- Following the derivation in Problem 6.12, derive an expression for the current associated with heat conduction through a metal/vacuum interface assuming the electrons impinging on the surface are confined to move only along the axis perpendicular to the interface, i.e., derive the heat conduction current in 1D. This problem is less academic than it seems since it can describe the heat current associated with thermionic emission from carbon nanotubes or other one-dimensional structures (e.g., ZnO nanowires) where carrier transport can be assumed to be along the direction of current flow along the main axis of the 1D structure.

References

[1] Bandyopadhyay S. and Cahay, M. (2015) *Introduction to Spintronics* 2nd edition, Chapter 18, CRC Press, Boca Raton, FL.

[2] Aurora, V. K. (1982) Onset of degeneracy in confined systems. *Physical Review B* 26, pp. 2247–2249.

[3] Delagebeaudeuf, D. and Linh, N. T. (1982) Metal-(n) AlGaAs–GaAs two-dimensional electron gas FET. *IEEE Trans. Electron. Devices* 29, pp. 955–960.

[4] Krantz, R. J. and Bloss, W. L. (1990) The role of acceptor density on the high channel carrier density I–V characteristics of AlGaAs/GaAs MODFETS. *Solid-State Electronics* 33, pp. 941–945.

[5] Mahan, G. D. (1994) Thermionic refrigeration. *Journal of Applied Physics* 76, pp. 4362–4366.

Suggested Reading

- Melvin, M. A. (1955) Blackbody radiation and Lambert's law. *American Journal of Physics* 23, pp. 508–510.

- Shanks, D. (1956) Monochromatic approximation of blackbody radiation. *American Journal of Physics* 24, pp. 244–246.

- Sherwin, C. W. (1957) On the average volume per photon in blackbody radiation. *American Journal of Physics* 25, pp. 117–118.

- Magram, S. J. (1957) Concentration and size of photons from particle interpretation of blackbody radiation. *American Journal of Physics* 25, pp. 283–284.

- Duley, W. W. (1972) Blackbody radiation in small cavities. *American Journal of Physics* 40, pp. 1337–1338.

- Baltes, H. P. (1974) Comment on blackbody radiation in small cavities. *American Journal of Physics* 42, pp. 505–507.

- Dahm, A. J. and Langenberg, D. N. (1975) Blackbody radiation from a single mode source: a demonstration. *American Journal of Physics* 43, pp. 1004–1006.

- Manikopoulos, C. N. and Aquirre, J. F. (1977) Determination of the blackbody radiation constant hc/k in the modern physics laboratory. *American Journal of Physics* 45, pp. 576–578.

- Yorke, E. D. (1983) Effect of blackbody radiation on the hydrogen spectrum. *American Journal of Physics* 51, pp. 16–19.

- Kelly, R. E. (1981) Thermodynamics of blackbody radiation. *American Journal of Physics* 49, pp. 714–719.

- Pearson, J. M. (1984) A note on the thermodynamics of blackbody radiation. *American Journal of Physics* 52, pp. 262–263.

- Dryzek, J. and Ruebenbaur, K. (1992) Planck's constant determination from blackbody radiation. *American Journal of Physics* 60, pp. 251–253.

- Heald, M. A. (2003) Where is the "Wien peak"? *American Journal of Physics* 71, pp. 1322–1323.

- Ribeiro, C. I. (2014) Blackbody radiation from an incandescent lamp. *The Physics Teacher* 52, pp. 371–372.

Chapter 7: Transfer Matrix

In this chapter, the transfer matrix formalism is introduced as a general approach for treating both bound state and tunneling problems [1–7]. It has been used extensively in the past to study bound states of quantum wells of arbitrary shape [8] and finite periodic potentials [9], and tunneling through finite repeated structures [10–12], among others.

The transfer matrix formalism can be used to show that the problem of finding the bound states of an arbitrary confined one-dimensional potential energy profile $E_c(z)$ can be reformulated as a tunneling problem (see Problem 7.12) [13]. The following theorem is proved:

For an electron confined to a quantum well of width W with an arbitrary potential profile $E_c(z)$ within the well and a constant potential V_0 outside the well, the bound state energies $(E_1, E_2, E_3, \dots)$ can be found by adding two barriers of width d and height V_0 on either side of the well and calculating the energies at which the transmission probability $T(E)$ through the resonant tunneling structure so formed reaches unity. The energies at which the transmission coefficient reaches unity converge toward the bound state energy levels when the thickness d tends to infinity.

The theorem is proved for the case of a spatially independent effective mass, but can be easily extended to the case of a spatially varying effective mass.

Preliminary: Concept of transfer matrix

The definition of the transfer matrix is based on the concepts of linearly independent solutions of the Schrödinger equation and the Wronskian introduced in Chapter 1. If we can find two linearly independent solutions, their linear combination is the general solution to the Schrödinger equation

$$\phi(z) = c_1\phi_1(z) + c_2\phi_2(z). \tag{7.1}$$

Suppose we seek two solutions $\phi_1(z)$ and $\phi_2(z)$ of the Schrödinger equation such that

$$\phi_1(0) = 0, \quad \dot{\phi}_1(0) = 1, \tag{7.2}$$

$$\phi_2(0) = 1, \quad \dot{\phi}_2(0) = 0. \tag{7.3}$$

Here, the dot symbol denotes the first derivative in space. In this book, we have intentionally used various notations—prime and dot, for example—to denote the spatial derivative since the literature in this field uses both conventions (plus, perhaps, a few others) and it is important that the reader is comfortable with all notations and conventions.

Problem Solving in Quantum Mechanics: From Basics to Real-World Applications for Materials Scientists, Applied Physicists, and Devices Engineers, First Edition.
Marc Cahay and Supriyo Bandyopadhyay.

The two solutions $\phi_1(z)$ and $\phi_2(z)$ are indeed linearly independent since their Wronskian, which is independent of z (see Chapter 1), is equal to

$$W(z) = W(0) = \dot{\phi}_1(0)\phi_2(0) - \phi_1(0)\dot{\phi}_2(0) = 1. \tag{7.4}$$

Since $\phi_1(z)$ and $\phi_2(z)$ satisfy the conditions (7.2) and (7.3), we have

$$c_1 = \dot{\phi}(0), \tag{7.5}$$

$$c_2 = \phi(0). \tag{7.6}$$

Hence, from Equation (7.1) and its first derivative with respect to z, we get

$$\phi(L) = \dot{\phi}(0)\phi_1(L) + \phi(0)\phi_2(L), \tag{7.7}$$

$$\dot{\phi}(L) = \dot{\phi}(0)\dot{\phi}_1(L) + \phi(0)\dot{\phi}_2(L). \tag{7.8}$$

These last two equations can be rewritten in a matrix form:

$$\begin{bmatrix} \dot{\phi}(L) \\ \phi(L) \end{bmatrix} = \begin{bmatrix} \dot{\phi}_1(L) & \dot{\phi}_2(L) \\ \phi_1(L) & \phi_2(L) \end{bmatrix} \begin{bmatrix} \dot{\phi}(0) \\ \phi(0) \end{bmatrix}. \tag{7.9}$$

The 2×2 matrix appearing on the right-hand side is called the transfer matrix because it relates the column vector $(\dot{\phi}(z), \phi(z))^\dagger$ (where the $\dagger$ stands for the transpose operation) at location $z = L$ to its value at location $z = 0$.

Cascading rule for transfer matrices: An arbitrary spatially varying potential energy profile $E_c(z)$ can always be approximated by a series of steps where the potential energy in each step is replaced by its average value over that interval. The accuracy of this approximation increases with decreasing interval size. Since the potential within each section is constant, the transfer matrix of each small section can be derived exactly (see Problem 7.2). Once the individual transfer matrix for each small segment is known, the overall transfer matrix W_{TOT} needed to relate the wave function on the right to the wave function on the left of the potential (i.e., describe the tunneling process of a particle incident from the left contact) is the product of the individual transfer matrices associated with each small segment, i.e.,

$$W_{\text{TOT}} = W_N W_{N-1} \cdots W_2 W_1, \tag{7.10}$$

where W_i is the transfer matrix associated with the ith segment counted from the left contact.

In Equation (7.10), it is important to mutiply the individual matrices from right to left (and not the other way around) to obtain the overall transfer matrix since the individual matrices do not commute in general.

---

* Problem 7.1: Transmission and reflection probabilities across an arbitrary potential energy profile

Consider a semiconductor device with arbitrary conduction band energy profile varying along the z-direction only, $E_c(z)$, sandwiched between two contacts (regions of constant or zero potential) at $z = 0$ and $z = L$ (see Figure 7.1). Consider the case of zero bias applied between the two contacts and assume a constant effective mass throughout. Derive an expression for the transmission amplitude t and transmission probability $T(E) = |t(E)|^2$ across the device as a function of the incident wavevector of the electron, k_0, and the elements of the overall transfer matrix across the region $[0, L]$.

Solution: The overall transfer matrix W_{TOT} relates the wave functions and their first derivatives at the left and right contacts (assumed to be at the same potential) according to

$$\begin{bmatrix} \frac{d\phi}{dz}(L^+) \\ \phi(L^+) \end{bmatrix} = W_{TOT} \begin{bmatrix} \frac{d\phi}{dz}(0^-) \\ \phi(0^-) \end{bmatrix}. \tag{7.11}$$

For an electron incident from the left, the electronic states in the left and right contacts are given by

$$\phi(z) = e^{ik_0 z} + r e^{-ik_0 z} \qquad\qquad (z < 0)$$
$$\phi(z) = t e^{ik_0(z-L)} \qquad\qquad (z > L), \tag{7.12}$$

where $k_0 = \frac{1}{\hbar}\sqrt{2m_c^* E_p}$ is the z-component of the electron's wavevector in the contact and r and t are the reflection and transmission amplitudes through the region $[0, L]$, respectively. The quantity E_p is the electron's kinetic energy component in the left contact associated with z-directed motion.

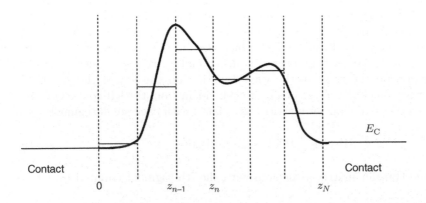

Figure 7.1: Approximation of an arbitrary conduction band energy profile $E_c(z)$ as a series of steps. The effective mass may be assumed to be different in each interval. There is no bias applied between the two contacts, i.e., $E_c(0) = E_c(z_N)$.

Using these scattering states for the wave functions at $z = 0^-$ and $z = L^+$, Equation (7.11) becomes

$$t \begin{bmatrix} ik_0 \\ 1 \end{bmatrix} = W_{\text{TOT}} \begin{bmatrix} ik_0(1-r) \\ 1+r \end{bmatrix}. \tag{7.13}$$

This is a system of two equations for the two unknowns t and r. We can solve these equations to find t and r. The explicit form of the transmission amplitude t in terms of the matrix elements of the total transfer matrix is given by

$$t = \frac{2ik_0 \left[W_{\text{TOT}}^{11} W_{\text{TOT}}^{22} - W_{\text{TOT}}^{12} W_{\text{TOT}}^{21} \right]}{ik_0 \left[W_{\text{TOT}}^{11} + W_{\text{TOT}}^{22} \right] + \left[W_{\text{TOT}}^{21} k_0^2 - W_{\text{TOT}}^{12} \right]}. \tag{7.14}$$

Since the Wronskian is independent of z, W_{TOT} is a unimodular matrix, i.e., $\det(W_{\text{TOT}}) = 1$. Hence, the term within the square brackets in the numerator of Equation (7.14) is unity. In addition, by approximating $E_c(z)$ by a series of steps, the W_{TOT}^{ij} are always purely real, and the transmission probability through the structure is given by

$$T(E) = |t|^2 = \frac{4k_0^2}{k_0^2 [W_{\text{TOT}}^{11} + W_{\text{TOT}}^{22}]^2 + [W_{\text{TOT}}^{21} k_0^2 - W_{\text{TOT}}^{12}]^2}. \tag{7.15}$$

This last equation shows that the transmission probability reaches unity when the following conditions are satisfied:

$$W_{\text{TOT}}^{11} + W_{\text{TOT}}^{22} = \pm 2, \quad W_{\text{TOT}}^{21} = W_{\text{TOT}}^{12} = 0. \tag{7.16}$$

---

* Problem 7.2: Transmission probability across a square barrier

Derive the analytical expressions for the four elements of the transfer matrix through a region with constant E_c and effective mass m^*.

Solution: Let E_p be the electron's total energy. If $E_p > E_c$ (so that the electron wave is a traveling wave and not an evanescent wave), then the two linearly independent solutions satisfying Equations (7.2) and (7.3) are given by (for a first-principles proof, see Chapter 3 of Ref. [14])

$$\phi_1(z) = \frac{\sin(kz)}{k}, \tag{7.17}$$

$$\phi_2(z) = \cos(kz), \tag{7.18}$$

where $k = \frac{1}{\hbar}\sqrt{2m^*(E_p - E_c)}$. Therefore the explicit form of the transfer matrix across a region of width W is given by

$$W(E_p > E_c) = \begin{pmatrix} \cos(kW) & -k\sin(kW) \\ \frac{1}{k}\sin(kW) & \cos(kW) \end{pmatrix}. \tag{7.19}$$

On the other hand, if $E_p < E_c$ (i.e., the electron wave in the region of interest is evanescent), then the two linearly independent solutions satisfying Equations (7.2) and (7.3) are given by (for a first-principles proof, see Chapter 3 of Ref. [14])

$$\phi_1(z) = \frac{\sinh(\kappa z)}{\kappa}, \tag{7.20}$$

$$\phi_2(z) = \cosh(\kappa z), \tag{7.21}$$

where $\kappa = \frac{1}{\hbar}\sqrt{2m^*(E_c - E)}$.

In this case, the explicit form of the transfer matrix is given by

$$W(E_p < E_c) = \begin{pmatrix} \cosh(\kappa W) & \kappa\sinh(\kappa W) \\ \frac{1}{\kappa}\sinh(\kappa W) & \cosh(\kappa W) \end{pmatrix}. \tag{7.22}$$

When the potential energy profile is approximated by a series of steps, the overall transfer matrix W_{TOT} is the product of matrices of the form (7.19) or (7.22), and therefore its matrix elements are real, as stated earlier.

---

* Problem 7.3: Transfer matrix across a region with varying $E_c(z)$ and $m^*(z)$.

Consider an electron with a total energy E moving in a region where the potential energy $E_c(z)$ and the effective mass $m^*(z)$ vary only in the z-direction.

Assume that both $E_c(z)$ and $m^*(z)$ profiles are segmented into sections, and that within each section the values of E_c and m^* are constant. Starting with results of Problem 1.1, write down the Schrödinger equation for the envelope of the wave function $\phi(z)$ in any of the steps.

Using the results of the previous step, find the analytical expression for the transfer matrix across a region where $E_c(z)$ and $m^*(z)$ are assumed to be constant. Across a region of length L (i.e., for $0 < z < L$), the transfer matrix is defined as follows:

$$\begin{bmatrix} \frac{\phi'}{\gamma}(L-\epsilon) \\ \phi(L-\epsilon) \end{bmatrix} = W \begin{bmatrix} \frac{\phi'}{\gamma}(0+\epsilon) \\ \phi(0+\epsilon) \end{bmatrix}, \tag{7.23}$$

where $\gamma = m^*/m_c^*$, and m_c^* is the effective mass in the immediate left of the region.

Calculate the explicit forms of the transfer matrix for the case where: (a) $E > E_t/\gamma + E_c$ and (b) $E < E_t/\gamma + E_c$, where E_t is the transverse component of the kinetic energy (see Problem 1.1).

Solution: In a region where both E_c and γ are constant (spatially invariant), the Schrödinger equation becomes (see Problem 1.1)

$$\frac{d}{dz}\left[\frac{1}{\gamma}\frac{d\phi}{dz}\right] + \frac{2m_c^*}{\hbar^2}\left(E - \frac{E_t}{\gamma} - E_c\right)\phi(z) = 0, \tag{7.24}$$

where $\phi(z)$ is the z-component of the wave function.

To derive the transfer matrix through a section of length L where both E_c and γ are constant, we look for solutions $\phi(z)$ of Equation (7.24) in the region $[0, L]$ which satisfy the boundary conditions (7.1) and (7.2).

Since the solutions $\phi_{1,2}(z)$ are linearly independent solutions (their Wronskian is unity), a general solution of Equation (7.24) can be written as

$$\phi(z) = A_1 \phi_1(z) + A_2 \phi_2(z). \tag{7.25}$$

Using this last result and Equation (7.23), which defines the transfer matrix, we obtain

$$W = \begin{bmatrix} \phi_1'(L) & \frac{1}{\gamma}\phi_2'(L) \\ \gamma\phi_1(L) & \phi_2(L) \end{bmatrix}. \tag{7.26}$$

The explicit forms for $\phi_{1,2}(z)$ are:

(a) If $E > \frac{E_t}{\gamma} + E_c$,

$$\phi_1(z) = \frac{\sin(\beta z)}{\beta}, \tag{7.27}$$

$$\phi_2(z) = \cos(\beta z), \tag{7.28}$$

where

$$\beta^2 = \frac{2m^*}{\hbar^2}\left[E - \frac{E_t}{\gamma} - E_c\right]. \tag{7.29}$$

(b) If $E < \frac{E_t}{\gamma} + E_c$,

$$\phi_1(z) = \frac{\sinh(\kappa z)}{\kappa}, \tag{7.30}$$

$$\phi_2(z) = \cosh(\kappa z), \tag{7.31}$$

where

$$\kappa^2 = \frac{2m^*}{\hbar^2}\left[\frac{E_t}{\gamma} + E_c - E\right]. \tag{7.32}$$

** Problem 7.4: Tunneling probability through a region with an arbitrary spatially varying conduction band energy profile $E_c(z)$ and effective mass $m^*(z)$

Derive an expression for the transmission probability of an electron tunneling through a region of finite spatial extent located in the interval $[0, L]$. The region is interposed between two contacts described by a constant potential energy $E_c = 0$. Within the region, the potential energy $E_c(z)$ and the effective mass $m^*(z)$ vary only in the z-direction. The electron's effective mass in the two contacts ($z < 0$ and $z > L$) are the same, spatially invariant, and equal to m_c^*.

Solution: The time-independent Schrödinger equation describing the steady-state (ballistic) motion of an electron through the potential energy profile described above is given by (see Problem 1.1)

$$-\frac{\hbar^2}{2m^*(z)}\frac{\partial^2 \psi}{\partial x^2} - \frac{\hbar^2}{2m^*(z)}\frac{\partial^2 \psi}{\partial y^2} - \frac{\hbar^2}{2}\frac{\partial}{\partial z}\left[\frac{1}{m^*(z)}\frac{\partial \psi}{\partial z}\right] + E_c(z)\psi = E\psi. \qquad (7.33)$$

Here, E is the total kinetic energy in the contact, where the bottom of the conduction band is taken as the zero of energy.

Because the Hamiltonian in Equation (7.33) is invariant in the x- and y-directions, the transverse wavevector $\vec{k}_t$ is a good quantum number. Furthermore, since the z-component of the electron's motion is decoupled from the transverse motion in the x–y plane, the wave function ψ can be written as

$$\psi(\vec{r}) = \phi(z)e^{i\vec{k}_t \cdot \vec{\rho}}, \qquad (7.34)$$

where $\vec{k}_t = (k_y, k_z)$ and $\vec{\rho} = (y, z)$. Plugging the result (7.34) in Equation (7.33), we get the effective Schrödinger equation for the z-component of the wave function $\phi(z)$:

$$\frac{\mathrm{d}}{\mathrm{d}z}\left[\frac{1}{\gamma(z)}\frac{\mathrm{d}\phi}{\mathrm{d}z}\right] + \frac{2m_c^*}{\hbar^2}\left[E_p + E_t(1 - \gamma(z)^{-1}) - E_c(z)\right]\phi(z) = 0, \qquad (7.35)$$

where m_c^* is the effective mass of the electrons in the contacts sandwiching the region of interest (m_c^* is spatially invariant within the contacts and isotropic), $\gamma(z)$ $= \frac{m^*(z)}{m_c^*}$, $E_t = \frac{\hbar^2 k_t^2}{2m_c^*}$, and E_p is the kinetic energy associated with the z-component of the motion in the contacts, $E_p = \frac{\hbar^2 k_z^2}{2m_c^*}$.

Equation (7.35) cannot be solved exactly for an arbitrary potential $E_c(z)$. However, an approximate solution can be found by approximating the potential profile by a series of potential steps (see Figure 7.1), or by using a piecewise linear approximation for the potential. Within each interval the potential and the effective mass are assumed to be *constant*. In that case, the wave function and its first derivative at the left and right edges of any interval are related by the transfer matrix, whose elements can be determined analytically (see Problem 7.3).

The transfer matrix for the nth interval $[z_{n-1}, z_n]$ is defined according to

$$\begin{bmatrix} \frac{1}{\gamma(z_n^-)}\frac{\mathrm{d}\phi}{\mathrm{d}z}(z_n^-) \\ \phi(z_n^-) \end{bmatrix} = \begin{pmatrix} W_{11}^{(n)} & W_{12}^{(n)} \\ W_{21}^{(n)} & W_{22}^{(n)} \end{pmatrix} \begin{bmatrix} \frac{1}{\gamma(z_{n-1}^+)}\frac{\mathrm{d}\phi}{\mathrm{d}z}(z_{n-1}^+) \\ \phi(z_{n-1}^+) \end{bmatrix}, \qquad (7.36)$$

where $W_{ij}^{(n)}$ are the elements of the transfer matrix, and z_{n-1}^+ and z_n^- stand for $z_{n-1} + \epsilon$ and $z_n - \epsilon$ respectively, with ϵ being a vanishingly small positive quantity.

Assuming continuity of $\phi(z)$ and $\frac{1}{\gamma(z)}\frac{\mathrm{d}\phi}{\mathrm{d}x}$ everywhere in the structure, the overall transfer matrix W_{TOT} describing the entire region $[0, L]$ is then found by cascading (multiplying) the individual transfer matrices for the individual intervals:

$$W_{\mathrm{TOT}} = W^{(N)} \cdots W^{(1)}, \qquad (7.37)$$

where $W^{(n)}$ is the transfer matrix for the nth interval, as given by Equation (7.36).

The overall transfer matrix W_{TOT} relates the wave functions and their first derivatives at the left and right contacts:

$$\left[\begin{array}{c} \frac{1}{\gamma(L^+)} \frac{d\phi}{dx}(L^+) \\ \phi(L^+) \end{array} \right] = W_{TOT} \left[\begin{array}{c} \frac{1}{\gamma(0^-)} \frac{d\phi}{dx}(0^-) \\ \phi(0^-) \end{array} \right]. \tag{7.38}$$

In Equation (7.38), $\phi(0^-)$ and $\phi(L^!)$ are the electronic states inside the left and right contacts. For an electron incident from the left contact, we have

$$\phi(z) = e^{ik_0 z} + r e^{-ik_0 z} \quad (z < 0)$$
$$\phi(z) = t e^{ik_0(z \; L)} \quad (z > L), \tag{7.39}$$

where $k_0 \; (= \frac{1}{\hbar} \sqrt{2 m_c^* E_p})$ is the z-component of the electron's wave vector in the contacts, and r and t are the overall reflection and transmission amplitudes through the region $[0, L]$, respectively. Using these scattering states for the wave functions at $z = 0^-$ and $z = L^+$ and noting that, by definition, $\gamma(L^+) = \gamma(0^-) = 1$, we obtain, from Equation (7.38),

$$t \left[\begin{array}{c} ik_0 \\ 1 \end{array} \right] = W_{TOT} \left[\begin{array}{c} ik_0(1-r) \\ 1+r \end{array} \right], \tag{7.40}$$

which are two equations for the two unknowns t and r. Eliminating r leads to

$$t = \frac{2ik_0 \left[W_{TOT}^{11} W_{TOT}^{22} - W_{TOT}^{12} W_{TOT}^{21} \right]}{ik_0 \left[W_{TOT}^{11} + W_{TOT}^{22} \right] + \left[W_{TOT}^{21} k_0^2 - W_{TOT}^{12} \right]}, \tag{7.41}$$

where W_{TOT}^{ij} are the elements of the matrix W_{TOT} that are found from Equation (7.37).

Since W_{TOT} is a unimodular matrix, the term within the square brackets in the numerator is unity. In addition, the W_{TOT}^{ij} are purely real (see Problem 7.3). Therefore, the general expression for the transmission probability is given by

$$T = |t|^2 = \frac{4k_0^2}{k_0^2 \left[W_{TOT}^{11} + W_{TOT}^{22} \right]^2 + \left[W_{TOT}^{21} k_0^2 - W_{TOT}^{12} \right]^2}. \tag{7.42}$$

---

* Problem 7.5: Reflection and transmission probabilities across a potential step

Consider a potential step as shown in Figure 7.2. The effective masses on the left and right side of the step are equal to $m_1{}^*$ and $m_2{}^*$, respectively. The step height is ΔE_c in eV.

Starting with the general time-independent Schrödinger equation for an electron moving in an arbitrary potential energy profile $E_c(z)$ and with a spatially varying effective mass $m^*(z)$ (see Equation (7.35) and Problem 1.1):

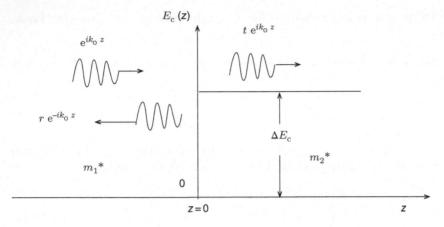

Figure 7.2: Scattering problem for an electron incident from the left on a potential energy step of height ΔE_c. The electron effective mass is assumed to be different on both sides of the step.

(a) Write down the Schrödinger equation for the z-component of the wave function $\phi(z)$ on the left and right sides of the potential step, assuming that the electron is incident from the left with a transverse kinetic energy $E_t = \frac{\hbar^2 k_t^2}{2m_1^*}$.

(b) Assume a plane wave is incident from the left and that the total energy of the incident electron is large enough so that it is transmitted on the other side. Write down the analytical form of the solution of the Schrödinger equation on either side of the potential step.

(c) By matching the wave function $\phi(z)$ at $z = 0$ and also $\frac{1}{m^*(z)} \frac{d\phi(z)}{dz}$ at $z = 0$, calculate the reflection and transmission amplitudes of the incident wave.

(d) Calculate the reflection and transmission probabilities across the step starting with the quantum mechanical expression for the current densities of the incident, reflected, and transmitted beams (see Chapter 5).

(e) Prove that the sum of the reflection and transmission probabilities is equal to unity.

Solution:
(a) Starting with the results of Problem 7.3 and using m_1^* as the effective mass in the region to the left of the potential step, the Schrödinger equation in the region $z < 0$ becomes

$$\frac{d^2\phi(z)}{dz^2} + \frac{2m_1^*}{\hbar^2}(E - E_t)\phi(z) = 0, \tag{7.43}$$

where E, E_t are the total and transverse components of the energy of the electron, respectively.

For the region $z > 0$, we have

$$\frac{d^2\phi(z)}{dz^2} + \frac{2m_2^*}{\hbar^2}\left(E - \frac{E_t}{\gamma} - \Delta E_c\right)\phi(z) = 0, \qquad (7.44)$$

where $\gamma = \frac{m_2^*}{m_1^*}$.

(b) For a plane wave incident from the left to be transmitted, we must have

$$E > \frac{E_t}{\gamma} + \Delta E_c. \qquad (7.45)$$

For $z < 0$, the solution of the Schrödinger equation is

$$\phi_I = e^{ik_1 z} + re^{-ik_1 z}, \qquad (7.46)$$

with

$$k_1 = \frac{1}{\hbar}\sqrt{2m_1^*(E - E_t)}. \qquad (7.47)$$

For $z > 0$, the solution of the Schrödinger equation is

$$\phi_{II} = te^{ik_2 z}, \qquad (7.48)$$

with

$$k_2 = \frac{1}{\hbar}\sqrt{2m_2^*\left(E - \frac{E_t}{\gamma} - \Delta E_c\right)}. \qquad (7.49)$$

(c) Continuity of the wave function at $z = 0$ mandates

$$1 + r = t. \qquad (7.50)$$

Continuity of $\frac{1}{m^*(z)}\frac{d\phi}{dz}$ requires

$$\frac{ik_1}{m_1^*}(1 - r) = \frac{ik_2 t}{m_2^*}, \qquad (7.51)$$

which can be rewritten as

$$1 - r = \left(\frac{k_2}{k_1}\frac{m_1}{m_2}\right)t. \qquad (7.52)$$

Adding Equations (7.50) and (7.52), we get

$$t = \frac{2}{\left[1 + \left(\frac{k_2}{k_1}\right)\left(\frac{m_1^*}{m_2^*}\right)\right]}. \qquad (7.53)$$

Substitution of this result in Equation (7.50) gives the reflection amplitude

$$r = \frac{1 - \frac{k_2}{k_1}\frac{m_1^*}{m_2^*}}{1 + \frac{k_2}{k_1}\frac{m_1^*}{m_2^*}}. \qquad (7.54)$$

(d,e) The proof that $|r|^2 + \frac{k_2}{k_1}\frac{m_1^*}{m_2^*}|t|^2 = 1$ by equating the incident current density to the sum of the reflected and transmitted current densities is left as an exercise. The quantities $|r|^2$ and $\frac{k_2}{k_1}\frac{m_1^*}{m_2^*}|t|^2$ are the reflection transmission probabilities across the potential step, respectively. Note the importance of the prefactor $\frac{k_2}{k_1}\frac{m_1^*}{m_2^*}$ in calculating the transmission probability.

---

* Problem 7.6: Tunneling probability across an arbitrary conduction band energy profile under bias

If a bias is applied across the two contacts sandwiching a device region with an arbitrary potential energy profile $E_c(z)$ (see Figure 7.1), derive the new general expression for the transmission probability of an electron incident from the left contact.

Solution: Following the approach used in Problem 7.4, the conduction band energy profile $E_c(z)$ in the interval $[0, L]$ (device region) is approximated as a series of steps in which both $E_c(z)$ and the electron effective mass are assumed to be constant. The transfer matrix across each individual section can then be determined and the overall transfer matrix W_{TOT} is found by multiplying the transfer matrices of the individual sections between the two contacts.

Following the approach described in Problem 7.4, the reflection and transmission amplitudes associated with an electron incident from the left contact are found to be solutions of the two equations

$$\begin{pmatrix} ik_{\mathrm{R}}t \\ t \end{pmatrix} = \begin{bmatrix} W_{\mathrm{TOT}}^{11} & W_{\mathrm{TOT}}^{12} \\ W_{\mathrm{TOT}}^{21} & W_{\mathrm{TOT}}^{22} \end{bmatrix} \begin{pmatrix} ik_{\mathrm{L}}(1-r) \\ 1+r \end{pmatrix}, \tag{7.55}$$

where the W_{TOT}^{ij} are the matrix elements of the total transfer matrix, k_{L} ($= \frac{1}{\hbar}\sqrt{2m_c^*E_p}$) and k_{R} ($= \frac{1}{\hbar}\sqrt{2m_c^*(E_p + qV_{\mathrm{bias}})}$) are the z-components of the electron's wave vector in the left and right contact, respectively. The quantity V_{bias} is the applied bias between the two contacts and q is the magnitude of the charge of the electron.

Starting with Equation (7.55), the reflection r and transmission t amplitudes must satisfy the following two equations:

$$ik_{\mathrm{R}}t = ik_{\mathrm{L}}(1-r)W_{\mathrm{TOT}}^{11} + W_{\mathrm{TOT}}^{12}(1+r), \tag{7.56}$$

$$t = ik_{\mathrm{L}}(1-r)W_{\mathrm{TOT}}^{21} + W_{\mathrm{TOT}}^{22}(1+r). \tag{7.57}$$

The last two equations are rewritten as

$$ik_{\mathrm{R}}t + (ik_{\mathrm{L}}W_{\mathrm{TOT}}^{11} - W_{\mathrm{TOT}}^{12})r = ik_{\mathrm{L}}W_{\mathrm{TOT}}^{11} + W_{\mathrm{TOT}}^{12}, \tag{7.58}$$

$$t + (ik_{\mathrm{L}}W_{\mathrm{TOT}}^{21} - W_{\mathrm{TOT}}^{22})r = ik_{\mathrm{L}}W_{\mathrm{TOT}}^{21} + W_{\mathrm{TOT}}^{22}. \tag{7.59}$$

Solving for t, we get

$$t = \frac{-2ik_L \left(W_{TOT}^{11} W_{TOT}^{22} - W_{TOT}^{12} W_{TOT}^{21} \right)}{\left(W_{TOT}^{12} - k_L k_R W_{TOT}^{21} \right) + i \left(k_R W_{TOT}^{22} + k_L W_{TOT}^{11} \right)}. \tag{7.60}$$

Since $W_{TOT}^{11} W_{TOT}^{22} - W_{TOT}^{12} W_{TOT}^{21} = 1$ and the elements of the total transfer matrix are real, we get the transmission probability as

$$|t|^2 = \frac{4k_L^2}{\left(W_{TOT}^{11} W_{TOT}^{12} - k_L k_R W_{TOT}^{11} W_{TOT}^{21} \right)^2 + \left(k_R W_{TOT}^{22} + k_L W_{TOT}^{11} \right)^2}. \tag{7.61}$$

As shown in Problem 5.3, the transmission probability associated with tunneling through an arbitrary potential under bias is given by $T = \frac{k_R}{k_L} |t|^2$, i.e.,

$$T = \frac{4k_R k_L}{\left(W_{TOT}^{11} W_{TOT}^{12} - k_L k_R W_{TOT}^{11} W_{TOT}^{21} \right)^2 + \left(k_R W_{TOT}^{22} + k_L W_{TOT}^{11} \right)^2}. \tag{7.62}$$

** **Problem 7.7: Tunneling and reflection probabilities through a one-dimensional delta scatterer**

Derive an analytical expression for the transfer matrix through a one-dimensional delta scatterer located at $z = 0$, for which the scattering potential is given by $\Gamma \delta(z)$ (the units of Γ are typically specified in eV-Å). Calculate the tunneling and reflection probabilities through the delta scatterer. Plot the tunneling and reflection probabilities as a function of the incident energy of the electron with effective mass $m^* = 0.067 m_0$ incident on a repulsive delta scatterer of strength $\Gamma = 5\,eV\text{-}Å$.

Solution: The Schrödinger equation describing propagation of an electron through a delta scatterer of strength Γ located at $z = 0$ is

$$-\frac{\hbar}{2m^*} \ddot{\phi}(z) + \Gamma \delta(z) \phi(z) = E_p \phi(z), \tag{7.63}$$

where E_p is the longitudinal component of the kinetic energy of the electron and $E_c(z)$ is assumed to be zero for both $z < 0$ and $z > 0$. Once again, the double dot superscript will represent the second derivative in space, and the single dot superscript will represent the first derivative.

Integrating the Schrödinger equation on both sides from $z = 0_-$ to $z = 0_+$, we get

$$-\frac{\hbar^2}{2m^*} \left[\dot{\phi}(0_+) - \dot{\phi}(0_-) \right] + \Gamma \phi(0) = 0, \tag{7.64}$$

which leads to

$$\dot{\phi}(0_+) = \dot{\phi}(0_-) + \frac{2m^* \Gamma}{\hbar^2} \phi(0_+). \tag{7.65}$$

Since the wave function is assumed to be continuous across the delta scatterer,

$$\phi(0_+) = \phi(0_-). \tag{7.66}$$

By regrouping the previous two equations in matrix form, we get

$$\begin{bmatrix} \dot{\phi}(0_+) \\ \phi(0_+) \end{bmatrix} = \begin{bmatrix} 1 & \frac{2m^*\Gamma}{\hbar^2} \\ 0 & 1 \end{bmatrix} \begin{bmatrix} \dot{\phi}(0_-) \\ \phi(0_-) \end{bmatrix}. \tag{7.67}$$

Therefore, the transfer matrix across a delta scatterer is given by

$$W_\delta = \begin{bmatrix} 1 & 2k_\delta \\ 0 & 1 \end{bmatrix}, \tag{7.68}$$

where we have introduced the quantity $k_\delta = \frac{m^*\Gamma}{\hbar^2}$.

The tunneling probability through the delta scatterer can be easily obtained from the general expression derived earlier:

$$T = |t|^2 = \frac{4k_0^2}{[k_0^2 W_{\text{TOT}}^{21} - W_{\text{TOT}}^{12}]^2 + k_0^2 [W_{\text{TOT}}^{11} + W_{\text{TOT}}^{22}]^2}, \tag{7.69}$$

where $k_0 = \frac{1}{\hbar}\sqrt{2m^* E_{\mathrm{p}}}$ is the wave vector of the incident electron and the W_{ij} are the elements of the transfer matrix.

Using Equation (7.68), the tunneling probability through the delta scatterer is found to be

$$T = \frac{k_0^2}{k_\delta^2 + k_0^2}. \tag{7.70}$$

The reflection probability is given by

$$R = 1 - T = \frac{k_\delta^2}{k_\delta^2 + k_0^2}. \tag{7.71}$$

Figure 7.3 is a plot of T and R versus the reduced wavevector k/k_δ. The Matlab code to generate this plot is given in Appendix G.

**** Problem 7.8: Floquet's theorem

Floquet's theorem states that the solution $\psi(z)$ of a homogeneous linear differential equation with periodic coefficients of period L can be written as

$$\psi(z) = e^{\sigma z}\phi(z), \tag{7.72}$$

where $\phi(z)$ is a periodic function of z, i.e.,

$$\phi(z) = \phi(z + L). \tag{7.73}$$

Prove this theorem for the case of the time-independent Schrödinger equation when the potential energy profile is periodic. Then, apply the results to the description of energy bands in one-dimensional infinite crystal considered as the infinite repetition of a unit cell. When generalized to the case of three-dimensional crystals, Floquet's theorem is referred to as Bloch's theorem.

This last equation is of the general form given in Equation (7.74), where functions $f_n(z)$ are given by

$$f_2(z) = -\frac{\hbar^2}{2m^*}, \tag{7.77}$$

$$f_1(z) = 0, \tag{7.78}$$

$$f_0(z) = E_c(z) - E_p. \tag{7.79}$$

Both $f_2(z)$ and $f_1(z)$ are constant and therefore they also automatically satisfy the periodic condition; i.e., $f_1(z) = f_1(z + L)$ and $f_2(z) = f_2(z + L)$. Moreover, since $E_c(z)$ is periodic, so is $f_0(z)$.

Suppose that $\psi_1(z)$ and $\psi_2(z)$ are two linearly independent solutions of Equation (7.76). Then, a general solution of this equation can be written as a linear superposition of these two solutions:

$$\psi(z) = A_1\psi_1(z) + A_2\psi_2(z). \tag{7.80}$$

Because the coefficients $f_n(z)$ for $n = 0, 1, 2$ are periodic, the functions $\psi_1(z + L)$ and $\psi_2(z + L)$ are also solutions of Equation (7.76). Hence, they can be written as linear combinations of the functions $\psi_1(z)$ and $\psi_2(z)$, i.e.,

$$\psi_1(z + L) = \alpha\psi_1(z) + \beta\psi_2(z), \tag{7.81}$$

$$\psi_2(z + L) = \gamma\psi_1(z) + \delta\psi_2(z). \tag{7.82}$$

Next, we show that we can write

$$\psi(z + L) = A_1'\psi_1(z) + A_2'\psi_2(z), \tag{7.83}$$

where the coefficients A_1' and A_2' are determined later.

Since $\psi(z + L)$ is also a solution of Equation (7.76), it can be written as a linear combination of $\psi_1(z)$ and $\psi_2(z)$ as well. Indeed, using Equations (7.80)–(7.82), we get that

$$\psi(z + L) = A_1\psi_1(z + L) + A_2\psi_2(z + L) = A_1\Big(\alpha\psi_1(z) + \beta\psi_2(z)\Big)$$
$$+ A_2\Big(\gamma\psi_1(z) + \delta\psi_2(z)\Big). \tag{7.84}$$

Hence,

$$\psi(z + L) = (A_1\alpha + A_2\gamma)\psi_1(z) + (A_1\beta + A_2\delta)\psi_2(z) = A_1'\psi_1(z) + A_2'\psi_2(z), \tag{7.85}$$

with $A_1' = A_1\alpha + A_2\gamma$ and $A_2' = A_1\beta + A_2\delta$.

If a value of k can be found such that $A_1' = kA_1$ and $A_2' = kA_2$, then the following two equations must be satisfied by k:

$$A_1\alpha + A_2\gamma = kA_1, \tag{7.86}$$

$$A_1\beta + A_2\delta = kA_2. \tag{7.87}$$

These can be written in matrix form:

$$\begin{bmatrix} \alpha & \gamma \\ \beta & \delta \end{bmatrix} \begin{bmatrix} A_1 \\ A_2 \end{bmatrix} = k \begin{bmatrix} A_1 \\ A_2 \end{bmatrix}, \tag{7.88}$$

which shows that the k are the eigenvalues of the matrix

$$A = \begin{bmatrix} \alpha & \gamma \\ \beta & \delta \end{bmatrix}. \tag{7.89}$$

As shown below, the k eigenvalues of this matrix exist, which means that the matrix A must have a non-zero determinant. Because the k eigenvalues exist, we can write Equation (7.85) as

$$\psi(z + L) = kA_1\psi_1(z) + kA_2\psi_2(z) = k\psi(z). \tag{7.90}$$

If we assume that the solution of the Schrödinger Equation (7.76) can be written in the form

$$\psi(z) = e^{\sigma z}\phi(z), \tag{7.91}$$

then using Equation (7.90), we get

$$\psi(z + L) = e^{\sigma(z+L)}\phi(z + L) = k\psi(z) = ke^{\sigma z}\phi(z). \tag{7.92}$$

If we select σ such that

$$k = e^{\sigma L}, \tag{7.93}$$

then from Equation (7.92), we obtain

$$\psi(z + L) = e^{\sigma(z+L)}\phi(z) = e^{\sigma z}e^{\sigma L}\phi(z) = e^{\sigma(z+L)}\phi(z + L), \tag{7.94}$$

which implies that the following equality must be satisfied:

$$\phi(z) = \phi(z + L). \tag{7.95}$$

This completes the proof of Floquet's theorem.

To show that the matrix A in Equation (7.89) has a non-zero determinant, we use the fact that

$$\psi_1(z + L) = \alpha\psi_1(z) + \beta\psi_2(z), \tag{7.96}$$
$$\psi_2(z + L) = \gamma\psi_1(z) + \delta\psi_2(z). \tag{7.97}$$

Since $\psi_1(z)$ and $\psi_2(z)$ are linearly independent, so are $\psi_1(z + L)$ and $\psi_2(z + L)$. Therefore, the only values of α' and β' that satisfy the condition

$$\alpha'\psi_1(z + L) + \beta'\psi_2(z + L) = 0 \tag{7.98}$$

are $\alpha' = \beta' = 0$. Hence,

$$\alpha'\psi_1(z+L) + \beta'\psi_2(z+L) = \alpha'\Big(\alpha\psi_1(z) + \beta\psi_2(z)\Big)$$

$$+ \beta'\Big(\gamma\psi_1(z) + \delta\psi_2(z)\Big) = (\alpha'\alpha + \beta'\gamma)\psi_1(z)$$

$$+ (\alpha'\beta + \beta'\delta)\psi_2(z) = 0. \tag{7.99}$$

Since ψ_1 and ψ_2 are linearly independent, we must have

$$\alpha'\alpha + \beta'\gamma = 0, \tag{7.100}$$

$$\alpha'\beta + \beta'\delta = 0. \tag{7.101}$$

These last two equations can be written in matrix form:

$$\begin{bmatrix} \alpha & \gamma \\ \beta & \delta \end{bmatrix} \begin{bmatrix} \alpha' \\ \beta' \end{bmatrix} = k \begin{bmatrix} 0 \\ 0 \end{bmatrix}. \tag{7.102}$$

This system admits the $(0,0)$ solution only if the determinant of the matrix on the left-hand side is non-zero. Hence, the matrix A in Equation (7.96) has a non-zero determinant.

**** Problem 7.9: Tunneling probability through N identical barriers

Starting with the one-dimensional time-independent Schrödinger equation, derive an expression for the tunneling probability T_N through a structure which consists of N repetitions of a unit cell, as shown in Figure 7.4. Determine the kinetic energy of the incident electron E_{p} for which T_N reaches unity [15].

Solution: In the first unit ($0 \le z \le L$), we write the solution of the Schrödinger equation

$$\left[-\frac{\hbar^2}{2m^*} \frac{d^2\psi}{dz^2} + E_{\mathrm{c}}(z)\psi \right] = E_{\mathrm{p}}\psi$$

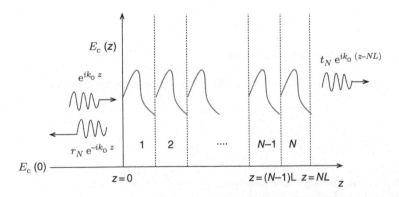

Figure 7.4: Scattering problem for an electron incident from the left on a periodic potential energy profile composed of N identical unit cells.

as a linear combination, $A_1\psi_1(z) + A_2\psi_2(z)$, of two linearly independent solutions $\psi_1(z)$ and $\psi_2(z)$ satisfying the boundary conditions

$$\begin{array}{ll} \psi_1(0) = 0 & \psi_2(0) = 1, \\ \psi_1'(0) = 1 & \psi_2'(0) = 0. \end{array} \tag{7.103}$$

These solutions are indeed independent since their Wronskian is unity.

The most general solution of the Schrödinger equation in the nth barrier can then be written as

$$\psi_n(z) = A_n\psi_1(z - (n-1)L) + B_n\psi_2(z - (n-1)L). \tag{7.104}$$

Matching the wave functions at the boundary between the nth and $(n+1)$th cells leads to

$$A_n\psi_1(L) + B_n\psi_2(L) = A_{n+1}\psi_1(0) + B_{n+1}\psi_2(0) = B_{n+1} \tag{7.105}$$

Furthermore, matching the wave function derivatives at the same boundary leads to

$$A_n\psi_1'(L) + B_n\psi_2'(L) = A_{n+1}\psi_1'(0) + B_{n+1}\psi_2'(0) = A_{n+1}. \tag{7.106}$$

Equations (7.105) and (7.106) can be written in matrix form:

$$\begin{pmatrix} A_{n+1} \\ B_{n+1} \end{pmatrix} = \begin{pmatrix} \psi_1'(L) & \psi_2'(L) \\ \psi_1(L) & \psi_2(L) \end{pmatrix} \begin{pmatrix} A_n \\ B_n \end{pmatrix} = \begin{pmatrix} W_{11} & W_{12} \\ W_{21} & W_{22} \end{pmatrix} \begin{pmatrix} A_n \\ B_n \end{pmatrix}, \tag{7.107}$$

where

$$W = \begin{pmatrix} \psi_1'(L) & \psi_2'(L) \\ \psi_1(L) & \psi_2(L) \end{pmatrix} \tag{7.108}$$

is the transfer matrix for each unit cell.

As shown in Figure 7.4, for an electron incident from the left, we have

$$\psi(z) = e^{ik_0 z} + r_N e^{-ik_0 z} \text{ for } z \leq 0 \tag{7.109}$$

and

$$\psi(z) = t_N e^{ik_0(z - NL)} \text{ for } z \geq NL, \tag{7.110}$$

where $k_0 = \frac{1}{\hbar}\sqrt{2m^* E_{\mathrm{p}}}$ is the wave vector of the incident electron and E_{p} its kinetic energy.

Enforcing the continuity of the wave function and its first derivative at $z = 0$ leads to:

$$1 + r_N = B_1, \tag{7.111}$$

$$ik_0(1 - r_N) = A_1. \tag{7.112}$$

Similarly, at $z = NL$, we get

$$t_N = A_N\psi_1(L) + B_N\psi_2(L), \tag{7.113}$$

$$ik_0 t_N = A_N\psi_1'(L) + B_N\psi_2'(L). \tag{7.114}$$

Furthermore, by induction,

$$\begin{pmatrix} A_2 \\ B_2 \end{pmatrix} = W^1 \begin{pmatrix} A_1 \\ B_1 \end{pmatrix}$$

$$\begin{pmatrix} A_3 \\ B_3 \end{pmatrix} = W^1 \begin{pmatrix} A_2 \\ B_2 \end{pmatrix} = W^2 \begin{pmatrix} A_1 \\ B_1 \end{pmatrix}$$

$$\begin{pmatrix} A_4 \\ B_4 \end{pmatrix} = W^1 \begin{pmatrix} A_3 \\ B_3 \end{pmatrix} = W^3 \begin{pmatrix} A_1 \\ B_1 \end{pmatrix}$$

$$\vdots \tag{7.115}$$

Hence,

$$\begin{pmatrix} A_N \\ B_N \end{pmatrix} = W^{N-1} \begin{pmatrix} A_1 \\ B_1 \end{pmatrix}. \tag{7.116}$$

Equations (7.111)–(7.115) form a system of six equations for the six unknowns r_N, t_N, A_1, B_1, A_N, and B_N.

Equations (7.113)–(7.115) can then be rewritten as

$$t_N \begin{pmatrix} ik_0 \\ 1 \end{pmatrix} = W \begin{pmatrix} A_N \\ B_N \end{pmatrix} = W^N \begin{pmatrix} A_1 \\ B_1 \end{pmatrix}. \tag{7.117}$$

Using the shorthand notation $W^N \equiv D$ for the transfer matrix through the N unit cells, Equations (7.111), (7.112), and (7.116) can be rewritten as

$$ik_0 t_N - D_{11} A_1 - D_{12} B_1 + 0 \cdot r_N = 0,$$
$$t_N - D_{21} A_1 - D_{22} B_1 + 0 \cdot r_N = 0,$$
$$0 \cdot t_N + 0 \cdot A_1 + B_1 - r_N = 1,$$
$$0 \cdot t + A_1 + 0 \cdot B_1 + ik_0 r_N = ik_0. \tag{7.118}$$

Eliminating A_1 and B_1 from these equations leads to two equations for the two unknowns r_N and t_N:

$$ik_0 t_N + r_N(ik_0 D_{11} - D_{12}) = D_{12} + ik_0 D_{11}, \tag{7.119}$$
$$t_N + r_N(ik_0 D_{21} - D_{22}) = D_{22} + ik_0 D_{21}. \tag{7.120}$$

Multiplying Equation (7.119) by $(ik_0 D_{21} - D_{22})$ and Equation (7.120) by $(ik_0 D_{11} - D_{12})$ and subtracting the resulting equations leads to the transmission amplitude t_N,

$$t_N = \frac{2ik_0(D_{11} D_{22} - D_{12} D_{21})}{(ik_0 D_{11} - D_{12}) - ik_0(ik_0 D_{21} - D_{22})}, \tag{7.121}$$

which can be further simplified since $D_{11} D_{22} - D_{12} D_{21} = \det D = \det W^N = [\det W]^N = 1$, giving the final expression:

$$t_N = \frac{2ik_0}{(ik_0 D_{11} - D_{12}) - ik_0(ik_0 D_{21} - D_{22})}. \tag{7.122}$$

Since the W_{ij} are real, so are the D_{ij}. The transmission probability $T_N = |t_N|^2$ is therefore given by

$$T_N = |t_N|^2 = \frac{4k_0^2}{\left[k_0^2 D_{21} - D_{12}\right]^2 + k_0^2 \left[D_{11} + D_{22}\right]^2}. \tag{7.123}$$

Next, we determine at which values of the incident kinetic energy T_N reaches unity. To do so, we rewrite T_N in Equation (7.123) as follows:

$$T_N = \frac{4k_0^2}{k_0^4 D_{21}^2 + k_0^2 \left[(D_{11} + D_{22})^2 - 2D_{12}D_{21}\right] + D_{12}^2}. \tag{7.124}$$

Since the numerator is a polynomial in k_0^2, T_N will reach unity when the D_{ij} satisfy the following conditions:

$$D_{21} = D_{12} = 0, \tag{7.125}$$

$$(D_{11} + D_{22})^2 - 2D_{12}D_{21} = 4. \tag{7.126}$$

Since $D_{12} = D_{21} = 0$, the last equation amounts to

$$(D_{11} + D_{22})^2 = 4. \tag{7.127}$$

If we call the eigenvalues of the matrix W $\lambda^{(1)}$ and $\lambda^{(2)}$, then the eigenvalues of the matrix $D = W^N$ are $\lambda^{(1)N}$ and $\lambda^{(2)N}$. Furthermore, we have

$$D_{11} + D_{22} = \text{Tr}(D) = \lambda^{(1)N} + \lambda^{(2)N}. \tag{7.128}$$

Since $\det W = 1$, we have $\lambda^{(2)} = \frac{1}{\lambda^{(1)}}$ and $\lambda^{(1)}$ must satisfy

$$\left(\lambda^{(1)N} + \frac{1}{\lambda^{(1)N}}\right)^2 = 4. \tag{7.129}$$

This last equation can be simplified to

$$\left[\lambda^{(1)N}\right]^4 - 2\left[\lambda^{(1)N}\right]^2 + 1 = 0 \quad \text{or} \quad \left(\lambda^{(1)2N} - 1\right)^2 = 0, \tag{7.130}$$

and therefore

$$\left[\lambda^{(1)}\right]^{2N} = 1, \tag{7.131}$$

$$\left[\lambda^{(2)}\right]^{2N} = 1. \tag{7.132}$$

So the $\lambda^{(1)}$ are the $2N$ square roots of unity, i.e.,

$$\lambda^{(1)} = \left(e^{i2\pi k}\right)^{\frac{1}{2N}}, \qquad k = 0, 1, 2, \ldots, 2N - 1. \tag{7.133}$$

The distinct solutions for $\lambda^{(1)}$ and $\lambda^{(2)}$ are given by (see suggested problems):

$$\begin{aligned} \lambda^{(1)} &= e^{i\frac{pik}{N}}, \\ \lambda^{(2)} &= e^{-i\frac{pik}{N}}, \qquad k = 1, 2, \ldots, N - 1. \end{aligned} \tag{7.134}$$

Therefore, the energies at which T_N is unity are the energies for which the following relation is satisfied:

$$\operatorname{Tr} W = e^{i\frac{\pi k}{N}} + e^{-i\frac{\pi k}{N}} = 2\cos\left(\frac{\pi k}{N}\right). \qquad (7.135)$$

** Problem 7.10: Relation between the band structure of an infinite periodic lattice and the transmission of an electron through a finite repeated structure

Prove the following theorem: The transmission coefficient of an electron through a periodic structure, formed by N repetitions of a basic subunit, as shown in Figure 7.4, reaches unity at the following energies: (a) energies at which the transmission through the basic subunit is unity, and (b) N − 1 energies in each energy band of the lattice formed by infinite periodic repetition of the basic subunit, where these N − 1 energies are given by $E = E_i(k = \pm\frac{n\pi}{NL})$ (n = 1, 2, 3, ..., N − 1) and L is the length of a subunit. Here, $E_i(k)$ is the energy–wavevector relationship (or dispersion relation) for the ith band of the infinite lattice [15].

Solution: In Equation (7.133), we reject the case $k = 0, k = N$. In this case,

$$\begin{cases} \lambda^{(1)} = \lambda^{(2)} = 1 \text{ are real and } \operatorname{Tr}(W) = 2 & \text{for } k = 0 \\ \lambda^{(1)} = \lambda^{(2)} = -1 \text{ are real and } \operatorname{Tr}(W) = -2 & \text{for } k = N. \end{cases}$$

Next, we show that these two cases correspond to the energies at the edges of the energy bands of the lattice obtained by infinite repetition of the unit cell.

We first derive the general expression for the eigenvalues of W associated with a subunit of the periodic lattice in terms of its matrix elements, starting with the relation

$$\det\begin{pmatrix} W_{11} - \lambda & W_{12} \\ W_{21} & W_{22} - \lambda \end{pmatrix} = 0,$$

whose solutions give the eigenvalues

$$\lambda^{(1),(2)} = \frac{\operatorname{Tr}(W)}{2} \pm \left[\left(\frac{\operatorname{Tr}(W)}{2}\right)^2 - 1\right]^{\frac{1}{2}}. \qquad (7.136)$$

This last equation shows that, if $|\operatorname{Tr}(W)| > 2$, the eigenvalues of W are real. The values of the energies for which it occurs are in the stop bands of the infinite superlattice, as we will show next. The values of the energy for which $|\operatorname{Tr}(W)| = 2$ then give the lower and upper limits of the energy of the pass band. If $|\operatorname{Tr}(W)| < 2$, the magnitudes of the eigenvalues $\lambda^{(1),(2)}$ are equal to unity.

For an infinite number of unit cells, we know from Bloch's theorem, proved in Problem 7.8, that the solution of the Schrödinger equation can be written as

$$\Psi(z) = e^{i\xi z}\phi(z), \qquad (7.137)$$

where $\phi(z)$ is a periodic function with the same period as the potential of the infinite superlattice, i.e., $\phi(z + L) = \phi(z)$, and ξ is purely real and is referred to as the electron Bloch wave vector of the infinite superlattice.

So we can rewrite Equation (7.107) as

$$\begin{pmatrix} A_{n+1} \\ B_{n+1} \end{pmatrix} = \begin{pmatrix} W_{11} & W_{12} \\ W_{21} & W_{22} \end{pmatrix} \begin{pmatrix} A_n \\ B_n \end{pmatrix} = e^{i\xi L} \begin{pmatrix} A_n \\ B_n \end{pmatrix} = \lambda \begin{pmatrix} A_n \\ B_n \end{pmatrix}. \qquad (7.138)$$

Using the periodicity of ϕ, we can write $\Psi(z + L) = e^{i\xi(z+L)}\phi(z + L) = e^{i\xi L}\Psi(z)$.

Equation (7.138) can be rewritten as

$$\begin{pmatrix} A_{n+1} \\ B_{n+1} \end{pmatrix} = \begin{pmatrix} W_{11} & W_{12} \\ W_{21} & W_{22} \end{pmatrix} \begin{pmatrix} A_n \\ B_n \end{pmatrix} = \begin{pmatrix} e^{i\xi L} & 0 \\ 0 & e^{i\xi L} \end{pmatrix} \begin{pmatrix} A_n \\ B_n \end{pmatrix}, \qquad (7.139)$$

which shows that ξ satisfies the equation

$$\det\left(W - e^{i\xi L}\delta_{ij}\right) = 0. \qquad (7.140)$$

The eigenvalues of this last equation are $\lambda^{(1)}, \lambda^{(2)} = e^{\pm i\xi L}$. We have a propagating wave in the infinite superlattice only if ξ is real, i.e., if the λ are complex.

If λ is real (ξ complex), then the wave function is either growing (if λ is positive) or decaying (if λ is negative). These correspond to forbidden energy bands (or stop bands) for the infinite superlattice.

The above analysis shows that the eigenvalues $\lambda^{(1)}$ and $\lambda^{(2)}$ satisfy the relation

$$\lambda^{(1)} + \lambda^{(2)} = \text{Tr}(W) = 2\cos(\xi L). \qquad (7.141)$$

One must solve this equation for a given energy to find the corresponding value of ξ. If that value turns out to be real, then the energy is in the pass band. On the other hand, if the value of ξ turns out to be complex, then the energy is in the gap between two pass bands, i.e., it is in the stop band. If ξ is real, then obviously $|t|^2 = 1$ in the infinite superlattice since this corresponds to the pass band.

In the previous problem, we found that the values of the energies E for which $|t_N|^2 = 1$ for a system of N barriers obey the relation

$$\text{Tr}(W) = 2\cos\left(\frac{\pi k(E)L}{N}\right) \quad \text{for } k = 1, \ldots, N - 1. \qquad (7.142)$$

This equation is formally identical to Equation (7.141) if we put $\xi L = \frac{k\pi}{N}$.

So, we arrive at the important conclusion that for a system of N barriers, the transmission coefficient is exactly equal to 1 whenever $\xi = \frac{k\pi}{NL}$ in the infinite model.

Stated otherwise, in a system of N barriers, $|t_N|^2 = 1$ at energies corresponding to different Bloch wave vectors

$$\xi = \frac{k\pi}{NL} \quad \text{for } k = 1, \ldots, N - 1. \qquad (7.143)$$

** Problem 7.11: The Krönig–Penney problem: energy dispersion relation of an infinitely repeated structure

Solve the one-dimensional time-independent Schrödinger equation (assuming constant effective mass) and derive the analytical expression for the energy dispersion (E versus k) relation in the structure formed by infinite repetition of the unit cell shown in Figure 7.5.

Solution: As shown in the previous problem, the energy dispersion relations in the various energy bands of a periodic potential are given by

$$\text{Tr}\, W_{\text{TOT}} = 2\cos(\xi L), \tag{7.144}$$

where ξ is the electron wave number and W_{TOT} is the transfer matrix of the unit cell.

Referring to Figure 7.5, we write the transfer matrix across a unit cell as

$$W_{\text{TOT}} = W_{\text{II}} \times W_\delta \times W_{\text{I}}, \tag{7.145}$$

where W_{I} and W_{II} are the transfer matrices associated with the free propagation regions on the left and right sides of the delta scatterer, respectively.

The transfer matrix across a delta scatterer was derived in Problem 7.7:

$$W_\delta = \begin{bmatrix} 1 & \frac{2m^*\Gamma}{\hbar^2} \\ 0 & 1 \end{bmatrix}. \tag{7.146}$$

To determine the transfer matrices in the free propagation regions I and II we use the results of Problem 7.2, where the transfer matrix through a region of

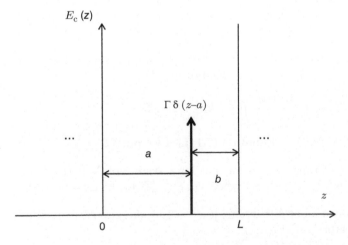

Figure 7.5: Basic unit cell used to calculate the energy dispersion relation of an infinite periodic lattice. The effective mass is assumed to be the same throughout.

width L, with constant potential E_c, was found to be

$$W(E_p \geq E_c) = \begin{pmatrix} \cos(kL) & -k\sin(kL) \\ \sin(kL)/k & \cos(kL) \end{pmatrix}, \tag{7.147}$$

where

$$k = \frac{1}{\hbar}\sqrt{2m^*(E_p - E_c)} \tag{7.148}$$

and E_p is the longitudinal component of the kinetic energy of the incident electron, i.e., the energy measured above E_c.

The transfer matrices through the free propagation regions have the form of Equation (7.147) with E_c equal to zero. We call k_0 the corresponding wave vector inside that region.

The overall transfer matrix W_{TOT} associated with the unit cell of the device composed of the two delta scatterers is

$$W_{TOT} = \begin{bmatrix} \cos(k_0 b) & -k_0 \sin(k_0 b) \\ \sin(k_0 b)/k_0 & \cos(k_0 b) \end{bmatrix} \times \begin{bmatrix} 1 & 2m^*\Gamma/\hbar^2 \\ 0 & 1 \end{bmatrix}$$
$$\times \begin{bmatrix} \cos(k_0 a) & -k_0 \sin(k_0 a) \\ \sin(k_0 a)/k_0 & \cos(k_0 a) \end{bmatrix}. \tag{7.149}$$

Performing the matrix multiplications, the following expressions for the matrix elements W_{TOT}^{11} and W_{TOT}^{22} are found:

$$W_{TOT}^{11} = \cos(k_0 b)\left[\cos(k_0 b) + \frac{2k_\delta}{k_0}\sin(k_0 a) - \sin(k_0 a)\sin(k_0 b)\right], \tag{7.150}$$

$$W_{TOT}^{22} = \frac{\sin(k_0 b)}{k_0}[-k_0\sin(k_0 a) + 2k_\delta \cos(k_0 a)] + \cos(k_0 a)\cos(k_0 b), \tag{7.151}$$

where $k_\delta = m^*\Gamma/\hbar^2$.

Therefore, Equation (7.144) leads to

$$2\cos(\xi L) = \mathrm{Tr}\, W_{TOT} = 2\cos(k_0 L) + \frac{2k_\delta}{k_0}\sin(k_0 L). \tag{7.152}$$

The energy dispersion relation for the infinite superlattice with the unit cell shown in Figure 7.5 is therefore given by

$$\cos(\xi L) = \cos(k_0 L) + \frac{k_\delta}{k_0}\sin(k_0 L). \tag{7.153}$$

**** Problem 7.12: Connection between bound state and tunneling problems

In this problem, the transfer matrix formalism is used to show that the problem of finding the bound states of an arbitrary confined one-dimensional potential energy profile $E_c(z)$ can be reformulated as a tunneling problem. More specifically, the following theorem is proved: For an electron confined to a region of width W with an arbitrary conduction band energy profile $E_c(z)$ [$E_c(z) = V_0$ for z outside the well], as shown in Figure 7.6, the bound state energies $(E_1, E_2, E_3, \ldots)$ can be found by adding two barriers of width d and height V_0 on two sides of the region and calculating the energies at which the transmission probability $T(E)$ through the quantum well structure so formed reaches unity. The energies at which the transmission coefficient reaches unity converge toward the bound state energy levels when the barrier thickness d tends to infinity. The theorem is proved for the case of a spatially independent effective mass but can be easily extended to the case of a spatially varying effective mass [13].

Solution: We first consider the tunneling through the quantum well structure shown in Figure 7.6 using the bottom of the quantum well as the zero of energy. Calling V_0 the maximum depth of the quantum well, the transfer matrix for each barrier on either side of the quantum well for $E \leq V_0$ is given by (see Problem 7.2)

$$W_{\rm B} = \begin{pmatrix} \cosh(\kappa d) & \kappa \sinh(\kappa d) \\ \sinh(\kappa d)/\kappa & \cosh(\kappa d) \end{pmatrix}, \tag{7.154}$$

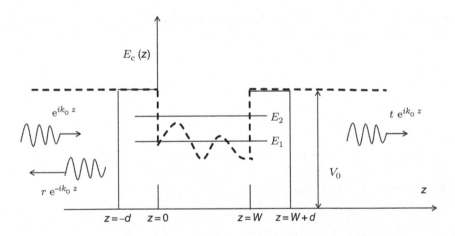

Figure 7.6: Schematic of a quantum well (dashed line) of width L with an arbitrary conduction band energy profile and maximum depth V_0. The zero of energy is selected to coincide with the bottom of the well. Also shown are the locations of the two lowest bound states E_1 and E_2 in the well. The latter coincide with the energies for unit transmission probability for an electron incident from the left barrier region. The quantities r and t are the reflection and transmission amplitudes, respectively, of the incident electron [13].

where $\kappa = \frac{1}{\hbar}\sqrt{2m^*(V_0 - E)}$ and m^* is the effective mass of the electron, assumed to be constant throughout.

The overall transfer matrix is given by the product of the following three matrices:

$$W_{\text{TOT}} = W_{\text{B}} \times W_{\text{well}} \times W_{\text{B}}, \tag{7.155}$$

where

$$W_{\text{well}} = \begin{pmatrix} \phi_1{}'(L) & \phi_2{}'(L) \\ \phi_1(L) & \phi_2(L) \end{pmatrix} \tag{7.156}$$

is the transfer matrix associated with the well region, the functions $\phi_1(z)$ and $\phi_2(z)$ being two linearly independent solutions of the Schrödinger equation satisfying the boundary conditions $\phi_1{}'(0) = 1$, $\phi_1(0) = 0$, $\phi_2{}'(0) = 0$, and $\phi_2(0) = 1$.

Performing the matrix multiplication, we obtain

$$W_{\text{TOT}} = \cosh^2(\kappa d) \begin{bmatrix} 1 & \kappa \tanh(\kappa d) \\ \tanh(\kappa d/\kappa) & 1 \end{bmatrix}$$
$$\times \begin{bmatrix} \phi_1{}' + \phi_2{}'(L)\tanh(\kappa d)/\kappa & \kappa\phi_1{}'(L)\tanh(\kappa d) + \phi_2{}'(L) \\ \phi_1(L) + \phi_2(L)\tanh(\kappa d)/\kappa & \kappa\phi_1(L)\tanh(\kappa d) + \phi_2(L) \end{bmatrix}. \tag{7.157}$$

In the limit $d \to \infty$, we have

$$W_{\text{TOT}} = \cosh^2(\kappa d) \begin{bmatrix} 1 & \kappa \\ \frac{1}{\kappa} & 1 \end{bmatrix} \begin{bmatrix} \phi_1{}'(L) + \phi_2{}'(L)/\kappa & \kappa\phi_1{}'(L) + \phi_2{}'(L) \\ \phi_1(L) + \phi_2(L)/\kappa & \kappa\phi_1(L) + \phi_2(L) \end{bmatrix}. \tag{7.158}$$

Multiplying the two matrices on the right-hand side, we get the elements of the matrix W_{TOT}:

$$W_{\text{TOT}}^{11} = \cosh^2(\kappa d)\left[\phi_1{}'(L) + \phi_2{}'(L)/\kappa + \kappa\phi_1(L) + \phi_2(L)\right], \tag{7.159}$$

$$W_{\text{TOT}}^{22} = W_{\text{TOT}}^{11}, \tag{7.160}$$

$$W_{\text{TOT}}^{12} = \cosh^2(\kappa d)\left[\kappa\phi_1{}'(L) + \phi_2{}'(L) + \kappa^2\phi_1(L) + \kappa\phi_2(L)\right], \tag{7.161}$$

$$W_{\text{TOT}}^{21} = \cosh^2(\kappa d)\left[\phi_1{}'(L)/\kappa + \phi_2{}'(L)/\kappa + \phi_1(L) + \phi_2(L)/\kappa\right]. \tag{7.162}$$

The transmission probability through the quantum well region depends on the elements of the transfer matrix, and reaches unity when the following two conditions are satisfied, as shown in Problem 7.1:

$$W_{12}^{\text{TOT}} = W_{21}^{\text{TOT}} = 0, \tag{7.163}$$

$$W_{11}^{\text{TOT}} + W_{22}^{\text{TOT}} = \pm 2. \tag{7.164}$$

Using Equations (7.159)–(7.162), these last two conditions amount to the following requirement:

$$\kappa\phi_1{}'(L) + \phi_2{}'(L) + \kappa^2\phi_1(L) + \kappa\phi_2(L) = 0. \tag{7.165}$$

Bound state problem: Next, we prove that Equation (7.165) is also the condition that must be satisfied to find the bound states in the well. With the zero of energy at the bottom of the well, the solutions of the Schrödinger equation for the bound state problem are given by: in region I ($z < 0$):

$$\psi_{\mathrm{I}} = A_1 e^{\kappa z} + B_1 e^{-\kappa z}; \tag{7.166}$$

in region II ($0 < z < L$):

$$\psi_{\mathrm{II}} = A_2 \phi_1(z) + B_2 \phi_2(z); \tag{7.167}$$

and in region III ($z > L$):

$$\psi_{\mathrm{III}} = A_3 e^{\kappa(z-L)} + B_3 e^{-\kappa(z-L)}; \tag{7.168}$$

where $\kappa = \frac{1}{\hbar}\sqrt{2m^*(V_0 - |E|)}$.

Matching the wave function and its derivative at $z = 0$, we get the following relations between the coefficients (A_1, B_1) and (A_2, B_2):

$$\begin{pmatrix} \kappa & -\kappa \\ 1 & 1 \end{pmatrix} \begin{bmatrix} A_1 \\ B_1 \end{bmatrix} = \begin{pmatrix} \phi_1'(0) & \phi_2'(0) \\ \phi_1(0) & \phi_2(0) \end{pmatrix} \begin{bmatrix} A_2 \\ B_2 \end{bmatrix}. \tag{7.169}$$

Similarly, at $z = W$ we get

$$\begin{pmatrix} \phi_1'(L) & \phi_2'(L) \\ \phi_1(L) & \phi_2(L) \end{pmatrix} \begin{bmatrix} A_2 \\ B_2 \end{bmatrix} = \begin{pmatrix} \kappa & -\kappa \\ 1 & 1 \end{pmatrix} \begin{bmatrix} A_3 \\ B_3 \end{bmatrix}. \tag{7.170}$$

The coefficient $B_1 = 0$ must be zero for the wave function to be well behaved for $z < 0$.

Since $\phi_1'(0) = 1$, $\phi_1(0) = 0$, $\phi_2'(0) = 0$, and $\phi_2(0) = 1$, Equation (7.169) leads to the following requirements:

$$A_2 = \kappa A_1, \tag{7.171}$$

$$B_2 = A_1. \tag{7.172}$$

Equation (7.170) can be expanded as follows:

$$A_2 \phi_1'(L) + B_2 \phi_2'(L) = \kappa A_3 - \kappa B_3, \tag{7.173}$$

$$A_2 \phi_1(L) + B_2 \phi_2(L) = A_3 + B_3. \tag{7.174}$$

Multiplying the last equation by κ and adding it to Equation (7.173), we obtain

$$A_3 = \frac{1}{2\kappa} \left[A_2 \left(\kappa \phi_1(L) + \phi_1'(L) \right) + B_2 \left(\kappa \phi_2(L) + \phi_2'(L) \right) \right]. \tag{7.175}$$

Taking into account Equations (7.171)–(7.172), this last equation becomes

$$A_3 = \frac{A_1}{2\kappa} \left[\kappa^2 \phi_1(L) + \kappa \phi_1'(L) + \kappa \phi_2(L) + \phi_2'(L) \right]. \tag{7.176}$$

For the wave function to be well behaved for $z > W$, we must have $A_3 = 0$. Since $A_1 \neq 0$, this leads to

$$g(E) = \kappa^2 \phi_1(L) + \kappa \phi_1'(L) + \kappa \phi_2(L) + \phi_2'(L) = 0, \tag{7.177}$$

which is the same as Equation (7.165) derived earlier for unity transmission through the tunneling structure as the width d of the two barriers on either side of the well approaches infinity. The left-hand side of Equation (7.177) is a function of energy $g(E)$ whose zeros correspond to the bound state energies of the quantum well. Numerical examples of the calculations of bound state energies for various quantum wells with different $E_c(z)$ using the results proven in this problem are given in Ref. [13].

** **Problem 7.13: Quantum mechanical wave impedance in terms of elements of the transfer matrix**

For the general conduction band energy profile shown in Figure 7.1 (with $z_N = L$), assuming a constant effective mass throughout, show that the quantum mechanical wave impedance $Z(0)$ is related to $Z_{QM}(L)$ by the general expression

$$Z_{QM}(0) = C\frac{Z_{QM}(L)\phi_2(L) - C\phi_2'(L)}{C\phi_1'(L) - Z_{QM}(L)\phi_1(L)}, \tag{7.178}$$

where $C = 2\hbar/(m^*i)$ and the ϕ_1, ϕ_2 functions are two linearly independent solutions of the Schrödinger equation in the interval $[0, L]$.

Solution: By definition, the quantum mechanical wave impedance [16, 17] is (see Chapter 1)

$$Z_{QM}(z) = C(d\phi/dz)/\phi, \tag{7.179}$$

where $\phi(z)$ is a solution of the time-independent Schrödinger equation

$$-\frac{\hbar^2}{2m^*}\ddot{\phi}(z) + E_c(z)\phi(z) = E_p\phi(z), \tag{7.180}$$

where all the quantities have their usual meaning.

The general solution of this second-order differential equation for $\phi(z)$ can be written as a linear combination of two linearly independent solutions $\phi(z) = A_1\phi_1(z) + A_2\phi_2(z)$ (see Chapter 1).

The quantum mechanical wave impedance can therefore be written as

$$Z_{QM}(z) = C\left[\frac{A_1\phi_1'(z) + A_2\phi_2'(z)}{A_1\phi_1(z) + A_2\phi_2(z)}\right]. \tag{7.181}$$

Introducing the quantity $\beta = \frac{A_2}{A_1}$, we get

$$Z_{QM}(z) = C\left[\frac{\phi_1'(z) + \beta\phi_2'(z)}{\phi_1(z) + \beta\phi_2(z)}\right]. \tag{7.182}$$

Writing this equation at $z = L$ and solving for β from the above equation leads to

$$\beta = \frac{C\phi_1'(L) - Z_{QM}(L)\phi_1(L)}{Z_{QM}(L)\phi_2(L) - C\phi_2'(L)}. \tag{7.183}$$

Plugging this value of β back into Equation (7.179) evaluated at $z = 0+$, we obtain

$$Z_{\text{QM}}(0_+) = C \left[\frac{Z_{\text{QM}}(L)\phi_2(L) - C\phi_2'(L)}{C\phi_1'(L) - Z_{\text{QM}}(L)\phi_1(L)} \right]. \tag{7.184}$$

If we consider a tunneling problem with an electron incident from the left, $Z_{\text{QM}}(L)$ is the characteristic load impedance of the right contact,

$$Z_{\text{QM}}(L) = \frac{2\hbar}{m^* i} k_{\text{F}}, \tag{7.185}$$

where k_{F} is the electron wave vector in the right contact.

Equation (7.184) is the quantum mechanical equivalent of the well-known formula for computing the impedance of a transmission line starting from the impedance of the load [18–21].

In the next problem, we show how to compute the reflection coefficient from an arbitrary conduction band energy profile using the quantum mechanical wave impedance concept.

---

* Problem 7.14: Reflection coefficient in terms of quantum mechanical wave impedance

For an electron incident from the left on an arbitrary conduction band energy profile $E_{\text{c}}(z)$ in the domain $[0, L]$, derive an expression for the reflection probability in terms of the quantum mechanical impedance at $z = 0+$ and the characteristic quantum mechanical impedance Z_0 of the contact $(z < 0)$.

Solution: The quantum mechanical wave impedance at $z = 0+$ can be calculated from the load impedance in the right contact using the approach described in the previous problem. Because the wave function and its derivative are continuous, so is the quantum mechanical wave impedance. For an electron incident from the left, $\phi(z) = e^{ik_0 z} + re^{-ik_0 z}$, where r is the reflection amplitude and k_0 is the z-component of its wave vector. As a result,

$$Z_{\text{QM}}(0_-) = \frac{2\hbar k_0}{m^*} \frac{(1-r)}{1+r} = Z_0 \left(\frac{1-r}{1+r} \right), \tag{7.186}$$

where $Z_0 = \frac{2\hbar k_0}{m^*}$ is the characteristic quantum mechanical wave impedance in the left contact. Since $Z_{\text{QM}}(0_+) = Z_{\text{QM}}(0_-)$, we get

$$(1 + r)Z_{\text{QM}}(0_+) = (1 - r)Z_0, \tag{7.187}$$

leading to the following expression for the reflection amplitude:

$$r = \frac{Z_0 - Z_{\text{QM}}(0_+)}{Z_0 + Z_{\text{QM}}(0_+)}. \tag{7.188}$$

This is the analog of the reflection amplitude formula in transmission line theory [19–21]. The reflection probability is therefore given by

$$R = |r|^2 = \left| \frac{Z_0 - Z_{\text{QM}}(0_+)}{Z_0 + Z_{\text{QM}}(0_+)} \right|^2. \tag{7.189}$$

**** Problem 7.15: Quantum mechanical wave impedance approach to tunneling through a square barrier

Using the concept of quantum mechanical wave impedance, derive the energy dependence of the transmission probability through a square barrier of height V_0 and width W. Assume that the electron is impinging from the left and the effective mass is constant throughout.

Solution: In the barrier region, if $E > V_0$, the solution of the one-dimensional Schrödinger Equation (7.180) is

$$\phi = A^+ e^{ikz} + A^- e^{-ikz}, \tag{7.190}$$

where $k = \frac{\sqrt{2m^*(E-V_0)}}{\hbar}$.

The quantum mechanical wave impedance in the barrier region is therefore given by

$$Z_{\text{QM}}(z) = \frac{2\hbar}{im^*} \frac{\phi'}{\phi} = \frac{2\hbar k}{m^*} \left(\frac{A^+ e^{ikz} - A^- e^{-ikz}}{A^+ e^{ikz} + A^- e^{-ikz}} \right), \tag{7.191}$$

which can be rewritten as

$$Z_{\text{QM}}(z) = \frac{2\hbar k}{m^*} \left[\frac{A^+ e^{ikW} e^{ik(z-W)} - A^- e^{-ikW} e^{-ik(z-W)}}{A^+ e^{ikW} e^{ik(z-W)} + A^- e^{-ikW} e^{-ik(z-W)}} \right]. \tag{7.192}$$

Introducing $\beta^+ = A^+ e^{ikW}$ and $\beta^- = A^- e^{-ikW}$, we obtain

$$Z_{\text{QM}}(z) = \frac{2\hbar k}{m^*} \left[\frac{\beta^+ e^{ik(z-W)} - \beta^- e^{-ik(z-W)}}{\beta^+ e^{ik(z-W)} + \beta^- e^{-ik(z-W)}} \right], \tag{7.193}$$

and therefore

$$Z_{\text{QM}}(z) = \frac{2\hbar k}{m^*} \left[\frac{(\beta^+ - \beta^-)\cos[k(z-W)] + j(\beta^+ + \beta^-)\sin[k(z-W)]}{(\beta^+ + \beta^-)\cos[k(z-W)] + j(\beta^+ - \beta^-)\sin[k(z-W)]} \right], \tag{7.194}$$

or

$$Z_{\text{QM}}(z) = \frac{2\hbar k}{m^*} \left[\frac{\frac{\beta^+ - \beta^-}{\beta^+ + \beta^-}\cos[k(z-W)] + j\sin[k(z-W)]}{\cos[k(z-W)] + j\frac{\beta^+ - \beta^-}{\beta^+ + \beta^-}\sin[k(z-W)]} \right], \tag{7.195}$$

and finally

$$Z_{\text{QM}}(z) = Z_0 \left[\frac{Z_W \cos[k(z-W)] + jZ_0 \sin[k(z-W)]}{Z_0 \cos[k(z-W)] + jZ_W \sin[k(z-W)]} \right], \tag{7.196}$$

where $Z_0 = \frac{2\hbar k}{m^*}$ and Z_W is the load impedance, i.e., Z_{QM}, at $z = W$. According to Equation (7.195), this is given by

$$Z_W = Z_{\mathrm{QM}}(z = W) = Z_0 \frac{\beta^+ - \beta^-}{\beta^+ + \beta^-}. \tag{7.197}$$

For $E > V_0$, an electron incident from the left will see a load impedance due to the barrier given by

$$Z_{\mathrm{i}} = Z_{\mathrm{QM}}(z = 0+) = Z_0 \left[\frac{Z_W \cos(kW) - jZ_0 \sin(kW)}{Z_0 \cos(kW) - jZ_W \sin(kW)} \right]. \tag{7.198}$$

Similarly, if $E < V_0$, the solution of the Schrödinger equation in region II is

$$\phi_{\mathrm{II}} = A^+ e^{\kappa z} + A^- e^{-\kappa z}, \tag{7.199}$$

where $\kappa = \frac{\sqrt{2m^*(V_0 - E)}}{\hbar}$. In this case, the quantum mechanical wave impedance is given by

$$Z_{\mathrm{QM}}(z) = \frac{2\hbar}{im^*} \frac{\phi'}{\phi} = \frac{2\hbar\kappa}{im^*} \frac{A^+ e^{\kappa z} + A^- e^{-\kappa z}}{a^+ e^{\kappa z} + A^- e^{-\kappa z}}, \tag{7.200}$$

which we rewrite as

$$Z_{\mathrm{QM}}(z) = \frac{2\hbar\kappa}{im^*} \left[\frac{A^+ e^{\kappa W} e^{\kappa(z-W)} - A^- e^{-\kappa W} e^{-\kappa(z-W)}}{A^+ e^{\kappa W} e^{\kappa(z-W)} + A^- e^{-\kappa W} e^{-\kappa(z-W)}} \right]. \tag{7.201}$$

This last equation can be recast as

$$Z_{\mathrm{QM}}(z) = Z_0 \left[\frac{Z_W \cosh[k(z - W)] + Z_0 \sinh[k(z - W)]}{Z_0 \cosh[k(z - W)] + Z_W \sinh[k(z - W)]} \right]. \tag{7.202}$$

The barrier provides the following load impedance to an electron incident from the left:

$$Z_{\mathrm{i}} = Z_{\mathrm{QM}}(z = 0) = Z_0 \left[\frac{Z_L \cosh(kL) - Z_0 \sinh(kL)}{Z_0 \cosh(kL) - Z_L \sinh(kL)} \right]. \tag{7.203}$$

If $E > V_0$, the reflection amplitude of an electron incident from the left is given by (see Problem 7.14)

$$r = \frac{Z_L - Z_{\mathrm{i}}}{Z_L + Z_{\mathrm{i}}} = \frac{Z_L - Z_{\mathrm{i}} \left[\frac{Z_L \cos(kL) - jZ_0 \sin(kL)}{Z_0 \cos(kL) - jZ_L \sin(kL)} \right]}{Z_L + Z_{\mathrm{i}} \left[\frac{Z_L \cos(kL) - jZ_0 \sin(kL)}{Z_0 \cos(kL) - jZ_L \sin(kL)} \right]}, \tag{7.204}$$

which becomes

$$r = \frac{j(Z_0^2 - Z_L^2) \sin(kL)}{2Z_0 Z_L \cos(kL) - j(Z_L^2 - Z_0^2) \sin(kL)}. \tag{7.205}$$

Therefore, the reflection probability is given by (see Equation (7.189))

$$R = |r|^2 = \frac{(Z_0^2 - Z_L^2)^2 \sin^2(kL)}{4Z_0^2 Z_L^2 + (Z_L^2 - Z_0^2)^2 \sin^2(kL)}$$

$$= \frac{(Z_0^2 - Z_L^2)^2 \sin^2(kL)}{4Z_0^2 Z_L^2 + (Z_L^2 - Z_0^2)^2 \sin^2(kL)}, \tag{7.206}$$

from which we easily derive the transmission probability

$$|T|^2 = 1 - |R|^2. \tag{7.207}$$

With $Z_0 = \frac{2\hbar k}{m^*}$ and $Z_L = \frac{2\hbar k_0}{m^*}$, we get

$$|T|^2 = \cfrac{1}{1 + \left(\frac{Z_L^2 - Z_0^2}{2Z_0 Z_L}\right)^2 \sin^2(kL)} = \cfrac{1}{1 + \left(\frac{k^2 - k_0^2}{2k_0 k}\right)^2 \sin^2(kL)}, \tag{7.208}$$

or

$$|T|^2 = \cfrac{1}{1 + \left(\frac{V_0^2}{2E(E-V_0)}\right) \sin^2(kL)}, \tag{7.209}$$

where $k = \frac{2m^*}{\hbar}\sqrt{E - V_0}$.

Similarly, if $E < V_0$,

$$r = \frac{Z_W - Z_i}{Z_W + Z_i} = \cfrac{Z_W - Z_0 \left[\frac{Z_W \cosh(\kappa W) - Z_0 \sinh(\kappa W)}{Z_0 \cosh(\kappa W) - Z_W \sinh(\kappa W)}\right]}{Z_W + Z_0 \left[\frac{Z_W \cosh(\kappa W) - Z_0 \sinh(\kappa W)}{Z_0 \cosh(\kappa W) - Z_W \sinh(\kappa W)}\right]}, \tag{7.210}$$

which reduces to

$$r = \frac{(Z_0^2 - Z_W^2)^2 \sinh(\kappa W)}{2Z_0 Z_W \cosh(\kappa W) + (Z_W^2 + Z_0^2)^2 \sinh(\kappa W)}. \tag{7.211}$$

Hence, the reflection probability is

$$R = |r|^2 = \frac{(Z_0^2 + Z_W^2)^2 \sinh^2(\kappa W)}{4Z_0^2 Z_W^2 + (Z_W^2 + Z_0^2)^2 \sinh^2(\kappa W)}. \tag{7.212}$$

The transmission probability is therefore given by

$$T = 1 - |R|^2 = \cfrac{1}{1 + \left(\frac{Z_W^2 + Z_0^2}{2Z_0 Z_W}\right)^2 \sinh^2(\kappa W)} = \cfrac{1}{1 + \left(\frac{\kappa^2 + k_0^2}{2\kappa k_0}\right)^2 \sinh^2(\kappa W)}, \tag{7.213}$$

or

$$T = \cfrac{1}{1 + \left(\frac{V_0^2}{2E(V_0 - E)}\right) \sinh^2(\kappa W)}. \tag{7.214}$$

** Problem 7.16: Bound state energies using the quantum mechanical wave impedance concept

Starting with Equation (7.178) of Problem 7.13, which relates to the quantum mechanical impedance on either side of an arbitrary conduction band energy profile $E_c(z)$ in a quantum well, derive a general equation to locate the bound state energies in that well.

Solution: In Problem 7.13, we showed that

$$Z_{\mathrm{QM}}(0) = C \left[\frac{Z(L)\phi_2(L) - C\phi_2{}'(L)}{C\phi_1{}'(L) - Z(L)\phi_1(L)} \right], \qquad (7.215)$$

where $C = 2\hbar/m^*i$ and $\phi_1(L)$, $\phi_1{}'(L)$, $\phi_2(L)$, and $\phi_2{}'(L)$ are the matrix elements of the transfer matrix across the quantum well, i.e., in the region $[0, L]$.

On the right side of the well, the solution of the Schrödinger equation is $Ae^{-\kappa z}$; hence,

$$Z_{\mathrm{QM}}(L) = -\frac{2\hbar\kappa}{m^*i}, \qquad (7.216)$$

where $\kappa = \frac{1}{\hbar}\sqrt{2m^*|E|}$.

On the left side of the well,

$$\phi = Ae^{\kappa z} \qquad (7.217)$$

and

$$Z_{\mathrm{QM}}(0) = \frac{2\hbar\kappa}{m^*i}. \qquad (7.218)$$

Plugging the expressions for $Z_{\mathrm{QM}}(0)$ and $Z_{\mathrm{QM}}(L)$ above in Equation (7.215) and simplifying, we get

$$k^2\phi_1(L) + k\phi_1{}'(L) + k\phi_2(L) + \phi_2{}'(L) = 0. \qquad (7.219)$$

This is exactly the general Equation (7.177) we derived in Problem 7.10 for the location of the bound states in an arbitrary $E_c(z)$ in the region $[0, L]$.

*** Problem 7.17: Connection between transmission and transfer matrices

As shown in Figure 7.7, the transmission matrix across a region of length d is the matrix T which relates the amplitudes (B^+, B^-) of the outgoing and incoming waves at $z = d$ to the amplitudes of the incoming and outgoing waves (A^+, A^-) at $z = 0$. Find the relation between the transmission matrix T and the transfer matrix W in the region between $z = 0$ and $z = d$.

Solution: The transfer matrix W is defined as

$$W = \begin{bmatrix} \phi_1{}'(d) & \phi_2{}'(d) \\ \phi_1(d) & \phi_2(d) \end{bmatrix}, \qquad (7.220)$$

where ϕ_1, ϕ_2 are the two linearly independent solutions of the Schrödinger equation such that

$$\phi_1{}'(0) = \phi_2(0) = 1, \qquad (7.221)$$

$$\phi_2{}'(0) = \phi_1(0) = 0. \qquad (7.222)$$

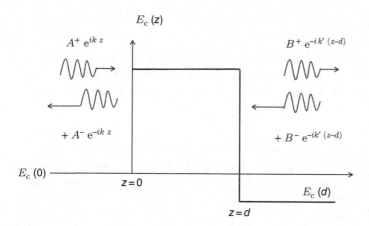

E_c (z)

$A^+\, e^{ik\, z}$

$+\, A^-\, e^{-ik\, z}$

$B^+\, e^{-ik'\, (z-d)}$

$+\, B^-\, e^{-ik'\, (z-d)}$

E_c (0)

z = 0

E_c (d)

z = d

Figure 7.7: Connections between the incoming and outgoing wave amplitudes across a region where the potential energy profile is approximated by a constant. The effective mass is assumed constant throughout.

In the interval $[0, d]$, we have

$$\phi(z) = A_1\phi_1(z) + A_2\phi_2(z). \tag{7.223}$$

Hence,

$$\phi(0) = A_2 = A^+ + A^- \tag{7.224}$$

and

$$\left.\frac{\mathrm{d}\phi}{\mathrm{d}z}\right|_{z=0} = A_1 = ik(A^+ - A^-). \tag{7.225}$$

Similarly, at $z = d$,

$$\phi(d) = A_1\phi_1(d) + A_2\phi_2(d) = B^+ + B^-, \tag{7.226}$$

$$\left.\frac{\mathrm{d}\phi}{\mathrm{d}z}\right|_{z=d} = A_1\phi_1{}'(d) + A_2\phi_2{}'(d) = B^+ - B^-. \tag{7.227}$$

Therefore,

$$\begin{bmatrix} ik' & -ik' \\ 1 & 1 \end{bmatrix} \begin{bmatrix} B^+ \\ B^- \end{bmatrix} = \begin{bmatrix} \phi_1{}'(d) & \phi_2{}'(d) \\ \phi_1(d) & \phi_2(d) \end{bmatrix} \begin{bmatrix} ik & -ik \\ 1 & 1 \end{bmatrix} \begin{bmatrix} A^+ \\ A^- \end{bmatrix}. \tag{7.228}$$

This last relation can be rewritten as

$$\begin{bmatrix} B^+ \\ B^- \end{bmatrix} = \begin{bmatrix} ik' & -ik' \\ 1 & 1 \end{bmatrix}^{-1} W \begin{bmatrix} ik & -ik \\ 1 & 1 \end{bmatrix} \begin{bmatrix} A^+ \\ A^- \end{bmatrix}, \tag{7.229}$$

and the transmission matrix T is

$$T = \frac{1}{2ik'} \begin{bmatrix} 1 & ik' \\ -1 & ik' \end{bmatrix} W \begin{bmatrix} ik & -ik \\ 1 & 1 \end{bmatrix}, \tag{7.230}$$

where W is the transfer matrix across the region of width d.

Suggested problems

- In Problem 7.1, start with Equation (7.13) and derive Equation (7.14) for the transmission amplitude. Derive an expression for the reflection amplitude and reflection probability in terms of the incident wave vector k_0 and the W_{TOT}^{ij} components of the overall transfer matrix.

- Starting with the results of Problem 7.3, write down the analytical expression for the tunneling probability of an electron incident from the left on a square barrier of width 50 Å and height $\Delta E_d = 0.3\,\text{eV}$. Plot the tunneling probability as a function of the z-component of the kinetic energy E_p for different values of the transverse energy E_t of the incident electron. Assume that the effective mass $m_1^* = 0.067m_0$ in the contacts and $m_2^* = 0.083m_0$ in the barrier region. Plot the tunneling probability versus E_p for E_t equal to 0, 25, 50, and 100 meV, respectively.

- Starting with the results of Problem 7.5 describing the tunneling across a potential step, use Equations (7.53) and (7.54) and show that the following relation is satisfied: $|r|^2 + \frac{k_2}{k_1}\frac{m_1^*}{m_2^*}|t|^2 = 1$.

- In Problem 7.6, derive an expression for the reflection probability in terms of the wave vectors of the electron on both sides of the device (k_L and k_R) and the W_{TOT}^{ij} components of the overall transfer matrix.

- Plot the transmission (T) and reflection (R) probabilities versus incident energy for an electron impinging on an attractive one-dimensional delta scatterer with strength $\Gamma = 5\,\text{eV-Å}$. Show that $R + T = 1$ for all energies.

- Using the results of the Problems 7.1 and 7.2, write the expressions for the tunneling probability through a square barrier of width W and height ΔE_c for the case of an electron incident with kinetic energy below and above ΔE_c. This problem illustrates the power of the transfer matrix formalism compared to the usual approach for this problem, which requires cumbersome algebra (see Problem 3.7).

- Starting with the result of the previous problem, show that the transmission probability above the square barrier reaches unity whenever the kinetic energy of the incident electron satisfies the relation

$$E_p = \Delta E_c + \frac{\hbar^2}{2m^*}\frac{n^2\pi^2}{W^2}.\tag{7.231}$$

- In Problem 7.9, it was shown that the energy dispersion relation through an infinite lattice is given by

$$\text{Tr}\,W = 2\cos(\xi L),\tag{7.232}$$

where ξ is the Bloch wave number, L the period of the unit cell, and W the transfer matrix of each unit cell.

Starting with Equation (7.133), show that the distinct solutions for $\lambda^{(1)}$ and $\lambda^{(2)}$ are given by

$$\lambda^{(1)} = e^{j\frac{pik}{N}},$$
$$\lambda^{(2)} = e^{-j\frac{pik}{N}}, \qquad k = 1, 2, \ldots, N - 1. \tag{7.233}$$

- The transfer matrix of a structure consisting of N repeated unit cells is given by $D = W^N$. Call $\theta = \xi L$ and using the results of Problem 7.9, show that

$$D = W \frac{\sin N\theta}{\sin \theta} - I \frac{\sin(N-1)\theta}{\sin \theta}, \tag{7.234}$$

where I is the 2×2 identity matrix.

- Use the results of the previous problem to derive the tunneling probability through a structure formed from N repeated units in terms of the wave vector of the incident electron, the element of the transfer matrix of a unit cell, and $\theta = \xi L$.

- Use the results of the previous problem to show that at the energies of unity transmission through a finite repeated structure with N periods, the following equality holds: $|T_{N_1}|^2 = |T_{N_2}|^2$ whenever $N_1 + N_2 = N$. Here, $|T_{N_1}|^2$ and $|T_{N_2}|^2$ are the transmission probabilities through two subsections with N_1 and N_2 periods respectively.

- Similary, using the results of the previous problems, show that at the energies of unity transmission ($|T_N|^2 = 1$) through a finite repeated structure with N periods, the following equality holds: $|T_{N+M}|^2 = |T_{N-M}|^2$ for all M such that $1 \leq M < N$.

- Starting with the results of Problem 7.11 and Figure 7.5, plot the energy dispersion relations for the two lowest energy bands in the first Brillouin zone, i.e., in the range of the electron wave number $[-\pi/L, \pi/L]$, L being the length of the unit cell. Use $L = 100\,\text{Å}$. Also, use $\Gamma = 5\,\text{eV-Å}$ and $z_0 = 25\,\text{Å}$ for the strength and location of the delta scatterer, respectively.

- Starting with the results of Problem 7.12, use Equation (7.177) and the results of Problem 7.2 to determine the transcendental equation for the bound states of a quantum well of width W and depth $-\Delta E_c$. Show that your result agrees with the transcendental equation derived in Problem 3.5.

- Starting with Equation (7.177) and the results of Problems 7.6 and 7.9, derive an analytical expression for the bound state of an attractive one-dimensional delta scatterer. Show that your result agrees with the result of Problem 3.1.

- Starting with the one-dimensional Schrödinger equation, derive an equation expressing the discontinuity of the quantum mechanical impedance across an attractive delta scatterer. Derive expressions for the reflection amplitude and probability associated with reflection of the electron from the delta scatterer in terms of the quantum mechanical impedance on both sides of the scatterer. Show that your results agree with the results of Problem 7.7.

- In Problem 7.15, starting with Equations (7.208) and (7.213), show that for $E = V_0$,

$$|T|^2(E = V_0) = \frac{1}{1 + \frac{m^* V_0 W^2}{\hbar^2}}. \tag{7.235}$$

- Derive an analytical expression for the transmission matrix across a one-dimensional delta scatterer.

- Derive an analytical expression for the transmission matrix across a potential step of height ΔE_c. Assume that the energy of the electron is above ΔE_c and that the effective mass is constant throughout.

References

[1] Walker, J. S. and Gathright, J. T. (1992) A transfer-matrix approach to one-dimensional quantum mechanics using Mathematica. *Computational Physics* 6, pp. 393–399.

[2] Walker, J. S. and Gathright, J. T. (1994) Exploring one-dimensional quantum mechanics with transfer matrices. *American Journal of Physics* 62, pp. 408–422.

[3] Kalotas, T. M. and Lee, A. R. (1991) A new approach to one-dimensional scattering. *American Journal of Physics* 59, pp. 48–52.

[4] Miller, D. A. B. (2008) *Quantum Mechanics for Scientists and Engineers*, Section 11.2, Cambridge University Press, New York.

[5] Sedrakian, D. M. and Khachatrian, A. Zh. (2003) Determination of bound state energies for a one-dimensional potential field, *Physica E* 19, p. 309.

[6] Mavromatis, H. A. (2000) Comment on "A single equation for finite rectangular well energy problems." *American Journal of Physics* 68, pp. 1151–1152.

[7] Barker, B. I., Rayborn, G. H., Ioup, J. W., and Ioup, G. E. (1991) Approximating the finite square well in an infinite well: energies and eigenfunctions. *American Journal of Physics* 59, pp. 1038–1042.

[8] Sprung, D. W. L., Wu, H., and Martorell, J. (1992) A new look at the square well potential. *European Journal of Physics* 13, pp. 21–25.

[9] Cameron Reed, B. (1990) A single equation for finite rectangular well energy eigenvalues. *American Journal of Physics* 58, p. 503.

[10] Pitkanen, P. H. (1955) Rectangular potential well problem in quantum mechanics. *American Journal of Physics* 23, pp. 111–113.

[11] Griffiths, D. J. and Steinke, C. A. (2001) Waves in locally periodic media. *American Journal of Physics* 69, pp. 137–154.

[12] Ghatak, A. K., Goyal, I. C., and Gallawa, R. L. (1990) Mean lifetime calculations of quantum well structures. *IEEE Journal of Quantum Electronics* 26, pp. 305–310.

[13] Bosken, M., Steller, A., Waring, B., and Cahay, M. (2014) Connection between bound state and tunneling problems. *Physica E* 64, pp. 141–145.

[14] Bandyopadhyay, S. (2012) *Physics of Nanostructured Solid State Devices*, Springer, New York.

[15] Vezzetti, D. J. and Cahay, M. (1986) Transmission resonances in finite, repeated structures. *Journal of Physics D—Applied Physics* 19, pp. L53–L55.

[16] Khondker, A. N., Rezwan Khan, M., and Anwar A. F. M. (1988) Transmission line analogy of resonance tunneling phenomena: the generalized impedance concept. *Journal of Applied Physics* 63(10), pp. 5191–5193.

[17] Fazlul Kalvi, S. M., Khan, M. R., and Alam, M. A. (1991) Application of quantum mechanical wave impedance in the solution of Schrödinger equation in quantum wells. *Solid-State Electronics*, 34(12), pp. 1466–1468.

[18] Jackson, J. D. (1975) *Classical Electrodynamics*, 2nd edition, Wiley, New York.

[19] Ramo, S., Whinnery, T. R., and van Duzer, T. (1965) *Fields and Waves in Communication Electronics*, John Wiley & Sons, New York.

[20] Pozar, D. M. (1990) *Microwave Engineering*, Addison-Wesley Publishing Company, New York.

[21] Gonzalez, G. (1997) *Microwave Transistor Amplifiers, Analysis and Design*, 2nd edition, Prentice Hall, Upper Saddle River, NJ.

Suggested Reading

- Olson, J. D. and Mace, J. L. (2003) Wave function confinement via transfer matrix methods. *Journal of Mathematical Physics* 44, pp. 1596–1624.

- Yu, K. W. (1990) Quantum transmission in periodic potentials: a transfer matrix approach. *Computers in Physics* 4, pp. 176–178.

- Cretu, N., Pop, I., and Rosca, I. (2012) Eigenvalues and eigenvectors of the transfer matrix. *American Institute of Physics Conference Proceedings* 1433, pp. 535–538.

- Erdos, P., Liviotti, E., and Herndon, R. C. (1997) Wave transmission through lattices, superlattices and layered media. *Journal of Physics D—Applied Physics* 30, pp. 338–345.

- Berman, P. R. (2013) Transmission resonances and Bloch states for a periodic array of delta function potentials. *American Journal of Physics* 81, pp. 190–201.

- Sanchez-Soto, L. L., Monzon, J. J., Alberto, A. G., et al. (2012) The transfer matrix: a geometrical perspective. *Physics Reports—Review Section of Physics Letters* 513, pp. 191–227.

- Jonsson, B. and Eng, S. T. (1990) Solving the Schrödinger equation in arbitrary quantum-well potential profiles using the transfer matrix. *IEEE Journal of Quantum Electronics* 26, pp. 2025–2035.

- Kalotas, T. M. and Lee, A. R. (1992) The bound states of a segmented potential. *American Journal of Physics* 59, pp. 1036–1038.

- Sprung, D. W. L., Sigetich, J. D., Wu, H., et al. (2000) Bound states of a finite periodic potential. *American Journal of Physics* 68, pp. 715–722.

- Sprung, D. W. L., Wu, H., and Martorell, J. (1993). Scattering by a finite periodic potential. *American Journal of Physics* 61, pp. 1118–1124.

- Griffiths, D. J. and Taussig, N. F. (1992) Scattering from a locally periodic potential. *American Journal of Physics* 60, pp. 883–888.

- Wu, H., Sprung, D. W. L., and Martorell, J. (1993) Periodic quantum wires and their quasi-one-dimensional nature. *Journal of Physics D—Applied Physics* 26, pp. 798–803.

Chapter 8: Scattering Matrix

The transfer matrix technique, discussed in Chapter 7, is a very useful method to treat both bound state and tunneling problems. However, the elements of an individual transfer matrix can end up being either very large or very small (when evanescent states are involved, i.e., the electron's kinetic energy is lower than the potential step and we are dealing with the wavevector κ rather than the wavevector k). When matrices with enormously large elements or vanishingly small elements are cascaded via matrix multiplication as described in Chapter 7, the resulting matrix can have elements that either blow up or become effectively zero, leading to numerical instabilities and truncation errors. For that reason, a *scattering matrix technique* is sometimes preferred over the transfer matrix technique, even though the rules for cascading scattering matrices are more complex. The cascading procedure for scattering matrices is not as simple as merely multiplying scattering matrices; on the other hand, the problem of handling extremely large or extremely small elements is ameliorated and the technique is numerically robust. This robustness is linked to the requirement that any scattering matrix used to describe coherent transport through nanoscale devices must be unitary, a property not shared by transfer matrices. The scattering matrix, however, has to be properly defined for unitarity, i.e., we must use the *current scattering matrix* as opposed to the more common *amplitude scattering matrix*. The unitarity follows from current conservation [1, 2].

We first describe the scattering matrix technique and how it can solve tunneling problems. Next, we present the cascading rule for cascading scattering matrices describing adjacent sections of an arbitrary potential profile (or conduction band energy profile in a semiconductor device) that has been partitioned into small sections each with a constant (spatially invariant) potential which is the spatial average of the potential within that section. Explicit analytical expressions for the elements of the scattering matrix for simple tunneling problems are then derived. Finally, we exemplify this approach with a derivation of the scattering matrix associated with tunneling through a two-dimensional delta scatterer in a two-dimensional quantum waveguide.

The scattering matrix relates the *current* amplitudes associated with incoming to outgoing electron waves on both sides of a region of width L containing an arbitrary potential energy profile [2–4]. If the latter depends only on the z-coordinate (the direction of current flow), as shown in Figure 8.1, then the scattering matrix associated with the section located in the interval $[0, L]$ is such that

$$\begin{bmatrix} \phi^+(L) \\ \phi^-(0) \end{bmatrix} = S \begin{bmatrix} \phi^+(0) \\ \phi^-(L) \end{bmatrix}, \tag{8.1}$$

where S is the scattering matrix relating the incoming and outgoing current density amplitudes $\phi^+(L)$, $\phi^-(0)$, $\phi^+(0)$, and $\phi^-(L)$ on both sides of region of width L (see Problem 5.4), as shown in Figure 8.1.

Problem Solving in Quantum Mechanics: From Basics to Real-World Applications for Materials Scientists, Applied Physicists, and Devices Engineers, First Edition.
Marc Cahay and Supriyo Bandyopadhyay.

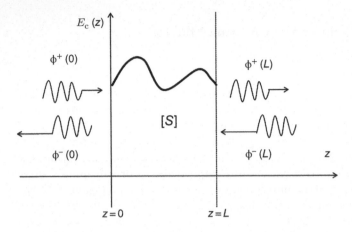

Figure 8.1: The scattering matrix relates the current amplitudes of incoming $[\phi^+(0), \phi^-(L)]$ to outgoing $[\phi^+(L), \phi^-(0)]$ waves on both sides of a region of width L containing an arbitrary potential energy profile. The scattering matrix S is defined such that Equation (8.1) is satisfied.

The scattering matrix is a 2×2 matrix whose elements are defined as follows:

$$S = \begin{bmatrix} t & r' \\ r & t' \end{bmatrix}, \tag{8.2}$$

where r and t are, respectively, the (current) reflection and transmission amplitudes for an electron incident from the left contact, and r' and t' are, respectively, the current reflection and transmission amplitudes for an electron incident from the right contact. The four elements of the scattering matrix are not equal to, but are related to, the wave function reflection and transmission amplitudes which were derived in Chapter 5.

The scattering matrix in Equation (8.2) can be easily extended to the case of *multi-moded* or *multi-channeled* transport that takes place when electron waves are impinging from multiple modes and reflecting back into multiple modes in the regions surrounding the scattering region. If there are M modes to consider, then each of the elements in the scattering matrix in Equation (8.2) will be an $M \times M$ matrix, each representing transmission or reflection from one mode to another, or unto itself. In that case, the scattering matrix will become a $2M \times 2M$ matrix.

---

* Problem 8.1: Scattering matrix describing a free propagation region

Calculate the four components of the scattering matrix describing free propagation (no reflection) through a device of width L as shown in the Figure 8.2.

Solution: Since the potential is spatially constant throughout, there is no force acting on the electron and hence the electron travels with a constant velocity.

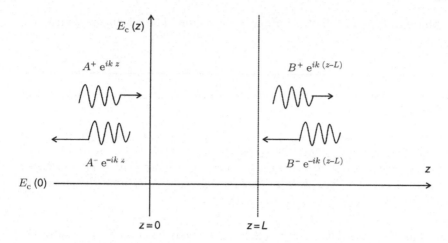

Figure 8.2: The only non-zero elements of the scattering matrix describing a free propagating region are simple phase shifts of the waves incident from the left and right contacts. The effective mass m^* is assumed to be constant throughout.

Consequently, the current amplitudes associated with the left- and right-going waves on both sides of the free propagation region are proportional to the corresponding wave function amplitudes. For $z < 0$, the solution of the Schrödinger equation is given by

$$A^+ e^{ikz} + A^- e^{-ikz}, \tag{8.3}$$

and, for $z > L$, we have

$$\phi = B^+ e^{ik(z-L)} + B^- e^{-ik(z-L)}, \tag{8.4}$$

where $k = \frac{1}{\hbar}\sqrt{2m^* E_p}$ is the z-component of the wavevector of the electrons and E_p is the longitudinal component of its kinetic energy.

Since there is no reflection for a wave incident from the left, we can use the two preceding equations to show that, at $z = L$,

$$A^+ e^{ikL} = B^+, \tag{8.5}$$

which shows that all that the wave experiences in traversing the region $[0, L]$ is a phase shift kL. Similarly, for an electron incident from the right, equating (8.3) and (8.4) at $z = 0$ leads to

$$A^- = B^- e^{ikL}. \tag{8.6}$$

From Equations (8.5) and (8.6), we obtain

$$\begin{bmatrix} \phi^+(L) \\ \phi^-(0_-) \end{bmatrix} = \begin{bmatrix} e^{ikL} & 0 \\ 0 & e^{ikL} \end{bmatrix} \begin{bmatrix} \phi^+(0_-) \\ \phi^-(L) \end{bmatrix}, \tag{8.7}$$

where the 2×2 matrix is the scattering matrix describing free propagation across the region of width L. Only its diagonal elements (t, t') are non-zero, and each is equal to a phase shift across the free propagation region. The off-diagonal elements (r, r') are zero.

*** Problem 8.2: Scattering matrix describing electron wave propagation across a potential step

Consider a potential step as shown in Figure 5.1. Assume a constant effective mass throughout and call k_1 and k_2 the z-components of the wavevector on the left and right sides of the potential step, respectively. Assume the incident energy is larger than the height of the step, i.e., both k_1 and k_2 are real. Determine the scattering matrix across the step and show that it is unitary.

Solution: The general solutions of the Schrödinger equation on both sides of the step are given by

$$A^+ e^{ik_1 z} + A^- e^{-ik_1 z} \text{ for } z < 0, \tag{8.8}$$

$$B^+ e^{ik_2 z} + B^- e^{-ik_2 z} \text{ for } z > 0. \tag{8.9}$$

The current density amplitudes associated with the left- and right-going plane waves on both sides of the step can then be calculated using the general expressions derived in Chapter 5 (see Problem 5.4):

$$\phi^+(z) = \frac{1}{2} \sqrt{\frac{\hbar k}{m^*}} \left[\phi(z) + \frac{1}{ik} \frac{d\phi(z)}{dz} \right], \tag{8.10}$$

$$\phi^-(z) = \frac{1}{2} \sqrt{\frac{\hbar k}{m^*}} \left[\phi(z) - \frac{1}{ik} \frac{d\phi(z)}{dz} \right], \tag{8.11}$$

where k is set to either k_1 or k_2 depending on whether the electron is on the left or right side of the step, respectively.

Using Equations (8.11) and (8.12), we easily obtain

$$\phi^+(0_+) = \sqrt{\frac{\hbar k_2}{m^*}} B^+, \tag{8.12}$$

$$\phi^-(0_+) = \sqrt{\frac{\hbar k_2}{m^*}} B^-, \tag{8.13}$$

$$\phi^+(0_-) = \sqrt{\frac{\hbar k_1}{m^*}} A^+, \tag{8.14}$$

$$\phi^-(0_-) = \sqrt{\frac{\hbar k_1}{m^*}} A^-. \tag{8.15}$$

Before deriving the scattering matrix across the potential step, we first derive the 2×2 matrix connecting the wave function amplitudes A^-, A^+, B^-, and B^+ as

$$\begin{bmatrix} B^+ \\ A^- \end{bmatrix} = M \begin{bmatrix} A^+ \\ B^- \end{bmatrix}. \tag{8.16}$$

Continuity of the wave function across the interface requires that

$$A^+ + A^- = B^+ + B^-. \tag{8.17}$$

Continuity of the wave function's derivative across the interface requires that

$$k_1 A^+ - k_1 A^- = k_2 B^+ - k_2 B^-. \tag{8.18}$$

Multiplying Equation (8.17) by k_1, we get that

$$k_1 A^+ + k_1 A^- = k_1 B^+ + k_1 B^-. \tag{8.19}$$

Adding the last two equations leads to

$$2k_1 A^+ = (k_1 + k_2)B^+ + (k_1 - k_2)B^-, \tag{8.20}$$

which we rewrite as

$$B^+ = \frac{2}{1 + \frac{k_2}{k_1}} A^+ - \frac{1 - \frac{k_2}{k_1}}{1 + \frac{k_2}{k_1}} B^-. \tag{8.21}$$

Similarly, multiplying Equation (8.17) by k_2, we get that

$$k_2 A^+ + k_2 A^- = k_2 B^+ + k_2 B^-. \tag{8.22}$$

Subtracting this last equation from Equation (8.18), we get

$$(k_1 - k_2)A^+ - (k_1 + k_2)A^- - 2k_2 B^-, \tag{8.23}$$

which we rewrite as

$$A^- = -\frac{1 - \frac{k_1}{k_2}}{1 + \frac{k_1}{k_2}} A^+ + \frac{2}{1 + \frac{k_1}{k_2}} B^-. \tag{8.24}$$

The matrix M defined in Equation (8.16) can then be inferred from Equations (8.22) and (8.25) written in the following form:

$$\begin{bmatrix} B^+ \\ A^- \end{bmatrix} = \begin{bmatrix} \frac{2k_1}{k_1 + k_2} & -\frac{k_1 - k_2}{k_1 + k_2} \\ -\frac{k_2 - k_1}{k_1 + k_2} & \frac{2k_2}{k_1 + k_2} \end{bmatrix} \begin{bmatrix} A^+ \\ B^- \end{bmatrix}. \tag{8.25}$$

It is then easy to derive the scattering matrix relating the incoming and outgoing current amplitudes from the defining relation

$$\begin{bmatrix} \sqrt{\frac{\hbar k_2}{m^*}} B^+ \\ \sqrt{\frac{\hbar k_1}{m^*}} A^- \end{bmatrix} = S \begin{bmatrix} \sqrt{\frac{\hbar k_1}{m^*}} A^+ \\ \sqrt{\frac{\hbar k_2}{m^*}} B^- \end{bmatrix}. \tag{8.26}$$

We get:

$$S = \begin{bmatrix} \frac{2\sqrt{k_1 k_2}}{k_1 + k_2} & -\left(\frac{k_1 - k_2}{k_1 + k_2}\right) \\ -\left(\frac{k_2 - k_1}{k_1 + k_2}\right) & \frac{2\sqrt{k_1 k_2}}{k_1 + k_2} \end{bmatrix}. \tag{8.27}$$

Next, we prove that S is unitary, i.e., $S^\dagger S = I$, where the $\dagger$ stands for the Hermitian conjugate and I is the identity matrix. Indeed,

$$S^\dagger = \begin{bmatrix} \frac{2\sqrt{k_1 k_2}}{k_1+k_2} & -\left(\frac{k_2-k_1}{k_1+k_2}\right) \\ -\left(\frac{k_1-k_2}{k_1+k_2}\right) & \frac{2\sqrt{k_1 k_2}}{k_1+k_2} \end{bmatrix}. \tag{8.28}$$

Hence,

$$S^\dagger S = \begin{bmatrix} \frac{2\sqrt{k_1 k_2}}{k_1+k_2} & -\left(\frac{k_2-k_1}{k_1+k_2}\right) \\ -\left(\frac{k_1-k_2}{k_1+k_2}\right) & \frac{2\sqrt{k_1 k_2}}{k_1+k_2} \end{bmatrix} \times \begin{bmatrix} \frac{2\sqrt{k_1 k_2}}{k_1+k_2} & -\left(\frac{k_1-k_2}{k_1+k_2}\right) \\ -\left(\frac{k_2-k_1}{k_1+k_2}\right) & \frac{2\sqrt{k_1 k_2}}{k_1+k_2} \end{bmatrix}. \tag{8.29}$$

Carrying out the matrix multiplication, we get

$$S^\dagger S = \begin{bmatrix} 1 & 0 \\ 0 & 1 \end{bmatrix} = I, \tag{8.30}$$

proving the unitarity of S.

** **Problem 8.3: Scattering matrix describing electron wave propagation across a one-dimensional delta scatterer**

Derive the expression for the scattering matrix describing propagation across a one-dimensional delta scatterer. Assume the potential energy $E_c(z)$ is equal to zero on both sides of the delta scatterer.

Solution: This tunneling problem was discussed in Problem 7.7 using the transfer matrix technique. Here, we treat it using the scattering matrix technique. Assuming a constant effective mass throughout, we have

$$\frac{-\hbar^2}{2m^*} \frac{d^2}{dz^2} \phi(z) + \Gamma \delta(z)\phi(z) = E\phi(z). \tag{8.31}$$

Integrating both sides of this equation from $-\epsilon$ to $+\epsilon$ and letting $\epsilon \to 0$, we get

$$\frac{d\phi}{dz}(0_+) - \frac{d}{dz}\phi(0_-) = \frac{2m^*}{\hbar^2}\Gamma\phi(0). \tag{8.32}$$

Using this boundary condition together with the continuity of ϕ at $z = 0$, we get, starting with Equations (8.8) and (8.9) for the description of the overall solutions of the Schrödinger equation on both sides of the delta scatterer,

$$A^+ + A^- = B^+ + B^-, \tag{8.33}$$

$$ik(B^+ - B^-) - ik(A^+ - A^-) = \frac{2m^*\Gamma}{\hbar^2}(A^+ + A^-). \tag{8.34}$$

We rewrite the last equation as

$$B^+ - B^- = \left[\frac{2m^*\Gamma}{ik\hbar^2} + 1\right] A^+ + \left[\frac{2m^*\Gamma}{ik\hbar^2} - 1\right] A^-. \tag{8.35}$$

Defining $\alpha = \frac{2m^*\Gamma}{ik\hbar^2}$, Equation (8.35) becomes

$$B^+ - B^- = (1 + \alpha)A^+ + (\alpha - 1)A^-. \tag{8.36}$$

Next, we rewrite Equations (8.33) and (8.36) as

$$B^+ - A^- = A^+ - B^-, \tag{8.37}$$

$$B^+ - (\alpha - 1)A^- - (1 + \alpha)A^+ + B^-. \tag{8.38}$$

These last two equations can be written in the matrix form

$$\begin{bmatrix} 1 & -1 \\ 1 & 1-\alpha \end{bmatrix} \begin{bmatrix} B^+ \\ A^- \end{bmatrix} = \begin{bmatrix} 1 & -1 \\ 1+\alpha & 1 \end{bmatrix} \begin{bmatrix} A^+ \\ B \end{bmatrix}. \tag{8.39}$$

A few extra steps of algebra then lead to

$$\begin{bmatrix} B^+ \\ A^- \end{bmatrix} = \begin{bmatrix} \frac{2}{2-\alpha} & \frac{\alpha}{2-\alpha} \\ \frac{\alpha}{2-\alpha} & \frac{2}{2-\alpha} \end{bmatrix} \begin{bmatrix} A^+ \\ B^- \end{bmatrix}. \tag{8.40}$$

Since the wavevector (and electron velocity) is the same on both sides of the delta scatterer, the 2×2 matrix on the right-hand side is the current scattering matrix. Hence, the current density transmission amplitudes for waves incident from the left or right are equal and given by

$$t = t' = \frac{2}{2 - \alpha}. \tag{8.41}$$

The current density reflection amplitudes for left- or right-incoming waves are also equal and given by

$$r = r' = \frac{\alpha}{2 - \alpha}. \tag{8.42}$$

It can be easily shown that the 2×2 scattering matrix is unitary, i.e., $S^\dagger S = S^{-1}S = I$, where I is the 2×2 identity matrix.

Furthermore, it can be easily checked that

$$|t|^2 + |r|^2 = 1, \tag{8.43}$$

which is an expression for current continuity.

** Problem 8.4: Cascading scattering matrices

Determine the cascading rule to derive the composite scattering matrix for a region consisting of two sections when the individual scattering matrix for each section is known.

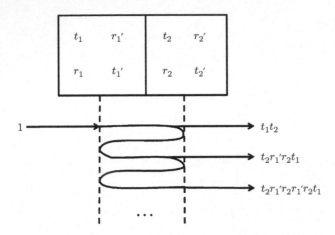

Figure 8.3: Summing the current transmission amplitudes of multiple Feynman paths to calculate the total current amplitude *transmitted* through two successive sections. The electron is incident from the left.

Solution: For any specific interval in a device, we have seen before that the scattering matrix relates incoming current density amplitudes to the outgoing current density amplitudes, as given by Equation (8.1). In a region composed of two adjoining intervals, an electron can suffer multiple reflections at the interface between the two adjacent regions, as illustrated in Figure 8.3. The elements of the composite scattering matrix can be derived by summing the amplitudes associated with the multiple paths leading to either reflection or transmission of electrons incident from either side of the composite region, as illustrated in Figures 8.3 through 8.6.

For instance, electrons incident from the left of the composite region can be transmitted straight through the structure or can experience an infinite number of multiple reflections, illustrated in Figure 8.3, before being transmitted. The total transmitted current density amplitude is the sum of the contributions from each possible path and is given by

$$t = t_2 \left[1 + r_1'r_2 + (r_1'r_2)^2 + \cdots \right] t_1, \qquad (8.44)$$

where the subscripts denote the two successive intervals with their respective current transmission and reflection amplitudes.

In Equation (8.44), the term within the square brackets forms an infinite geometric series and it results from the multiple reflections illustrated in Figure 8.3. The series is easily summed, yielding the current density amplitude of the beam transmitted across two successive regions:

$$t = t_2 \left[1 - r_1'r_2 \right]^{-1} t_1. \qquad (8.45)$$

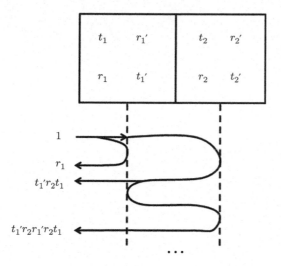

Figure 8.4: Summing the current reflection amplitudes of multiple Feynman paths to calculate the total current amplitude *reflected* by two successive sections. The electron is incident from the left.

Following a similar derivation and making use of Figures 8.4, 8.5, and 8.6, the remaining elements of the composite scattering matrix are found as

$$r = r_1 + t_1'r_2 \left[1 - r_1'r_2\right]^{-1} t_1, \tag{8.46}$$

$$t' = t_1' \left[1 + r_2[1 - r_1'r_2]^{-1}r_1'\right] t_2', \tag{8.47}$$

$$r' = r_2' + t_2 \left[1 - r_1'r_2\right]^{-1} r_1't_2'. \tag{8.48}$$

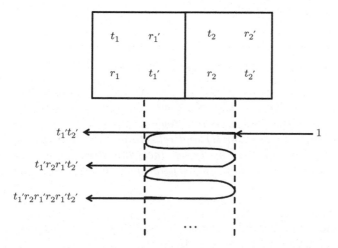

Figure 8.5: Summing the current transmission amplitudes of multiple Feynman paths to calculate the total current amplitude *transmitted* through two successive sections. The electron is incident from the right.

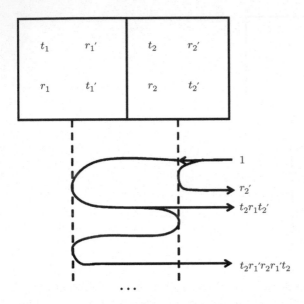

Figure 8.6: Summing the current reflection amplitudes of multiple Feynman paths to calculate the total current amplitude *reflected* by two successive sections. The electron is incident from the right.

Remark

Even if the product $r_1'r_2$ were to approach unity, making the inverse of $[1 - r_1'r_2]$ large, transmitted amplitudes t_1' and t_2 would approach zero, owing to current conservation. Since many computer algorithms allow small numbers to underflow gracefully to zero, a scattering matrix approach to tunneling problems can be implemented with relatively little error checking.

** Problem 8.5: Scattering matrix across a resonant tunneling device

A resonant tunneling device can be approximated as two repulsive delta scatterers separated by a distance L. Using the results of Problems 8.1 and 8.3 and the rule for cascading scattering matrices described in Problem 8.4, derive the analytical expression for the current transmission (t) and reflection (r) amplitudes as a function of the wavevector of the electron in the contact, the strength of each delta scatterer (assumed to be the same), and the separation L between the two scatterers.

Solution: Using the cascading rule derived in the previous problem, the scattering matrix across the resonant tunneling device is given by

$$S_{\text{tot}} = \begin{bmatrix} t_\delta & r_\delta \\ r_\delta & t_\delta \end{bmatrix} \otimes \begin{bmatrix} e^{ikL} & 0 \\ 0 & e^{ikL} \end{bmatrix} \otimes \begin{bmatrix} t_\delta & r_\delta \\ r_\delta & t_\delta \end{bmatrix}, \tag{8.49}$$

where the $\otimes$ symbol stands for the cascading rule applied to the scattering matrices describing adjacent regions. We use this symbol to distinguish it from mere multiplication.

The first and last 2×2 matrices are the scattering matrices for the two delta scatterers. The explicit expressions for t_δ and r_δ are given in Problem 8.3. The middle 2×2 matrix in Equation (8.49) is the scattering matrix for the free propagation region S_{free} between the two scatterers.

Using the cascading rules derived in the previous problem, we first calculate the composite scattering matrix describing propagation through the free region and the right delta scatterer. This leads to

$$S_{\text{free}} \otimes S_\delta = \begin{bmatrix} e^{ikL} & 0 \\ 0 & e^{ikL} \end{bmatrix} \otimes \begin{bmatrix} t_\delta & r_\delta \\ r_\delta & t_\delta \end{bmatrix} = \begin{bmatrix} t_\delta e^{ikL} & r_\delta \\ r_\delta e^{2ikL} & t_\delta e^{ikL} \end{bmatrix}. \tag{8.50}$$

Next, we calculate the overall scattering matrix S_{tot} by cascading the scattering matrix of the left delta scatterer with the last equation, to yield

$$S_{\text{tot}} = \begin{bmatrix} t_\delta & r_\delta \\ r_\delta & t_\delta \end{bmatrix} \otimes \begin{bmatrix} t_\delta e^{ikL} & r_\delta \\ r_\delta e^{2ikL} & t_\delta e^{ikL} \end{bmatrix}. \tag{8.51}$$

Using the cascading rules derived in Problem 8.4, the current density transmission amplitude for an electron incident from the left is found to be

$$t = \frac{t_\delta^2 e^{ikL}}{1 - r_\delta^2 e^{2ikL}}, \tag{8.52}$$

while the current reflection amplitude for an electron incident from the left is given by

$$r = r_\delta + \frac{t_\delta^2 r_\delta e^{2ikL}}{\left[1 - r_\delta^2 e^{2ikL}\right]}. \tag{8.53}$$

Using Equations (8.52) and (8.53), it can be shown that $|t|^2 + |r|^2 = 1$.

**** Problem 8.6: The law of summing series resistances derived using the rule for cascading scattering matrices

This problem will show that the semi-classical conductance (or resistance) of a random array of scatterers can be deduced by replacing the current amplitude scattering matrix S with the current probability scattering matrix Σ. The matrix Σ is obtained from S by replacing each of its elements with the square of its magnitude. This substitution leads to the semi-classical result for the conductance of a random array of scatterers, leading to the "series resistance law," which states that the total resistance of a number of resistors placed in series is the sum of their individual resistances.

To prove this law, consider an array of scatterers, all characterized by the same amplitude scattering matrix, but with different (random) spacing d_n between them.

For a two-dimensional channel with M propagating modes, the scattering matrix has dimensions $2M \times 2M$. Using the results of Problem 2.7, select the amplitude scattering matrix as

$$U = e^A, \tag{8.54}$$

where A is a $2M \times 2M$ matrix of the form $A = i\alpha a$, where α is a real parameter and a is a $2M \times 2M$ matrix with all its elements equal to unity. In this case, as shown in Problem 2.7, the amplitude scattering matrix for the mth scatterer is given by

$$S_m = I + \frac{(e^{2i\alpha M} - 1)}{2M} a, \tag{8.55}$$

where I is the $M \times M$ identity matrix.

Starting with the scattering matrix amplitude in Equation (8.55), show that the probability scattering matrix for the mth scatterer is given by

$$\Sigma_m = \begin{pmatrix} T & R \\ R & T \end{pmatrix}, \tag{8.56}$$

where the matrix $R = \delta u$ and u is the $N \times N$ matrix whose matrix elements are all equal to unity, i.e.

$$u = \begin{bmatrix} 1 & \cdots & 1 \\ \vdots & \ddots & \vdots \\ 1 & \cdots & 1 \end{bmatrix}, \tag{8.57}$$

and the matrix T is given by

$$T = \begin{bmatrix} 1 - (2M-1)\delta & \delta & \cdots & \delta \\ \delta & 1 - (2M-1)\delta & \cdots & \delta \\ \vdots & \vdots & \ddots & \vdots \\ \delta & \delta & \cdots & 1 - (2M-1)\delta \end{bmatrix}, \tag{8.58}$$

with

$$\delta = \left| \frac{e^{2iM\alpha} - 1}{2M} \right|^2. \tag{8.59}$$

For propagation between scatterers, the probability scattering matrix is given by

$$P_m = \begin{pmatrix} I & 0 \\ 0 & I \end{pmatrix}. \tag{8.60}$$

I is the identity matrix, because the probability associated with the phase shift between scatterers is equal to unity.

It is easy to show that the probability scattering matrices cascade in the same way as the amplitude scattering matrices. It is also easy to show, using the rule for cascading scattering matrices derived in Problem 8.4, that cascading any P_m with *any matrix Σ_m yields back Σ_m. In other words, the spacing between scatterers d_n will not appear in the final expression for the overall scattering matrix describing propagation across a sample (as expected in a semi-classical treatment). Therefore, the overall scattering matrix across N scatterers is obtained by cascading N identical scattering matrices Σ_m.*

Use the rule for cascading probability scattering matrices and show that the matrix giving the reflection probabilities through an array of N scatterers is given by

$$R_N = \delta_N u, \tag{8.61}$$

where δ_N satisfies the relation

$$\frac{M\delta_N}{1 - M\delta_N} = N\left(\frac{M\delta}{1 - M\delta}\right). \tag{8.62}$$

Use this result to show that the zero temperature Landauer conductance G_N through an array of N scatterers is given by [2]

$$G = \frac{2e^2}{h}\sum_i^M\sum_j^M T_{ij}, \tag{8.63}$$

where the T_{ij} are the matrix elements of the transmission probabilities associated with the ith mode incident from the left transmitting into the jth mode on the right of the array of scatterers.

Show that for large N, G_N is inversely proportional to N, and hence the resistance $R_N = 1/G_N$ is proportional to N. This is the series law of resistance.

Solution: Consider first the result of cascading the probability scattering matrices of two adjacent scatterers. In the resulting composite scattering matrix, the reflection probability for M propogating modes is given by

$$R_{2\ \text{scatterers}} = R + TR(1 - R^2)^{-1}T. \tag{8.64}$$

The matrices R and T can be rewritten as

$$R = \delta u \tag{8.65}$$
$$T = (1 - 2M\delta)I + \delta u. \tag{8.66}$$

Equation (8.64) can then be simplified to yield

$$R_{2\ \text{scatterers}} = R\frac{1}{1 - \delta^2 M^2}TRT. \tag{8.67}$$

Here, we have used the relation

$$R^2 = \delta M R. \tag{8.68}$$

Equation (8.67) can be simplified further using Equation (8.64):

$$TRT = T(1 - M\delta)R = (1 - M\delta)^2 R. \tag{8.69}$$

Hence,

$$R_{2 \text{ scatterers}} = \frac{2}{1 + M\delta}R. \tag{8.70}$$

Thus, R_2 can be written as

$$R_2 = \delta_2 u, \tag{8.71}$$

where

$$\delta_2 = \frac{2\delta}{1 + \delta}. \tag{8.72}$$

Note that

$$\frac{M\delta_2}{1 - M\delta_2} = 2\left[\frac{M\delta}{1 - M\delta}\right]. \tag{8.73}$$

Similarly, by cascading two sections, each having two scatterers, we can show that

$$R_{4 \text{ scatterers}} = \delta_4 u, \tag{8.74}$$

where

$$\frac{M\delta_4}{1 - M\delta_4} = 2\frac{M\delta_2}{1 - M\delta_2} = 4\frac{M\delta}{1 - M\delta}. \tag{8.75}$$

We can continue this process indefinitely to get

$$R_{\text{N scatterers}} = \delta_N u, \tag{8.76}$$

where

$$\frac{M\delta_N}{1 - M\delta_N} = N\left(\frac{M\delta}{1 - M\delta}\right). \tag{8.77}$$

Next, we rewrite the Landauer conductance (8.63) in terms of the reflection probabilities, taking into account the fact that

$$\sum_j^M (T_{ij} + R_{ij}) = 1. \tag{8.78}$$

This leads to

$$G_N = \frac{2e^2}{h}\left(M - \sum_i^M \sum_j^M R_{ij}\right) = \frac{2e^2}{h}M\left(1 - \frac{MN\delta}{1 + (N - 1)M\delta}\right), \tag{8.79}$$

which can be written as

$$G_N = \frac{2e^2}{h}M\frac{\Lambda}{\Lambda + N}, \tag{8.80}$$

where

$$\Lambda = \frac{1 - M\delta}{M\delta}. \tag{8.81}$$

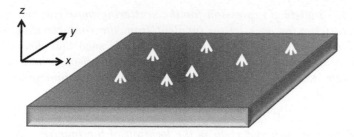

Figure 8.7: A quasi two-dimensional electron waveguide of finite width W containing randomly placed δ scatterers that cause elastic scattering.

We can rewrite Equation (8.80) as

$$G_N = \frac{2e^2}{h} M \frac{\Lambda_{el}}{\Lambda_{el} + N\langle d \rangle} = \frac{2e^2}{h} M \frac{\Lambda_{el}}{\Lambda_{el} + L}, \qquad (8.82)$$

where $\Lambda_{el} = \Lambda \langle d \rangle$, $\langle d \rangle$ is the average distance between scatterers, and L is the total length of the resistor. The last equation can be rewritten as

$$G_N = \frac{2e^2}{h} M \langle |T|^2 \rangle, \qquad (8.83)$$

where $\langle |T|^2 \rangle = \frac{\Lambda_{el}}{\Lambda_{el} + L}$ can be interpreted as the average transmission probability of a mode. The quantity Λ_{el} can be interpreted as the *elastic mean free path* [2]. The transmission probability $\langle |T|^2 \rangle$ approaches unity when $\Lambda_{el} \gg L$, which is the limit of ballistic transport.

In the limit of ballistic transport, the conductance (and hence also the resistance) of the device is independent of the length and is simply $2e^2/h$ times the number of modes M. It is independent of the number of scatterers (and hence the length) since the scatterers are "ineffective" in ballistic transport (no scattering). The resistance $h/(2e^2)$ has been interpreted as the contact resistance [5].

However, in the limit of severely diffusive transport, when $\Lambda_{el} \ll L$, the scatterers are very effective. In this case, Equation (8.82) shows that the conductance of the sample is inversely proportional to the number of scatterers, and therefore inversely proportional to the length, which means that the resistance is directly proportional to the length. This can be viewed as implying that the total resistance of several sections in series is the sum of the resistances of individual sections, which is the series law of resistance.

**** Problem 8.7: Scattering matrix of a two-dimensional δ scatterer

Consider an electron moving in an electron waveguide formed in a 2DEG with finite width W in the y-direction, which is perpendicular to the direction of current flow (x-axis), as shown in Figure 8.7. Assuming a particle-in-a-box confinement *in the y-direction, derive the analytical expression for the scattering matrix across a two-dimensional delta scatterer for an electron incident from the contacts with an energy equal to the Fermi level energy E_F. Assume that E_F is such that there are M propagating modes in the quantum wire. Assume a constant effective mass throughout and model the potential of the impurity as a two-dimensional δ function, i.e.*

$$V_\gamma(x,y) = \gamma\delta(x - x_i)\delta(y - y_i), \tag{8.84}$$

where γ is the strength of the impurity and x_i, y_i is the location of the impurity in the electron waveguide.

Solution: Our starting point is the two-dimensional one-electron effective mass Schrödinger equation

$$\left[-\frac{\hbar^2}{2m^*}\nabla^2 + E_c(x,y) + V_\gamma(x,y)\right]\psi(x,y,t) = i\hbar\frac{\partial\psi(x,y,t)}{\partial t}. \tag{8.85}$$

In a quantum wire with particle-in-a-box confinement along the y-axis, we write

$$E_c(x,y) = E_{c,x}(x) + E_{c,y}(y), \tag{8.86}$$

where we assume $E_{c,x}(x) \approx 0$ (i.e., a negligible bias electric field exists between the contacts to the quantum waveguide) and $E_{c,y} = 0$ for $0 < y < W$ and ∞ otherwise.

Since impurity scattering is elastic in nature (because an impurity is a time-independent scatterer), the electron's total energy is conserved while traversing the waveguide and therefore the time-dependent part of the wave function can be eliminated using the ansatz

$$\psi(x,y,t) = \phi(x,y)e^{-\frac{iEt}{\hbar}}. \tag{8.87}$$

The problem is then reduced to solving the time-independent two-dimensional Schrödinger equation

$$\left[-\frac{\hbar^2}{2m^*}\nabla^2 + E_c(x,y) + V_\gamma(x,y)\right]\phi(x,y) = E\phi(x,y), \tag{8.88}$$

where E is the energy of the electron in the waveguide.

Far from an impurity, ballistic motion in the quantum wire can be described by writing the solution of the Schrödinger Equation (8.88) as

$$\phi(x,y) = \sum_n C_n(x)|n\rangle, \tag{8.89}$$

where $|n\rangle$ are the eigenstates of the particle in a box confined in the y-direction (see Problem 3.5), and they form a complete set.

For a fixed energy of the electron incident from the contacts (taken as the Fermi energy), there will be M propagating states along the x-direction, i.e., states for which there exists a real value of the wavevector describing plane wave motion along the x-axis. The wavevectors corresponding to the Fermi energy in each of the propagating subbands in the waveguide are given by

$$E_{\mathrm{F}} = \epsilon_n^y + \frac{\hbar^2 k_{x,n}^2}{2m^*},$$ (8.90)

where

$$\epsilon_n^y = \frac{n^2\pi^2\hbar^2}{2m^*W_y^2}$$ (8.91)

are the energy eigenvalues due to the particle-in-a-box confinement along the y-axis.

For the particle-in-a-box confinement, the electron eigenstates in the y-direction are given by (see Problem 3.5)

$$\xi_n(y) = \sqrt{\frac{2}{W_y}} \sin\left(\frac{n\pi y}{W_y}\right).$$ (8.92)

Using the orthogonality property of the complete set (8.92), the current density $J(x)$ associated with the wave function (8.89) is given by

$$J(x) = c \sum_n \frac{i\hbar}{2m^*} \left[\left(\frac{\mathrm{d}C_n^*}{\mathrm{d}x}\right) C_n - C_n^* \left(\frac{\mathrm{d}C_n}{\mathrm{d}x}\right) \right].$$ (8.93)

Using the expansion (8.89) in Equation (8.88), and making use of the orthogonality condition of the eigenstates (8.91), the Schrödinger Equation (8.88) can be rewritten as an infinite set of coupled differential equations for the coefficients $C_n(z)$:

$$\frac{\mathrm{d}^2 C_n}{\mathrm{d}x^2} + k_n^2 C_n = \sum_m \Gamma_{nm}(x)C_m(x),$$ (8.94)

where

$$k_n^2 = \frac{2m^*}{\hbar^2}\left(E - \epsilon_n^y\right),$$ (8.95)

$$\Gamma_{nm}(x) = \frac{2m^*}{\hbar^2} \int \mathrm{d}y\,\xi_n^*(y)V(x,y)\xi_m(y).$$ (8.96)

The scattering problem is then reduced to the calculation of the coefficients Γ_{nm} and to the solution of the system of coupled differential equations (8.94). These equations must be written for the infinite set of modes given in Equation (8.91). Note that the sum over m in Equation (8.94) includes the term $m = n$.

In practice, the set of coupled differential equations in Equation (8.94) is reduced to M, the number of propagating modes in the quantum waveguide, since they alone carry current. The remaining modes, which are evanescent, do not carry current but still have a non-trivial effect in that they renormalize the values of Γ_{nm}. In other words, if we ignore the evanescent modes, then we must also alter the values of Γ_{nm}

from what is given by Equation (8.96) to compensate for neglecting the evanescent modes. The procedure for renormalization is not mentioned here but is described in [6].

In the set of coupled differential equations (8.94), both the k_n^2 and Γ_{nm} are functions of the variable x and depend respectively on the exact shape of the conduction band energy profile (assumed to be flat along the x-axis in our case) and the interacting potential.

As shown earlier, the scattering matrix relates the modes incident on an obstacle from either direction to the modes leaving the obstacle in either direction. Generalizing the definition of the current amplitude introduced in Problem 5.4 for a single-moded structure to the multi-moded case considered here, we introduce the following new set of variables for each of the propagating modes in the quantum wire:

$$C_n^{\pm}(x) = \frac{1}{2}\sqrt{\frac{\hbar k_n}{m^*}}\left(C_n \pm \frac{1}{ik_n}\frac{dC_n}{dx}\right), \tag{8.97}$$

which can easily be inverted to give

$$C_n = \frac{1}{2\kappa_n}\left(C_n^+ + C_n^-\right), \tag{8.98}$$

$$\dot{C}_n = \frac{dC_n}{dx} = \frac{ik_n}{2\kappa_n}\left(C_n^+ - C_n^-\right), \tag{8.99}$$

where, by definition,

$$\kappa_n = \frac{1}{2}\sqrt{\frac{\hbar k_n}{m^*}}. \tag{8.100}$$

The $C_n^{+,-}$ represent the amplitude of the current density in mode n traveling along the positive and negative x-direction, respectively.

Indeed, one can easily show using Equation (8.93) and the definitions (8.98)–(8.99) that the total current density in the electron waveguide is given by

$$J(x) = \sum_n (J_n^+ - J_n^-), \tag{8.101}$$

where $J_n^+ = (C_n^+)^* C_n^+$ and $J_n^- = (C_n^-)^* C_n^-$.

If we model the impurity scattering in a 2D sample (x–y plane) by a δ impurity interaction,

$$V(x,y) = \gamma\delta(x - x_i)\delta(y - y_i), \tag{8.102}$$

then the set of differential equations (8.94) can be written as

$$\frac{d^2C_m}{dx^2} + k_m^2 C_m = \sum_n \Gamma_{m,n}\delta(x - x_i)C_n, \tag{8.103}$$

where $\Gamma_{m,n}$ is given by

$$\Gamma_{m,n} = \frac{4m^*}{\hbar^2}\frac{1}{W}\sin\left(\frac{\alpha\pi y_i}{W}\right)\sin\left(\frac{\beta\pi y_i}{W}\right). \tag{8.104}$$

Integrating both sides of Equation (8.103) from $x_i - \epsilon$ to $x_i + \epsilon$ (ϵ being a small positive quantity), and taking into account the assumed continuity of the C_n, we obtain

$$\dot{C}_n(x_i + \epsilon) - \dot{C}_n(x_i - \epsilon) = \sum_m \Gamma_{nm} C_m(x_i + \epsilon), \qquad (8.105)$$

which we rewrite as

$$\dot{C}_n(x_i + \epsilon) = \dot{C}_n(x_i - \epsilon) + \sum_m \Gamma_{nm} C_m(x_i + \epsilon). \qquad (8.106)$$

Since all the C_n are continuous, we have

$$C_n(x_i + \epsilon) = C_n(x_i - \epsilon). \qquad (8.107)$$

Dividing Equation (8.106) by ik_α and adding the obtained result to Equation (8.107), we get

$$C_n(x_i + \epsilon) + \frac{1}{ik_n}\dot{C}_n(x_i + \epsilon) = \frac{1}{\kappa_n}C_n^+(x_i + \epsilon)$$
$$= \frac{1}{\kappa_n}C_n^+(x_i - \epsilon) - \sum_m \frac{\Gamma_{nm}}{ik_n}C_m(x_i). \qquad (8.108)$$

Now, using Equations (8.98) and (8.99), we have

$$C_m(x_i + \epsilon) = \frac{1}{2\kappa_m}\left[C_m^+(x_i + \epsilon) + C_m^-(x_i - \epsilon)\right]. \qquad (8.109)$$

Plugging this last result into Equation (8.108), we finally derive

$$C_n^+(x_i + \epsilon) = C_n^+(x_i - \epsilon) + \frac{\kappa_n}{2ik_n}\sum_m \Gamma_{nm}\frac{1}{\kappa_m}\left[C_m^+(x_i + \epsilon) + C_m^-(x_i - \epsilon)\right], \quad (8.110)$$

or, equivalently,

$$C_n^+(x_i + \epsilon) - \sum_m \frac{1}{2ik_n}\Gamma_{nm}\left(\frac{\kappa_n}{\kappa_m}\right)C_m^+(x_i + \epsilon)$$
$$= C_n^+(x_i - \epsilon) + \sum_m \frac{\Gamma_{nm}}{2ik_n}\left(\frac{\kappa_n}{\kappa_m}\right)C_m^-(x_i + \epsilon), \qquad (8.111)$$

valid for all modes n.

The equations for the different modes can be written in matrix form (which is done here for the case of two modes for the sake of simplicity):

$$\left[\begin{array}{cc} \left(1 - \frac{\Gamma_{11}}{2ik_1}\right) & \frac{-1}{2ik_2}\Gamma_{12} \end{array}\right]\left[\begin{array}{c} c_1^+(x_i + \epsilon) \\ c_2^+(x_i + \epsilon) \end{array}\right]$$
$$= \left[\begin{array}{cccc} 1 & 0 & \frac{\Gamma_{11}}{2ik_1} & \frac{\Gamma_{12}}{2ik_1} \\ 0 & 1 & \frac{\Gamma_{21}}{2ik_2}\frac{\kappa_2}{\kappa_1} & \frac{\Gamma_{22}}{2ik_2} \end{array}\right]\left[\begin{array}{c} c_1^+(x_i - \epsilon) \\ c_2^+(x_i - \epsilon) \\ c_1^-(x_i + \epsilon) \\ c_2^-(x_i + \epsilon) \end{array}\right], \qquad (8.112)$$

which we write more simply as

$$[\ I + ia_{++}\ ] \begin{bmatrix} c_1^+(x_i + \epsilon) \\ c_2^+(x_i + \epsilon) \end{bmatrix} = [\ I - ia_{++}\ ] \begin{bmatrix} c_1^+(x_i - \epsilon) \\ c_2^+(x_i - \epsilon) \\ c_1^-(x_i + \epsilon) \\ c_2^-(x_i + \epsilon) \end{bmatrix}, \qquad (8.113)$$

I being the 2×2 identity matrix; the matrix a_{++} is given by

$$a_{++} = \begin{bmatrix} \frac{1}{2k_1}\Gamma_{11} & \frac{1}{2k_1}\Gamma_{12}\frac{\kappa_1}{\kappa_2} \\ \frac{1}{2k_2}\Gamma_{21}\kappa_2\kappa_1 & \Gamma_{22}/(2k_2) \end{bmatrix}. \qquad (8.114)$$

From Equation (8.113), we then deduce

$$\begin{bmatrix} c_1^+(x_i + \epsilon) \\ c_2^+(x_i + \epsilon) \end{bmatrix} = [\ [I + ia_{++}]^{-1}\ ] [\ I \quad (-ia_{++})\ ] \begin{bmatrix} c_1^+(x_i - \epsilon) \\ c_2^+(x_i - \epsilon) \\ c_1^-(x_i + \epsilon) \\ c_2^-(x_i + \epsilon) \end{bmatrix}, \qquad (8.115)$$

or, equivalently,

$$\begin{bmatrix} c_1^+(x_i + \epsilon) \\ c_2^+(x_i + \epsilon) \end{bmatrix} = [\ (I + ia_{++})^{-1} \quad -(I + ia_{++})^{-1}ia_{++}\ ]$$

$$\times \begin{bmatrix} c_1^+(x_i - \epsilon) \\ c_2^+(x_i - \epsilon) \\ c_1^-(x_i + \epsilon) \\ c_2^-(x_i + \epsilon) \end{bmatrix}. \qquad (8.116)$$

Similarly, using Equations (8.109) and (8.110) and following a similar derivation, we obtain

$$\begin{bmatrix} c_1^-(x_i - \epsilon) \\ c_2^-(x_i - \epsilon) \end{bmatrix} = [\ -(I + ia_{++})^{-1}ia_{++} \quad (I + ia_{++})^{-1}\ ]$$

$$\times \begin{bmatrix} c_1^+(x_i - \epsilon) \\ c_2^+(x_i - \epsilon) \\ c_1^-(x_i + \epsilon) \\ c_2^-(x_i + \epsilon) \end{bmatrix}. \qquad (8.117)$$

Grouping the results (8.116) and (8.117), we obtain the final relation

$$
\begin{bmatrix} c_1^-(x_i - \epsilon) \\ c_2^-(x_i - \epsilon) \\ c_1^+(x_i + \epsilon) \\ c_2^+(x_i + \epsilon) \end{bmatrix} = \begin{bmatrix} -(I + ia_{++})^{-1}ia_{++} & (I + ia_{++})^{-1} \\ (I + ia_{++})^{-1} & -(I + ia_{++})^{-1}ia_{++} \end{bmatrix}
$$

$$
\times \begin{bmatrix} c_1^+(x_i - \epsilon) \\ c_2^+(x_i - \epsilon) \\ c_1^-(x_i + \epsilon) \\ c_2^-(x_i + \epsilon) \end{bmatrix}, \tag{8.118}
$$

where the 4×4 square matrix is the required scattering matrix.

For the general case of M modes, we can easily generalize the 2×2 matrix given in Equation (8.114). The general expression for the n, mth element of the matrix a_{++} can be written as

$$
a_{++,nm} = \frac{1}{2} \frac{\Gamma_{nm}}{\sqrt{k_n k_m}}. \tag{8.119}
$$

The analysis above only takes into consideration the propagating modes, i.e., the modes for which the bottom of their energy dispersion relationship is below the Fermi level in the contacts. Actually, it has been shown that evanescent modes (whose bottom of the energy dispersion relation is above the Fermi level in the contacts) can have a drastic influence on the conductance through mesoscopic systems [6–8].

Suggested problems

- The goal of this problem is to establish the connection between transfer (see Chapter 7) and scattering matrices. For a single-moded structure, show that the different elements of the transfer matrix W_{ij} $(i, j = 1, 2)$ and the corresponding scattering matrix (t, r, t', r') are related as follows:

$$
W_{11} = t - r't'^{-1}r, \tag{8.120}
$$

$$
W_{12} = r't'^{-1}, \tag{8.121}
$$

$$
W_{21} = t'^{-1}r, \tag{8.122}
$$

$$
W_{22} = t^{-1}. \tag{8.123}
$$

- Conversely, show that the various elements (t, r, t', r') of the scattering matrix can be expressed in terms of the various elements of the transfer matrix W_{ij} $(i, j = 1, 2)$ as

$$r = -W_{22}^{-1}W_{21}, \tag{8.124}$$

$$r' = W_{12}W_{22}^{-1}, \tag{8.125}$$

$$t = W_{11} - W_{12}W_{22}^{-1}W_{21}, \tag{8.126}$$

$$t' = W_{22}^{-1}. \tag{8.127}$$

- By using the multiple path approach described in Problem 8.4 to calculate the current transmission amplitude through two adjacent sections of a device, use Figures 8.4, 8.5, and 8.6 to show that the current amplitudes r, t', and r' of the composite scattering matrix are given by Equations (8.46), (8.47), and (8.48), respectively.

- Starting with the results of Problem 5.5 and following the derivation outlined in Problem 8.2, derive the expression for the scattering matrix across a potential step, noting that the electron's effective mass is not the same on the two sides of the step. Verify whether this scattering matrix is unitary.

- Starting with Equations (8.52) and (8.53) for the current transmission t and reflection r amplitudes derived in Problem 8.5 for a resonant tunneling device composed of two delta scatterers separated by a distance L, show that $|t|^2 + |r|^2 = 1$.

- Repeat Problem 8.5 to study the energy dependence of the transmission probability through a resonant tunneling structure composed of two one-dimensional delta scatterers of different strengths and separated by a distance L. Write Matlab code to study how the transmission probability through the resonant tunneling structure varies with the strength of the two delta scatterers.

- Starting with the results of Problem 8.6, show that the sum of the matrix elements in any row or column of the probability scattering matrix Σ_i of a scatterer is equal to unity. What is the physical significance of this result?

References

[1] Bandyopadhyay, S. (2012) *Physics of Nanostructured Solid State Devices*, Springer, New York.

[2] Datta, S. (1995) *Electronic Transport in Mesoscopic Systems*, Cambridge University Press, Cambridge.

[3] Datta, S. (1989) *Quantum Phenomena*, Modular Series on Solid State Devices, Vol. VIII, Addison-Wesley Publishing Company, Reading, MA.

[4] Kroemer, H. (1994) *Quantum Mechanics for Engineering, Materials Science, and Applied Physics*, Chapter 5, Prentice Hall, Englewood Cliffs, NJ.

[5] Imry, Y. (1997) in *Introduction to Mesoscopic Physics*, eds. H. G. Craighead, et al., Oxford University Press, Oxford, pp. 89–120.

[6] Bagwell, P. F. (1990) Evanescent modes and scattering in quasi-one-dimensional wires. *Physical Review B* 41, p. 10354.

[7] Datta, S., Cahay, M., and McLennan, M. (1987) Scatter-matrix approach to quantum transport. *Physical Review B, Rapid Communications* 36, p. 5655.

[8] Cahay, M., McLennan, M., and Datta, S. (1988) Conductance of an array of elastic scatterers: a scattering matrix approach. *Physical Review B* 37, p. 10125.

Chapter 9: Perturbation Theory

In many quantum mechanical problems, the solutions to the Schrödinger equation with a particular Hamiltonian are known, so that one knows the eigenstates (energies of the allowed states and their wave functions) for that Hamiltonian. However, if the system is perturbed and the Hamiltonian changes, the new solutions may not be easily found. In such circumstances, one is able to find the new solutions by a technique known as *perturbation theory* as long as the perturbation is weak and the change in the Hamiltonian due to the perturbation is small. If the perturbation is time independent, then one uses time-independent perturbation theory, whereas if the perturbation is time dependent, then one uses time-dependent perturbation theory. We give the results of time-independent perturbation theory below. Time-dependent perturbation theory has similar results, and can be found in [1]. However, very often, one is interested in the transitions between energy states of the unperturbed system caused by a time-dependent perturbation. This can be found using an important rule of quantum mechanics, *Fermi's Golden Rule*, which is derived from time-dependent perturbation theory [1]. Fermi's Golden Rule can yield the transition rates for transitions between states that are either degenerate or non-degenerate in energy.

In this chapter, we apply time-independent perturbation theory to solve some important problems that have applications in electro-optic modulators and band-structure calculations. We then use Fermi's Golden Rule to solve some problems involving the rates with which electrons transition between states due to either time-dependent or time-independent perturbations. In the former case, the electron will transition between states that are not degenerate in energy (inelastic transitions), whereas in the latter case the electron will transition between states that are degenerate in energy (elastic transitions).

Brief tutorial: First-order time-independent non-degenerate perturbation theory

Imagine a time-independent static system described by the time-independent Hamiltonian $H_0(\vec{r})$. An electron in the nth eigenstate of this system will have a time-independent wave function $\phi_n(\vec{r})$ satisfying the time-independent Schrödinger equation $H_0(\vec{r})\phi_n(\vec{r}) = E_n\phi_n(\vec{r})$, where E_n is the energy of the nth eigenstate.

Suppose now that the system is perturbed by a time-independent perturbation that changes the Hamiltonian to $H_0(\vec{r}) + H'(\vec{r})$. The new wave function and the new eigenenergy in the mth eigenstate can be approximately written as

$$\psi_m(\vec{r}) = C_m \left[\phi_m(\vec{r}) + \sum_{p=1, p\neq m}^{n} \frac{H_{pm}'}{E_m + H_{mm}' - E_p - H_{pp}'} \phi_p(\vec{r}) \right], \tag{9.1}$$

Problem Solving in Quantum Mechanics: From Basics to Real-World Applications for Materials Scientists, Applied Physicists, and Devices Engineers, First Edition.
Marc Cahay and Supriyo Bandyopadhyay.

$$\bar{E}_m = E_m + H_{mm}' + \sum_{p=1, p \neq m}^{n} \frac{|H_{pm}'|^2}{E_m + H_{mm}' - E_p - H_{pp}'}, \qquad (9.2)$$

where

$$H_{ij}' = \int \mathrm{d}^3 \vec{r} \phi_i^*(\vec{r}) H'(\vec{r}) \phi_j(\vec{r}), \qquad (9.3)$$

as long as no two states are degenerate in energy. The coefficient C_m is found by normalizing the wave function such that $\int \mathrm{d}^3 \vec{r} |\psi_m(\vec{r})|^2 = 1$. For a derivation of this result, see, for example, Refs. [1–6].

** Problem 9.1: Quantum-confined DC Stark effect

Consider a 10 nm layer of GaAs sandwiched between two ZnSe layers. The bulk bandgaps of ZnSe and GaAs are 2.83 eV and 1.42 eV, respectively, so that for all practical purposes the electrons and holes in GaAs can be considered to be confined in an infinite potential well. Suppose that the carrier concentration and temperature are such that only the lowest electron subband and heavy hole subband in the GaAs quantum well are occupied. The effective masses of electrons and heavy holes in GaAs are 0.067 and 0.45 times the free electron mass, m_0, respectively.

Assume that an electric field of 100 kV/cm is applied transverse to the heterointerfaces, which tilts the energy band diagram and skews the electron and hole wave functions, as shown in Figure 9.1.

(a) Use first-order time-independent non-degenerate perturbation theory to derive an expression for the skewed wave functions of electrons and heavy holes in

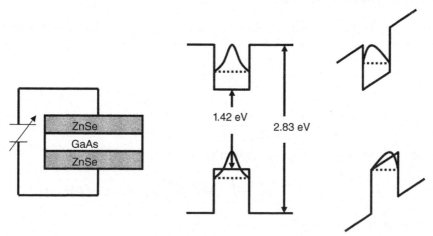

Figure 9.1: The quantum-confined DC Stark effect. Application of an electric field transverse to the heterointerfaces of a quantum well skews the electron and hole wave functions in opposite directions. That decreases the overlap between them and partially quenches photoluminescence intensity. The band bending within the quantum well due to the electric field also reduces the effective bandgap and causes a red-shift of the suppressed peak in the photoluminescence spectrum.

the lowest subbands. Remember that the unperturbed wave functions in the direction of confinement would have been $\Psi_e^0(z) = \Psi_{hh}^0(z) = \sqrt{\frac{2}{W}}\sin\left(\frac{n\pi z}{W}\right)$ *in the nth subband, where W is the width of the well (see Problem 3.5). When the electric field is turned on, the electrons see a perturbation which changes the Hamiltonian by an amount $-eEz$, and the holes experience a perturbation of $+eEz$, where E is the electric field.*

(b) Which wave function is skewed more, the electron or the heavy hole? Why?

(c) The device emits light when the electrons and holes recombine to emit photons. The intensity of the emitted light is given by $I \sim \left|\int_0^W \Psi_e^*(z)\Psi_{hh}(z)dz\right|^2$. Find the percentage decrease in the intensity when the electric field is turned on.

(d) The energy of emitted photons is the difference in the energy between the lowest-energy electron and the highest-energy heavy hole states. This energy is red-shifted by the electric field. Find the amount of red-shift. The red-shift is obviously the algebraic sum of the changes in the electron and hole subband energies caused by the applied electric field.

Solution:

(a) The skewed wave function is the perturbed wave function given by

$$\Psi_{\text{skewed}}(z) = \sum_n C_n \phi_n(z), \qquad (9.4)$$

where $\phi_n(z) = \sqrt{\frac{2}{W}}\sin\left(\frac{n\pi z}{W}\right)$. The perturbation is due to the electric field applied in the z-direction. Hence, $H' = -eEz$.

Since the unperturbed wave functions $\phi_n(z)$ are orthogonal to each other, the normalized skewed wave function is

$$\Psi_{\text{skewed}}(z) = \frac{\sum_n C_n \phi_n(z)}{\sqrt{\sum_n |C_n|^2}}, \qquad (9.5)$$

where

$$\frac{C_n}{C_1} = \frac{H_{1n}'}{E_1 + H_{11}' - E_n - H_{nn}'}. \qquad (9.6)$$

The so-called self-energy terms H_{mm}' are given by

$$H_{mm}' = -eE \int_0^W \phi_m^2(z)z\,dz = -\frac{2eE}{W}\int_0^W \sin^2\left(\frac{m\pi z}{W}\right)z\,dz = -eEW, \qquad (9.7)$$

which is independent of m.

Next, we evaluate the term H_{1n}':

$$H_{1n}' = -eE \int_0^W \phi_1(z)z\phi_n(z)dz = -\frac{2eE}{W}\int_0^W \sin\left(\frac{\pi z}{W}\right)\sin\left(\frac{n\pi z}{W}\right)dz$$

$$= -\frac{eE}{W}\int_0^W z\left[\cos\left(\frac{(n-1)\pi z}{W}\right) - \cos\left(\frac{(n+1)\pi z}{W}\right)\right]dz. \qquad (9.8)$$

Integrate by parts to obtain

$$
\begin{aligned}
H_{1n}' = & -\frac{eE}{W}\left[\frac{W}{[n-1]\pi}z\sin\left(\frac{[n-1]\pi z}{W}\right)\Big|_{z=0}^{z=W}\right.\\
& \left. -\int_0^W \frac{W}{[n-1]\pi}\sin\left(\frac{[n-1]\pi z}{W}\right)dz\right]\\
& +\frac{eE}{W}\left[\frac{W}{[n+1]\pi}z\sin\left(\frac{[n+1]\pi z}{W}\right)\Big|_{z=0}^{z=W}\right.\\
& \left. -\int_0^W \frac{W}{[n+1]\pi}\sin\left(\frac{[n+1]\pi z}{W}\right)dz\right]\\
= & -\frac{eE}{W}\left(\frac{W}{[n-1]\pi}\right)^2\cos\left(\frac{[n-1]\pi z}{W}\right)\Big|_{z=0}^{z=W}\\
& +\frac{eE}{W}\left(\frac{W}{[n+1]\pi}\right)^2\cos\left(\frac{[n+1]\pi z}{W}\right)\Big|_{z=0}^{z=W}.
\end{aligned}
\tag{9.9}
$$

For odd values of n, H_{1n}' vanishes. For even values of n, it is given by

$$
H_{1n}' = \frac{2eEW}{\pi^2}\left[\frac{1}{(n-1)^2}-\frac{1}{(n+1)^2}\right]\quad(n\text{ even}).
\tag{9.10}
$$

There is elegant physics behind the fact that the perturbation term vanishes for odd values of n. Note that the unperturbed wave function is symmetric about the center of the well. The perturbed (or skewed) wave function is not symmetric. Perturbation caused by the electric field mixes the wave functions of the higher subbands into the wave function of the lowest subband to skew the wave function. You cannot make a symmetric (unperturbed) wave function asymmetric (perturbed) by mixing in other symmetric wave functions. You can do that only by mixing in antisymmetric wave functions. Therefore, you would have gained nothing by mixing in wave functions of states with odd subband index (n) since those wave functions are symmetric. Only states with even subband index will help since these wave functions are antisymmetric. That is why the perturbation term vanishes for odd n, but not for even n.

Finally,

$$
\frac{C_{2p}}{C_1} = -\frac{\frac{2eEW}{\pi^2}\left[\frac{1}{(2p-1)^2}-\frac{1}{(2p+1)^2}\right]}{\frac{\hbar^2}{2m^*}\left[\left(\frac{2p\pi}{W}\right)^2-\left(\frac{\pi}{W}\right)^2\right]}.
\tag{9.11}
$$

Next, we find the skewed wave function or perturbed wave function:

$$
\Psi_{\text{skewed}}(z) = \frac{\sum_n C_n\phi_n(z)}{\sqrt{\sum_n |C_n|^2}} = \frac{\sum_p C_{2p}\phi_{2p}(z)}{\sqrt{\sum_p |C_{2p}|^2}}\frac{num}{denom},
\tag{9.12}
$$

where

$$num = \sqrt{\frac{2}{W}} \left\{ \sin\left(\frac{\pi z}{W}\right) - \sum_p \frac{\frac{2eEW}{\pi^2}\left[\frac{1}{(2p-1)^2} - \frac{1}{(2p+1)^2}\right]}{\frac{\hbar^2}{2m^*}\left[\frac{1}{(2p-1)^2} - \frac{1}{(2p+1)^2}\right]^2} \sin\left(\frac{2p\pi z}{W}\right) \right\}, \quad (9.13)$$

$$denom = \sqrt{1 + \sum_p \left(\frac{\frac{2eEW}{\pi^2}\left[\frac{1}{(2p-1)^2} - \frac{1}{(2p+1)^2}\right]}{\frac{\hbar^2}{2m^*}\left[\left(\frac{2p\pi}{W}\right)^2 - \left(\frac{\pi}{W}\right)^2\right]} \right)^2 }. \quad (9.14)$$

Let us get a sense of how much skewing is caused by the electric field. The first term in the summation above is (for $p = 1$)

$$\frac{\frac{2eEW}{\pi^2}\left[1 - \frac{1}{9}\right]}{\frac{\hbar^2}{2m^*}\left[\left(\frac{2\pi}{W}\right)^2 - \left(\frac{\pi}{W}\right)^2\right]} = \frac{128 m^* eEW^3}{27\hbar^2\pi^2}. \quad (9.15)$$

Plugging in the values of the electric field, well width, and effective mass, we find that the first term in the electron wave function skewing is 0.01. That means the leading term skews the electron wave function by a mere 1%. For holes, which are heavier, the leading term contributing to skewing is 0.067, or 6.7%. The hole wave function is therefore skewed more.

(b) It is easily seen that the only difference between the electron and hole is in the effective mass. The wave function of the heavier particle is more flexible and skews more. The physics behind this is the following. The subbands of the heavier particle are spaced more closely in energy since the subband separation in energy is $\frac{\hbar^2}{2m^*}(n^2 - 1)\left(\frac{\pi}{W}\right)^2$, which is inversely proportional to the effective mass. Hence the wave functions of the higher subbands are more easily mixed into the wave function of the lowest subband in the case of the heavier particle when perturbation is on, making the wave function of the heavier particle more easily skewed, or more flexible. This happens because when states are spaced closer in energy, they intermix more effectively.

(c) The skewed wave functions of the electron and heavy hole in the presence of the electric field are given by

$$\Phi_e^{\text{skewed}}(z) = \frac{\sqrt{\frac{2}{W}}\left[\sin\left(\frac{\pi z}{W}\right) - 0.01 \sin\left(\frac{2\pi z}{W}\right) + \cdots\right]}{\sqrt{1 + (0.01)^2 + \cdots}}$$

$$= \sqrt{\frac{2}{W}}\left[\sin\left(\frac{\pi z}{W}\right) - 0.01 \sin\left(\frac{2\pi z}{W}\right) + \cdots\right], \quad (9.16)$$

$$\Phi_{hh}^{\text{skewed}} = \frac{\sqrt{\frac{2}{W}}\left[\sin\left(\frac{\pi z}{W}\right) + 0.067 \sin\left(\frac{2\pi z}{W}\right) + \cdots\right]}{\sqrt{1 + (0.067)^2 + \text{small terms}}}$$

$$= 0.997\sqrt{\frac{2}{W}}\left[\sin\left(\frac{\pi z}{W}\right) + 0.067 \sin\left(\frac{2\pi z}{W}\right) + \cdots\right]. \quad (9.17)$$

In Equations (9.16) and (9.17), the three dots stand for negligible terms.

Note that the electron and hole wave functions are skewed in opposite directions since the signs of the second terms within the square brackets are opposite.

Therefore, when the electric field is on, the photoluminescence intensity is

$$
I_{\text{perturbed}} \sim \left| 0.997 \left(1 - 0.057 \frac{2}{W} \int_0^W \sin \left(\frac{\pi z}{W} \right) \sin \left(\frac{2\pi z}{W} \right) \, dz \right. \right.
$$

$$
\left. \left. - 0.00067 \frac{2}{W} \int_0^W \sin^2 \left(\frac{2\pi z}{W} \right) dz \right) \right|^2 . \tag{9.18}
$$

Hence,

$$
I_{\text{perturbed}} \sim |0.997 \times 0.99933|^2 = 0.9926. \tag{9.19}
$$

Therefore, the percentage quenching of photoluminescence is $(1 - 0.9926) \times 100 = 0.73\%$.

(d) The renormalization of the electron energy is

$$
\Delta_{\text{electron}} = \sum_{p=1}^{n} \left(\frac{\left\{ \frac{2eEW}{\pi^2} \left[\frac{1}{(2p-1)^2} - \frac{1}{(2p+1)^2} \right] \right\}^2}{\frac{\hbar^2}{2m_e^*} \left[\left(\frac{2p\pi}{W} \right)^2 - \left(\frac{\pi}{W} \right)^2 \right]} \sim \frac{\left[\frac{2eEW}{\pi^2} \frac{8}{9} \right]^2}{\frac{\hbar^2}{2m_e^*} \left[3 \left(\frac{\pi}{W} \right)^2 \right]} \right). \tag{9.20}
$$

This amounts to $\Delta_{\text{electron}} = 19.48 \, \mu\text{eV}$. The renormalization of the heavy hole energy is found from the same expression after replacing m_e^* with m_{hh}^*, leading to $\Delta_{\text{heavy hole}} = 130 \, \mu\text{eV}$. Therefore, the red-shift in the photon energy, which is the sum $\Delta_{\text{electron}} + \Delta_{\text{heavy hole}}$, is a mere $149.5 \, \mu\text{eV}$.

---

* Problem 9.2: Nearly free electron theory of crystal band structure

In this problem we will use perturbation theory to find the energy versus momentum (or wavevector) relation for an electron in a periodic potential such as a crystal. For a free electron, the energy (E) versus wavevector (k) relation is parabolic: $E = \frac{\hbar^2 k^2}{2m_0}$, where m_0 is the free electron mass.

A free electron does not experience any force and therefore sees a spatially invariant (constant) potential, which we can always assume to be zero since potential is undefined to the extent of an arbitrary constant. However, in a crystal, an electron sees a spatially varying potential which is periodic in space. This potential will be treated as a perturbation.

The Schrödinger equation describing an electron in a crystal is

$$
-\frac{\hbar^2}{2m} \nabla^2 \Psi(\vec{r}) + V_{\text{L}} \psi(\vec{r}) = E \Psi(\vec{r}), \tag{9.21}
$$

where m_0 is the free electron mass and $V_L(\vec{r})$ is the spatially periodic crystal or lattice potential. Assume that the crystal potential is weak and treat it like a perturbation within the ambit of perturbation theory. The unperturbed wave functions and eigenenergies can be written as

$$\phi_n(\vec{r}) = \frac{1}{\sqrt{a^3}} e^{i(\vec{k}+n\vec{G})\cdot\vec{r}}, \tag{9.22}$$

$$E_n = \frac{\hbar^2}{2m^*}|\vec{k}+n\vec{G}|^2, \tag{9.23}$$

where a^3 is the volume of a unit cell, $\vec{k}$ is the electron wavevector, and $\vec{G} = 2\pi/\vec{a}$ is the reciprocal lattice vector.

Find an expression for the electron wave function in the crystal and the energy–wavevector relation. Since the crystal potential is weak, the electron is nearly free and hence the calculated energy–wavevector relation is often called the "nearly free electron" (NFE) model of band structure.

Solution: Treating the crystal potential as the perturbation Hamiltonian, we get

$$\begin{aligned} H_{mn}{}' &= \frac{1}{a^3} \int \mathrm{d}^3\vec{r}\, e^{-i(\vec{k}+m\vec{G})\cdot\vec{r}} V_L(\vec{r}) e^{i(\vec{k}+n\vec{G})\cdot\vec{r}} \\ &= \frac{1}{a^3} \int \mathrm{d}^3\vec{r}\, V_L(\vec{r}) e^{i(n-m)\vec{G}\cdot\vec{r}} = \Pi_{(n-m)\vec{G}}, \end{aligned} \tag{9.24}$$

where the right-hand side is nothing but the Fourier component of the periodic lattice potential at $(n-m)\vec{G}$. The self-energy terms H_{mm} are the DC component of the periodic crystal potential and are independent of m or n.

Therefore, the wave function in the crystal lattice can be written (using perturbation theory) as

$$\begin{aligned} \phi_{\mathrm{NFE}}(\vec{r}) &= \frac{1}{\sqrt{a^3}} e^{i\vec{k}\cdot\vec{r}} + \sum_{p=1}^{n} \frac{\Pi_{p\vec{G}}}{E_0 + H_{00}{}' - E_p - H_{pp}{}'} \frac{1}{\sqrt{a^3}} e^{i(\vec{k}+p\vec{G})\cdot\vec{r}} \\ &= \frac{1}{\sqrt{(2\pi/G)^3}} e^{i\vec{k}\cdot\vec{r}} \left[1 + \sum_{p=1}^{n} \frac{\Pi_{p\vec{G}}}{\frac{\hbar^2|\vec{k}|^2}{2m} - \frac{\hbar^2|\vec{k}+p\vec{G}|^2}{2m}} e^{ip\vec{G}\cdot\vec{r}} \right], \end{aligned} \tag{9.25}$$

where we made use of the fact that the unperturbed energies are the free electron energies, i.e.,

$$E_0 = \frac{\hbar^2|\vec{k}|^2}{2m_0}, \tag{9.26}$$

$$E_p = \frac{\hbar^2|\vec{k}+p\vec{G}|^2}{2m_0}. \tag{9.27}$$

Hence, according to perturbation theory, the energy versus wavevector dispersion relation (or crystal band structure) is given by

$$E(\vec{k}) = E_0 + H_{00}' + \sum_{p=1}^{n} \frac{\left| \Pi_{p\vec{G}} \right|^2}{\frac{\hbar^2 |\vec{k}|^2}{2m_0} - \frac{\hbar^2 |\vec{k}+p\vec{G}|^2}{2m_0}}$$

$$= \frac{\hbar^2 |\vec{k}|^2}{2m_0} + V_0 + \sum_{p=1}^{n} \frac{\left| \Pi_{p\vec{G}} \right|^2}{\frac{\hbar^2 |\vec{k}|^2}{2m_0} - \frac{\hbar^2 |\vec{k}+p\vec{G}|^2}{2m_0}}, \tag{9.28}$$

where V_0 is a constant potential and can always be set equal to zero since potential is always undefined to the extent of an arbitrary constant.

Note that

$$V_0 = H_{nn}' = \frac{1}{a^3} \int d^3\vec{r} V_{\mathrm{L}}(\vec{r}) \tag{9.29}$$

is the DC component of the lattice potential $V_{\mathrm{L}}(\vec{r})$.

The above relations are not valid at the *Brillouin zone edges*, i.e., for $\vec{k} = \pm p\vec{G}/2$, since states at energies E_0 and E_p become "degenerate" at those wavevectors. Bandgaps open up at these wavevectors.

Brief tutorial: Electronic transitions and Fermi's Golden Rule

Many physical processes such as absorption and emission of light in a solid, or scattering of an electron due to impurities and phonons, involve the electron transitioning from one state to another. In a solid, the state of an electron is usually labeled by the wavevector $\vec{k}$. Transition from one state to another is caused by a time-independent (e.g., collisions with impurities) or time-dependent (e.g., collision with phonons) perturbation. Time-independent perturbations conserve energy, i.e., the energies of the electron in the initial and final states are same. Time-dependent perturbations do not conserve energy, meaning that the electron's energies in the initial and final states are different. The time rate of transition $S(\vec{k}, \vec{k}')$ from a wavevector state $\vec{k}$ to a wavevector state $\vec{k}'$ is given by the so-called Fermi's Golden Rule:

$$S(\vec{k}, \vec{k}') = \frac{2\pi}{\hbar} \left| M_{\vec{k},\vec{k}'}^{\pm} \right|^2 \delta(E_{\vec{k}'} - E_{\vec{k}} \pm \hbar\omega_0), \tag{9.30}$$

where $M_{\vec{k},\vec{k}'}^{\pm}$ is the so-called "matrix element" associated with a specific scattering potential, $E_{\vec{k}}$ is the energy of an electron in the wavevector state $\vec{k}$, and $\hbar\omega_0$ is the energy change in a non-energy-conserving (also known as inelastic) transition. In an energy-conserving (or elastic) transition, $\hbar\omega_0 = 0$. For a derivation of the above relation from time-dependent perturbation theory, see Ref. [2].

The matrix element $M_{\vec{k},\vec{k'}}$ is given by

$$M_{\vec{k},\vec{k'}} = \langle \Psi_{\vec{k'}}^*(\vec{r},t) | H_{\text{int}}(\vec{r},t) | \Psi_{\vec{k}}(\vec{r},t) \rangle$$

$$= \int \mathrm{d}^3 \vec{r}\, \Psi_{\vec{k'}}^*(\vec{r},t) H_{\text{int}}(\vec{r},t)\, \Psi_{\vec{k}}(\vec{r},t), \qquad (9.31)$$

where $H_{\text{int}}(\vec{r},t)$ is the operator for the interaction potential causing the transition and $\Psi(\vec{r},t)$ is the wave function of the electron in the wavevector state $\vec{k}$. The interaction Hamiltonian will, of course, be time independent for a time-independent perturbation such as interaction with an ionized impurity, but will be time dependent for a time-dependent perturbation such as a photon or phonon interaction.

**** Problem 9.3: Total electron scattering rate due to a screened ionized impurity

When an electron collides with a screened ionized impurity, the collision is elastic and the electron's kinetic energy is conserved. However, its momentum changes and the electron transitions from a wavevector state $\vec{k}$ to a wavevector state $\vec{k'}$. The interaction potential for this transition has an operator

$$H_{\text{int}}(\vec{r}) = \frac{-Zq^2 e^{-\lambda r}}{4\pi\epsilon |\vec{r}|}, \qquad (9.32)$$

where q is the magnitude of the electronic charge, Zq is the charge on the ionized impurity, λ is the inverse screening length, and ϵ is the dielectric constant of the material in which the electron and impurity are resident. The reader will recognize this as the screened Coulomb potential (sometimes referred to as the Yukawa potential).

Find the total electron scattering rate $\frac{1}{\tau(E_{\vec{k}})} = \sum_{\vec{k}} S(\vec{k},\vec{k'})$ and the momentum relaxation rate defined as $\frac{1}{\tau_m(E_{\vec{k}})} = \sum_{\vec{k}} S(\vec{k},\vec{k'})(\frac{k_z - k'_z}{k_z})$, where $E_{\vec{k}}$ is the energy of an electron in wavevector state $\vec{k}$ and k_z is the wavevector's component in an arbitrary direction that we call the z-direction. We assume that the initial and final electron wave functions are plane wave states. This would be an appropriate assumption in vacuum or in a solid whose band structure is approximately parabolic.

Solution: Since the initial and final wave functions are plane wave states,

$$\Psi_{\text{i}}(\vec{r},t) = \frac{1}{\sqrt{\Omega}} e^{i\vec{k}\cdot\vec{r}} e^{-iE_{\vec{k}}t/\hbar}, \qquad (9.33)$$

$$\Psi_{\text{f}}(\vec{r}) = \frac{1}{\sqrt{\Omega}} e^{i\vec{k'}\cdot\vec{r}} e^{-iE_{\vec{k'}}t/\hbar}, \qquad (9.34)$$

where Ω is the normalizing volume.

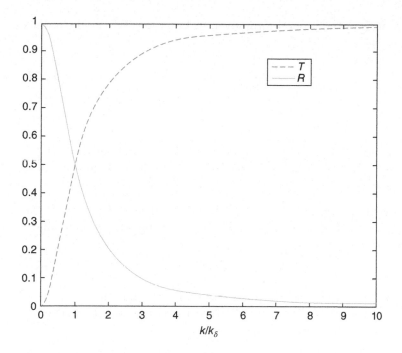

Figure 7.3: Plot of the transmission (T) and reflection (R) probabilities given by Equations (7.70) and (7.71), respectively, as a function of the reduced wavevector k/k_δ. Notice that $R = T = 0.5$ when $k/k_\delta = 1$.

Solution: The following proof is based on the transfer matrix formalism introduced earlier.

A general homogeneous linear differential equation of order n can be written as

$$\sum_n f_n(z)\frac{\mathrm{d}^n}{\mathrm{d}z^n}\psi(z) = 0. \tag{7.74}$$

If $f_n(z)$ is constant, it is well known that the solution of Equation (7.74) can be found by using the ansatz

$$\psi(z) = \mathrm{e}^{\sigma z} \tag{7.75}$$

and solving the resulting characteristic equation for the n σ-roots of the polynomial obtained by substituting Equation (7.75) into Equation (7.74). Floquet's theorem states that, if the functions $f_n(z)$ are periodic with period L, then the solutions given in Equation (7.74) are modulated by a periodic function of L. Our starting point is the one-dimensional effective mass Schrödinger equation for an electron moving in a periodic potential energy profile, i.e.,

$$-\frac{\hbar^2}{2m^*}\ddot{\psi}(z) + E_{\mathrm{c}}(z)\psi(z) = E_{\mathrm{p}}\psi(z), \tag{7.76}$$

where $E_{\mathrm{c}}(z) = E_{\mathrm{c}}(z+L)$ and E_{p} is the longitudinal component of the electron energy due to motion in the z-direction.

Because the collision is elastic, $E_{\vec{k}} = E_{\vec{k}'}$. Therefore, the matrix element is given by

$$M_{\vec{k},\vec{k}'} = \frac{1}{\Omega} \int \mathrm{d}^3\vec{r}\, e^{-i\vec{k}'\cdot\vec{r}} H_{\text{int}}\left(\vec{r}\right) e^{i\vec{k}\cdot\vec{r}} = \frac{1}{\Omega} \int \mathrm{d}^3\vec{r} H_{\text{int}}\left(\vec{r}\right) e^{i\left(\vec{k}-\vec{k}'\right)\cdot\vec{r}}, \qquad (9.35)$$

which is the Fourier component of the scattering potential at $\vec{k} - \vec{k}'$. The Fourier component of the screened Coulomb potential in Equation (9.32) is

$$M_{\vec{k},\vec{k}'} = -\frac{1}{\Omega} \frac{Zq^2}{4\pi\epsilon \left[\left|\vec{k} - \vec{k}'\right|^2 + \lambda^2\right]}. \qquad (9.36)$$

The total scattering rate is the sum of $S\left(\vec{k},\vec{k}'\right)$ over all final wavevector states $\vec{k}'$.

Summations over wavevector states can be converted to integrals over wavevector states if the latter form a continuum (as in bulk systems). We accomplish this by multiplying the summand with the density of states, which is $\Omega/(4\pi^3)$ (see Chapter 6), and then integrating over all $\vec{k}'$ states. Converting the sum into an integral in the calculation of the scattering rate $1/\tau(\vec{k})$, we get

$$\frac{1}{\tau(E_{\vec{k}})} = \sum_{\vec{k}'} S\left(\vec{k},\vec{k}'\right)$$

$$= \frac{2\pi}{\hbar} \sum_{\vec{k}'} \left|M_{\vec{k},\vec{k}'}\right|^2 \delta\left(E_{\vec{k}} - E_{\vec{k}'}\right)$$

$$= \frac{\Omega}{4\pi^3} \frac{2\pi}{\hbar} \left(\frac{Zq^2}{4\pi\epsilon}\right)^2 \frac{1}{\Omega} \int \mathrm{d}^3\vec{k}' \left[\frac{1}{\left|\vec{k}-\vec{k}'\right|^2 + \lambda^2}\right]^2 \delta\left(E_{\vec{k}} - E_{\vec{k}'}\right)$$

$$= \frac{Z^2 q^4}{32\hbar\left(\pi^2\epsilon\right)^2} \int \mathrm{d}^3\vec{q} \left[\frac{1}{q^2 + \lambda^2}\right]^2 \delta\left(E_{\vec{k}} - E_{\vec{k}\mp\vec{q}}\right), \qquad (9.37)$$

where $\vec{q} = \pm\left(\vec{k} - \vec{k}'\right)$. Note that we made a variable substitution from $\vec{k}'$ to $\vec{q}$ in the last line of the above equation. The upper sign corresponds to momentum loss and the lower sign to momentum gain.

The energy difference can be written as $E_{\vec{k}} - E_{\vec{k}'} = \frac{\hbar^2}{2m^*}\left(|\vec{k}'|^2 - |\vec{k}|^2\right) = \frac{\hbar^2}{2m^*}\left(q^2 \pm 2kq\cos\theta\right)$, where θ is the angle between the vectors $\vec{k}$ and $\vec{k}'$. Therefore, substituting this in the argument of the delta function above, we get

$$\frac{1}{\tau(E_{\vec{k}})} = \frac{Z^2 q^4}{32\hbar(\pi^2\epsilon)^2} \int\int\int q^2 \mathrm{d}q\, \mathrm{d}(\cos\theta)\mathrm{d}\phi \left[\frac{1}{q^2 + \lambda^2}\right]^2 \frac{2m^*}{\hbar^2}\delta(q^2 \pm 2kq\cos\theta)$$

$$= \frac{Zq^4}{32\hbar(\pi^2\epsilon)^2} \int\int\int q^2 \mathrm{d}q\, \mathrm{d}(cos\theta)\mathrm{d}\phi \left[\frac{1}{q^2 + \lambda^2}\right]^2 \frac{2m^*}{\hbar^2}\frac{1}{2kq}\delta\left(\frac{q}{2k} \pm \cos\theta\right)$$

$$= \frac{m^* Z^2 q^4 2\pi}{32\hbar^3 (\pi^2 \epsilon)^2 k} \int \int q dq [\frac{1}{q^2 + \lambda^2}]^2 d(\cos\theta)\delta\left(\frac{q}{2k} \pm \cos\theta\right)$$

$$= \frac{m^* Z^2 q^4 2\pi}{32\hbar^3 (\pi^2 \epsilon)^2 k} \int_{q_{min}}^{q_{max}} \frac{q}{(q^2 + \lambda^2)^2} dq, \qquad (9.38)$$

where $q_{min} = 0$ and $q_{max} = 2k$. It is easy to see that since the scattering process is elastic, the maximum momentum change is $\hbar \vec{q}_{max} = \hbar(2\vec{k})$.

Therefore,

$$\frac{1}{\tau\left(E_{\vec{k}}\right)} = \frac{m^* Z^2 q^4 2\pi}{32\hbar^3 (\pi^2 \epsilon)^2 k} \int_0^{2k} \frac{q}{(q^2 + \lambda^2)^2} dq$$

$$= -\frac{m^* Z^2 q^4 2\pi}{32\hbar^3 (\pi^2 \epsilon)^2 2k} \frac{1}{q^2 + \lambda^2}\Big|_0^{2k}$$

$$= \frac{m^* Z^2 q^4 \pi}{32\hbar^3 (\pi^2 \epsilon)^2 k} \left[\frac{1}{\lambda^2} - \frac{1}{4k^2 + \lambda^2}\right]$$

$$= \frac{\sqrt{m^*} Z^2 q^4}{32\sqrt{2}\hbar^2 \pi^3 \epsilon^2 \sqrt{E_k}} \left[\frac{1}{\lambda^2} - \frac{1}{8m^* E_k/\hbar^2 + \lambda^2}\right], \qquad (9.39)$$

where we have assumed a parabolic energy–wavevector relation $E_k = \frac{\hbar^2 k^2}{2m^*}$.

Clearly, the rate would diverge at $E_k = 0$ if $\lambda = 0$. For the momentum relaxation rate, we would get

$$1/\tau_m\left(E_{\vec{k}}\right) = \frac{m^* Z q^4 2\pi}{32\hbar^3 (\pi^2 \epsilon)^2 k} \int \int q dq d(\cos\theta) \left[\frac{1}{q^2 + \lambda^2}\right]^2$$

$$\times (1 - \cos\theta)\delta\left(\cos\theta \pm \frac{q}{2k}\right)$$

$$= \frac{m^* Z^2 q^4 2\pi}{32\hbar^3 (\pi^2 \epsilon)^2 k} \int_0^{2k} \frac{q}{(q^2 + \lambda^2)^2} \left[1 \pm \frac{q}{2k}\right] dq$$

$$= -\frac{m^* Z^2 q^4 2\pi}{32\hbar^3 (\pi^2 \epsilon)^2 2k} \frac{1}{q^2 + \lambda^2}\Big|_{q=0}^{q=2k}$$

$$\pm \frac{m^* Z^2 q^4 2\pi}{32\hbar^3 (\pi^2 \epsilon)^2 2\lambda k^2} \tan^{-1}\left(\frac{q}{\lambda}\right)\Big|_{q=0}^{q=2k}$$

$$\mp \frac{m^* Z^2 q^4 2\pi}{32\hbar^3 (\pi^2 \epsilon)^2 k^2} \frac{q}{q^2 + \lambda^2}\Big|_{q=0}^{q=2k}$$

$$= \frac{m^* Z^2 q^4 \pi}{32\hbar^3 (\pi^2 \epsilon)^2 k^3} \left[\frac{k^2}{\lambda^2} - \frac{k^2}{4k^2 + \lambda^2}\right]$$

$$+ \frac{m^* Z^2 q^4 \pi}{32\hbar^3 (\pi^2 \epsilon)^2 k^3} \left[\pm \frac{k}{2\lambda} \tan^{-1}(2k\lambda) \mp \frac{2k^2}{4k^2 + \lambda^2}\right]$$

$$= \frac{m^* Z^2 q^4 \pi}{32\hbar^3 (\pi^2 \epsilon)^2 \, (2m^* E_k)^{3/2}} \left\{ \left[\frac{2m^* E_k}{\hbar^2 \lambda^2} - \frac{1}{4 + \frac{\hbar^2 \lambda^2}{2m^* E_k}} \right] \right.$$

$$\left. \pm \frac{\sqrt{2m^* E_k}}{2\hbar\lambda} \tan^{-1} \left(2\sqrt{2m^* E_k} \hbar\lambda \right) \mp \frac{1}{2 + \frac{\hbar^2 \lambda^2}{4m^* E_k}} \right\}. \qquad (9.40)$$

The two signs correspond to momentum loss and momentum gain from the scattering process.

** Problem 9.4: Absorption coefficient of light

When light is absorbed in a semiconductor, an electron in the valence band absorbs a photon and is excited to the conduction band. Of course, the photon energy has to exceed the bandgap; otherwise, no absorption can occur via this process. We will use Fermi's Golden Rule to calculate the absorption coefficient of light. The absorption coefficient is defined by Beer's relation,

$$I(z) = I_0 e^{-\alpha z}, \qquad (9.41)$$

where $I(0)$ is the intensity of light impinging on the surface of a semiconductor and $I(z)$ is the intensity at some depth z beneath the surface. The intensity decays exponentially with distance into the semiconductor because photons from the incident beam are increasingly absorbed as the beam travels deep into the semiconductor. The coefficient of decay is the absorption coefficient α. Find an expression for it.

Solution: We seek an expression for α using Fermi's Golden Rule. We can view the absorption process as an electronic transition from the valence to the conduction band governed by Fermi's Golden Rule, whose *time* rate will be given by

$$S\left(\vec{k}_v, \vec{k}_c \right) = \frac{2\pi}{\hbar} \left| M_{\vec{k}_v, \vec{k}_c} \right|^2 \delta\left(E_{\vec{k}_c} - E_{\vec{k}_v} - \hbar\omega_l \right), \qquad (9.42)$$

where $\vec{k}_v$ is the electron's wavevector in the valence band, $\vec{k}_c$ is the wavevector in the conduction band, $\hbar\omega_l$ is the photon's energy, and $M_{\vec{k}_v, \vec{k}_c}$ is the matrix element for transition from the valence to the conduction band. The matrix element is given by the quantity

$$\left| M^-{}_{\vec{k}_v, \vec{k}_c} \right|^2 = \frac{\Gamma}{\hbar\omega_l \Omega} \delta_{\vec{k}_v, \vec{k}_c}, \qquad (9.43)$$

where Γ is a material constant that also depends on the band structure of the material, and Ω is the normalizing volume. The delta is the Krönecker delta implying momentum conservation (or the k-selection rule) in the absorption process, i.e., the wavevector of the electron in the valence band and in the conduction band are the same. That is why absorption is strong in direct gap semiconductors and ideally should be absent in indirect gap semiconductors.

Equation (9.43) gives the time rate of transition. However, the absorption coefficient deals with the spatial decay and therefore involves the spatial rate, which we can find easily by dividing the time rate by the velocity of light in the medium, which will be c/η , where c is the speed of light in vacuum and η is the refractive index of the material. Therefore, using Equations (9.41) and (9.43), we get

$$\alpha = \frac{\eta}{c} \frac{\Gamma}{\hbar\omega_1\Omega} \frac{2\pi}{\hbar} \sum_k \delta\left(E_c(k) - E_v(k) - \hbar\omega_1\right), \qquad (9.44)$$

where $E_c(k)$ and $E_v(k)$ are the energies of an electron with wavevector $\vec{k}$ in the conduction and valence bands, respectively. We will convert the summation to an integral by multiplying by the density of states, as usual. This yields

$$\alpha = \frac{\eta}{c} \frac{\Gamma}{\hbar\omega_1\Omega} \frac{2\pi}{\hbar} \frac{\Omega}{4\pi^3} \int d^3k \,\delta\left(E_c(k) - E_v(k) - \hbar\omega_1\right). \qquad (9.45)$$

The conduction and valence band energies in a direct gap semiconductor are given by

$$E_c(k) = E_c(0) + \frac{\hbar^2 k^2}{2m_c}$$

$$E_v(k) = E_v(0) - \frac{\hbar^2 k^2}{2m_v}, \qquad (9.46)$$

with $E_c(0) - E_v(0) = E_g$, where E_g is the bandgap of the semiconductor, m_c is the effective mass in the conduction band, and m_v is the effective mass in the valence band. Using the energy dispersion relations (9.46) in Equation (9.45), we obtain

$$\alpha = \frac{\eta}{c} \frac{\Gamma}{\hbar\omega_1\Omega} \frac{2\pi}{\hbar} \frac{\Omega}{4\pi^3} \int d^3k \,\delta\left(E_g + \frac{\hbar^2 k^2}{2\mu} - \hbar\omega_1\right), \qquad (9.47)$$

where μ is the "reduced mass," defined as $\frac{1}{\mu} = \frac{1}{m_c} + \frac{1}{m_v}$.

Equation (9.47) can be simplified to

$$\begin{aligned}
\alpha &= \frac{\eta}{c} \frac{\Gamma}{\hbar\omega_1\Omega} \frac{2\pi}{\hbar} \frac{\Omega}{4\pi^3} \int_0^{+\infty} 4\pi k^2 dk \,\delta\left(E_g + \frac{\hbar^2 k^2}{2\mu} - \hbar\omega_1\right) \\
&= \frac{\eta}{c} \frac{\Gamma}{\hbar\omega_1} \frac{1}{h} \int_0^{+\infty} k\,d\left(k^2\right) \delta\left(E_g + \frac{\hbar^2 k^2}{2\mu} - \hbar\omega_1\right) \\
&= \frac{\eta}{c} \frac{\Gamma}{\hbar\omega_1} \frac{1}{h} \int_0^{+\infty} k\,d\left(k^2\right) \delta\left(\frac{\hbar^2}{2\mu}\left[k^2 - \frac{2\mu}{\hbar^2}\{\hbar\omega_1 - E_g\}\right]\right) \\
&= \frac{\eta}{c} \frac{\Gamma}{\hbar\omega_1} \frac{1}{h} \frac{2\mu}{\hbar^2} \int_0^{+\infty} k\,d\left(k^2\right) \delta\left(k^2 - \frac{2\mu}{\hbar^2}\{\hbar\omega_1 - E_g\}\right) \\
&= \frac{\eta}{c} \frac{\Gamma}{\hbar\omega_1} \frac{1}{h} \left(\frac{2\mu}{\hbar^2}\right)^{3/2} \sqrt{\hbar\omega_1 - E_g}. \qquad (9.48)
\end{aligned}$$

Note that the above expression shows that no absorption can occur if the photon energy is less than the bandgap, since then the absorption coefficient becomes imaginary.

Suggested problems

- Consider an electron in a one-dimensional harmonic oscillator potential. It is described by the Hamiltonian

$$H_0 = \frac{p_z^2}{2m^*} + \frac{1}{2}m^*\omega^2 z^2.$$

  In the presence of a constant electric field E, the Hamiltonian changes to $H_0 + V(z)$, where $V(z) = -qEz$ and q is the charge of the electron.

  Using second-order perturbation theory, find the eigenvalues of the harmonic oscillator in the electric field and compare your results to the exact solution of this problem (see Problem 2.12).

- Consider the problem of a one-dimensional particle in a box in the presence of a repulsive delta scatterer modeled as $V(z) = \Gamma\delta(z - z_0)$, where $0 < z_0 < W$ and W is the width of the box. If the strength of the scatterer is assumed to be small, use second-order perturbation theory to calculate the energy of the ground state.

- Use second-order perturbation theory to calculate the energy of the ground state for the one-dimensional particle in a box in the presence of a constant force of strength F.

- Consider the one-dimensional Schrödinger equation

$$-\frac{\hbar^2}{2m^*}\frac{d^2\psi(z)}{dz^2} + [V_0(z) + gV_1(z)]\,\psi(z) = E\psi(z), \qquad (9.49)$$

  where g is a parameter characterizing the strength of the perturbing potential $V_1(z)$.

  Transform the previous equation by introducing the function $S(z)$ such that $\Psi(z) = e^{S(z)}$. Show that $S(z)$ satisfies the following non-linear second-order differential equation

$$-\frac{\hbar^2}{2m^*}\left[\frac{d^2S(z)}{dz^2} + S'(z)^2\right] + V_0(z) + gV_1(z) - E = 0. \qquad (9.50)$$

  Next, expand E and $S'(z)$ as follows:

$$E = E_0 + gE_1 + g^2E_2 + \cdots, \qquad (9.51)$$

$$S'(z) = C_0(z) + gC_1(z) + g^2C_2(z) + \cdots. \qquad (9.52)$$

  Substitute these last two expansions in Equation (9.50) and show that, by grouping terms of the same order in g, the $C_i{'}$ obey the following set of coupled differential equations:

$$C_0{'}(z) + C_0{}^2(z) = \frac{2m^*}{\hbar^2}[V_0(z) - E_0],$$

$$C_1'(z) + 2C_0(z)C_1(z) = \frac{2m^*}{\hbar^2}[V_1(z) - E_1],$$

$$C_2'(z) + 2C_0(z)C_2(z) = -\frac{2m^*}{\hbar^2}E_2 - C_1{}^2(z),$$

$$\vdots$$

$$C_n'(z) + 2C_0(z)C_n(z) = -\frac{2m^*}{\hbar^2}E_n - \sum_{k=1}^{n-1} C_k(z)C_{n-k}(z). \qquad (9.53)$$

The first of these equations is the Schrödinger equation of the unperturbed problem.

- Starting with the set of differential equations (9.53) for the C_i' derived in the previous problem, show that for bound state solutions of the Schrödinger equation, the first- (E_1) and second-order (E_2) corrections to the ground state energy E_0 are given by

$$E_1 = \int_{-\infty}^{+\infty} V_1(z)|\psi_0(z)|^2 dz, \qquad (9.54)$$

$$E_1 = -\frac{\hbar^2}{2m^*} \int_{-\infty}^{+\infty} C_1{}^2(z)|\psi_0(z)|^2 dz, \qquad (9.55)$$

where $\psi_0(z)$ is the normalized ground state wave function of the unperturbed problem.

- The Lennard-Jones potential is sometimes used to model the binding of two atoms into a molecule. Its functional form in one dimension is

$$V(z) = \frac{C_1}{z^{12}} - \frac{C_2}{z^6}.$$

This potential is plotted as a function of the coordinate z in Figure 9.2.

(1) Derive an expression for the location of the potential minimum z_0 in terms of C_1 and C_2.

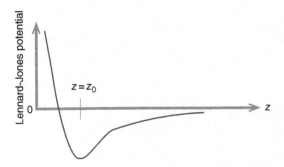

Figure 9.2: The Lennard-Jones potential plotted in one dimension.

(2) Show that in the neighborhood of z_0, the potential profile is approximately parabolic, like a simple harmonic motion oscillator's potential.

(3) Find the energy of the lowest bound state in this potential using lowest-order perturbation theory. You will need the following information [7]:

$$\int_{-\infty}^{\infty} u_0(z)^* x^3 u_1(z)\mathrm{d}z \approx 1.06 \left(m\omega/\hbar^2\right)^{-3/4},\qquad(9.56)$$

where $u_n(z)$ is the wave function of the nth excited state of the one-dimensional harmonic oscillator (see Appendix B).

References

[1] Bandyopadhyay, S. (2012) *Physics of Nanostructured Solid State Devices*, Springer, New York.

[2] Schiff, L. (1968) *Quantum Mechanics*, McGraw-Hill, New York.

[3] Shankar, R. (1980) *Principles of Quantum Mechanics*, Plenum, New York.

[4] Levi, A. F. J. (2006) *Applied Quantum Mechanics*, 2nd edition, Chapter 10, Cambridge University Press, Cambridge.

[5] Kroemer, H. (1994) *Quantum Mechanics for Engineering, Materials Science, and Applied Physics*, Chapter 15, Prentice Hall, Englewood Cliffs, NJ.

[6] Ohanian, H. C. (1990) *Principles of Quantum Mechanics*, Chapter 10, Prentice Hall, Upper Saddle River, NJ.

[7] Joshi, B. D., LaGrou, S. E., and Spooner, D. W. (1981) Integral of ζ^ν over harmonic oscillator wave functions. *Journal of Chemical Education*, 58, pp. 39–41.

Suggested Reading

- Imbo, T. and Sukhatme, U. (1984) Logarithmic perturbation expansions in nonrelativistic quantum mechanics. *American Journal of Physics* 52, pp. 140–146.

- Lapidus, I. R. (1987) Particle in a square well with a δ-function. *American Journal of Physics* 55, pp. 172–174.

- Mavromatis, H. A. (1991) The Dalgarno–Lewis summation technique: some comments and examples. *American Journal of Physics* 59, pp. 738–744.

- Maize, M. A., Antonacci, M. A., and Marsiglio, F. (2011) The static electric polarizability of a particle bound by a finite potential well. *American Journal of Physics* 79, pp. 222–225.

- Bera, N., Bhattacharyya, K., and Bhattacharjee, J. K. (2008) Perturbative and non-perturbative studies with the delta function potential. *American Journal of Physics* 76, pp. 250–257.

- Niblack, W. K. and Nigam, B. P. (1970) An alternative to perturbation theory. *American Journal of Physics* 38, pp. 101–108.

- Epstein, S. T. (1960) Application of the Rayleigh–Schrödinger perturbation theory to the delta function. *American Journal of Physics* 28, pp. 495–496.

- Epstein, S. T. (1954) Note on the perturbation theory. *American Journal of Physics* 22, pp. 613–614.

- Mavromatis, H. A. (2000) A Hamiltonian decomposition technique for improving the convergence of perturbation series. *American Journal of Physics* 68, pp. 1134–1138.

- Yao, D. and Shi, J. (2000) Projection operator approach to time-independent perturbation theory in quantum mechanics. *American Journal of Physics* 68, pp. 278–281.

- Reittu, H. J. (1995) Fermi's golden rule and Bardeen's tunneling theory. *American Journal of Physics* 63, pp. 940–944.

- Tang, A. Z., Lieber, M., and Chan, F. T. (1985) Simple example in second-order perturbation theory. *American Journal of Physics* 53, pp. 595–596.

- Adler, C. and Rose, O. (1979) Generalized bound-state perturbation theory. *American Journal of Physics* 47, pp. 822–823.

- Gottdiener, L. (1978) Derivation of the perturbation formulas for the energy in quantum mechanics. *American Journal of Physics* 46, pp. 893–895.

- Ipsen, A. C. and Splittorff, K. (2015) The van der Waals interaction in one, two, and three dimensions. *American Journal of Physics* 83, pp. 150–155.

- Holstein, B. R. (2001) The van der Waals interaction. *American Journal of Physics* 69, pp. 441–449.

- Dauphinee, T. and Marsiglio, F. (2015) Asymmetric wave functions from tiny perturbations. *American Journal of Physics* 83, pp. 861–866.

Chapter 10: Variational Approach

An important approach to finding approximate solutions to the Schrödinger equation is based on the variational principle known as the *Rayleigh–Ritz variational principle* [1]. For a specific problem, if the wave function associated with the ground or the first excited state of a Hamiltonian cannot be calculated exactly, a suitable guess for the general shape of the wave functions associated with these states can be inferred using some symmetry properties of the system and the general properties of the Schrödinger equation studied in Chapter 1. In this chapter, we first briefly describe the Rayleigh–Ritz variational procedure and apply it to the calculation of the energy of the ground and first excited states of problems for which an exact solution is known. Next, some general criteria for the existence of a bound state in a one-dimensional potential with finite range are derived.

Preliminary: The Rayleigh–Ritz variational procedure

The Rayleigh–Ritz variational principle is often used when the wave functions of the ground state and the first few excited states of a Hamiltonian cannot be found analytically, and when perturbation theory is too poor an approximation to calculate the energies of the lower eigenstates of the Hamiltonian. In that case, physical arguments are invoked to *guess* an analytical form for the eigenstates of the Hamiltonian in terms of some variational parameters. The latter are then varied until the expectation value of the Hamiltonian (or the energy eigenvalue) is minimized. More precisely, the Rayleigh–Ritz variational procedure is based on the premise that if $|\phi\rangle$ is a guess for the ground state wave function, then the actual ground state energy will satisfy the inequality

$$E_0 \leq E_{\min} = \min \left(\frac{\langle \phi | H | \phi \rangle}{\langle \phi | \phi \rangle} \right), \tag{10.1}$$

where $|\phi\rangle$ is the trial wave function that contains variational parameters. The denominator appearing on the right-hand side ensures that the trial wave function is normalized. The variational parameters are varied until the expression on the right-hand side is minimized. This yields the best guess for $|\phi\rangle$.

The variational method can also be applied to obtain the eigenfunctions of the higher excited states but, in that case, the trial function for the excited state must be selected such that it is orthogonal to the trial eigenfunctions selected to describe the wave functions of the lower-energy eigenstates. This obviously follows from the requirement that the eigenfunctions of a Hermitian operator are orthonormal.

A more thorough discussion of the Rayleigh–Ritz variational principle can be found in some quantum mechanics textbooks [1]. In this chapter, we illustrate this principle with a few simple problems, including an estimation of the ground state energy in a triangular well, which approximates the potential energy profile in a high electron mobility transistor (HEMT) close to the heterointerface (see Appendix F).

Problem Solving in Quantum Mechanics: From Basics to Real-World Applications for Materials Scientists, Applied Physicists, and Devices Engineers, First Edition.
Marc Cahay and Supriyo Bandyopadhyay.

We also discuss sufficient conditions for a one-dimensional confined potential to possess a bound state.

---

* Problem 10.1: Ground state of a particle in an infinite square well

The eigenfunctions and corresponding eigenvalues of the 1D particle in a box of width W can be found exactly, as shown in Problem 3.5. The following problem illustrates how an approximate wave function for the ground state leads to an average energy that is larger than the exact value $E_1 = \frac{\hbar^2 \pi^2}{2m^* W^2}$ derived in Problem 3.5.

Suppose that the ground state of a particle in a box is approximated by the following wave function:

$$\psi(z) = N(z^2 - Wz). \tag{10.2}$$

(a) Show that the normalization coefficient is given by $N = \sqrt{\frac{30}{L^5}}$.

(b) Calculate the average kinetic energy in the state (10.2) and determine how much larger it is than its exact value.

(c) Using the wave function (10.2), determine the uncertainty in the position Δz and momentum Δp_z of the particle. Show that the product satisfies the Heisenberg uncertainty principle. Show that this product is larger that its value derived for the true ground state (see Equation (4.47)).

Solution:
(a) Normalization requires that

$$\int_0^W \psi^*(z)\psi(z)\mathrm{d}z = 1. \tag{10.3}$$

Expanding the integral leads to

$$N^2 \int_0^W (z^2 - Wz)^2 \mathrm{d}z = N^2 \left(\frac{1}{5}L^5 - \frac{1}{2}L^5 + \frac{1}{3}L^5 \right) = 1. \tag{10.4}$$

Solving for N, we get

$$N = \sqrt{\frac{30}{L^5}}. \tag{10.5}$$

(b) The expectation value of the kinetic energy in the state (10.2) is given by

$$\bar{E} = \int_0^W \psi^* \left(-\frac{\hbar^2}{2m^*} \frac{\mathrm{d}^2\psi(z)}{\mathrm{d}z^2} \right) \mathrm{d}z = -\frac{\hbar^2 N^2}{m^*} \int_0^W (z^2 - Wz)\mathrm{d}z$$

$$= -\frac{\hbar^2 N^2}{m} \left(-\frac{1}{6}W^3 \right) = \frac{\hbar^2 (\sqrt{\frac{30}{W^5}})^2 W^3}{6m^*} = \frac{5\hbar^2}{m^* W^2}. \tag{10.6}$$

This last result is 1.03% larger than the exact energy of the ground state of the particle in a box, $E_1 = \frac{\pi^2 \hbar^2}{2m^* W^2}$ (see Problem 3.5).

(c) The expectation value of the position of the particle is given by

$$
\begin{aligned}
\bar{z} &= \int_0^W z|\psi(z)|^2 \mathrm{d}z = N^2 \int_0^W z\left(z^2 - Wz\right)^2 \mathrm{d}z \\
&= N^2 \left(\frac{1}{6}W^6 - \frac{2}{5}W^6 + \frac{1}{4}W^6\right) = \frac{1}{2}L.
\end{aligned}
\tag{10.7}
$$

This result makes sense. The expectation value of the particle's position is at the center of the well since the probability distribution given by (10.2) is symmetric with respect to the center of the well.

The uncertainty Δz in the position of the particle is given by

$$
\Delta z = \sqrt{\bar{z^2} - (\bar{z})^2}.
\tag{10.8}
$$

We first calculate

$$
\bar{z^2} = \int_0^W z^2|\psi(z)|^2 \mathrm{d}z = N^2 \int_0^W \left(z^6 - 2z^5 W + z^4 W^2\right) \mathrm{d}z.
\tag{10.9}
$$

Carrying out the integration leads to

$$
\bar{z^2} = N^2 \left(\frac{W^7}{105}\right) = \left(\sqrt{\frac{30}{W^5}}\right)^2 \frac{W^7}{105} = \frac{2W^2}{7}.
\tag{10.10}
$$

Hence the standard deviation of the position of the particle in the wave function (10.2) is given by

$$
\Delta z = \sqrt{\bar{z^2} - (\bar{z})^2} = \sqrt{\frac{2W^2}{7} - \left(\frac{W}{2}\right)^2} = 0.1898W.
\tag{10.11}
$$

Likewise, the uncertainty Δp in the momentum is given by

$$
\Delta p_z = \sqrt{\bar{p_z^2} - (\bar{p_z})^2}.
\tag{10.12}
$$

The average value of the momentum, $\bar{p_z}$, is given by

$$
\bar{p_z} = \int_0^W \psi(z)^* \left(-\frac{\hbar}{i}\frac{\mathrm{d}\psi(z)}{\mathrm{d}z}\right) \mathrm{d}z,
\tag{10.13}
$$

which, upon substituting the wave function (10.2) is found to be

$$
\bar{p_z} = \frac{\hbar N^2}{i} \int_0^W (z^2 - Wz)(2z - W)\mathrm{d}z = \frac{\hbar N^2}{i}\left(\frac{1}{2}W^4 - W^4 + \frac{1}{2}W^4\right) = 0.
\tag{10.14}
$$

This is to be expected, since the integral in Equation (10.14) is antisymmetric with respect to the center of the box. Next, we calculate $\bar{p^2}$ starting with Equation (10.6), which leads to

$$\bar{p_z^2} = \frac{10\hbar^2}{W^2}. \tag{10.15}$$

Hence, the standard deviation of the momentum of the particle in the state (10.2) is given by

$$\Delta p_z = \sqrt{\bar{p_z^2} - (\bar{p_z})^2} = \sqrt{\frac{10\hbar^2}{W^2} - (0)^2} = 3.16\frac{\hbar}{W}. \tag{10.16}$$

Using the previous results, the product $\Delta z \Delta p_z$ is therefore given by

$$\Delta z \Delta p_z = 0.1898W \times 3.16\frac{\hbar}{W} = 1.2\frac{\hbar}{2} \geq \frac{\hbar}{2}, \tag{10.17}$$

in agreement with the Heisenberg uncertainty principle for position and momentum.

---

* Problem 10.2: Excited state in an infinite square well

This problem is an application of the variational principle to approximate the wave function of the first excited state of the particle in a box. Suppose that the wave function associated with the first excited state of a particle in a box of width w is approximated as

$$\psi(z) = Nz\left(z - \frac{w}{2}\right)(z - w). \tag{10.18}$$

(a) Explain why this last expression is a good approximation for the wave function associated with the first excited state.

(b) Determine the normalization constant N.

(c) Calculate the average kinetic energy in the state (10.18) and determine how much larger it is than its exact value.

(d) Using the wave function (10.18), determine the uncertainty in the position Δz and momentum Δp_z of the particle. Show that the product satisfies the Heisenberg uncertainty principle. Show that this product is larger that its value derived for the true first excited state (see Equation (4.47)).

Solution:
(a) The trial wave function (10.18) is equal to zero at the edges of the box, as required by the principles of quantum mechanics (see Appendix A), since the box has infinite walls on both sides and therefore the wave function must be zero outside the box and equal to zero at the edges of the box, as required by the continuity of the wave function. Furthermore, the trial wave function (10.18) has only one additional node located at the center of the box, i.e., for $z = w/2$, and is odd with respect to the center of the box. This property is in agreement with the general properties of the solutions of the one-dimensional time-independent Schrödinger equation derived

in Problem 1.3. Finally, we note that the trial wave function associated with the first excited state is orthogonal to the trial wave function associated with the ground state considered in the previous problem.

(b) The normalization constant N is found by enforcing that

$$\int_0^w \psi^* \psi dz = 1. \tag{10.19}$$

Using the trial wave function (10.18), we get

$$\int_0^w \psi(z)^2 dz = N^2 \int_0^w \left[z \left(z - \frac{w}{2} \right) (z - w) \right]^2 dz = 1. \tag{10.20}$$

Carrying out the integration leads to

$$N^2 \left(\frac{w^7}{12} - \frac{3w^7}{8} + \frac{13w^7}{20} - \frac{w^7}{2} + \frac{w^7}{7} \right) = 1. \tag{10.21}$$

Therefore the normalization coefficient is given by

$$N = \sqrt{\frac{840}{w^7}}. \tag{10.22}$$

(c) Next, we determine the expectation value and standard deviation of the position of the particle:

$$\bar{z} = \int_0^w z |\psi(z)|^2 dz. \tag{10.23}$$

This leads to

$$\bar{z} = N^2 \int_0^w z \left[z \left(z - \frac{w}{2} \right) (z - w) \right]^2 dz. \tag{10.24}$$

Expanding the integrand, we get

$$\bar{z} = N^2 \int_0^w z \left(\frac{w^4 z^2}{4} - \frac{3w^3 z^3}{2} + \frac{13w^2 z^4}{4} - 3w z^5 + z^6 \right) dz. \tag{10.25}$$

Performing the integrations leads to

$$\bar{z} = N^2 \left(\frac{w^8}{16} - \frac{3w^8}{10} + \frac{13w^8}{24} - \frac{3w^8}{7} + \frac{w^8}{8} \right). \tag{10.26}$$

Simplifying and using the expression for the normalization coefficient (10.22), we obtain

$$\bar{z} = \left(\frac{840}{w^7} \right) \left(\frac{w^8}{1680} \right) = \frac{w}{2}. \tag{10.27}$$

This result is not too surprising since the probability density associated with the trial wave function (10.18) is symmetric with respect to the center of the well and the average location of the particle should therefore be at the center of the well.

(d) Next, we calculate the uncertainty Δz in the position of the particle in the state (10.18). We first calculate

$$\overline{z^2} = \int_0^w z^2 |\psi(z)|^2 dz. \tag{10.28}$$

With the explicit expression of the trial wave function (10.18), we get

$$\overline{z^2} = N^2 \int_0^w z^2 \left[z \left(z - \frac{w}{2} \right) (z - W) \right]^2 dz. \tag{10.29}$$

Expanding the integrand leads to

$$\overline{z^2} = N^2 \int_0^w z^2 \left(\frac{w^4 z^2}{4} - \frac{3w^3 z^3}{2} + \frac{13w^2 z^4}{4} - 3wz^5 + z^6 \right) dz. \tag{10.30}$$

Performing the integration of the different terms gives

$$\overline{z^2} = N^2 \left(\frac{w^9}{20} - \frac{3w^9}{4} + \frac{13w^9}{28} - \frac{3w^9}{8} + \frac{w^9}{9} \right). \tag{10.31}$$

Simplifying and using the expression of the normalization coefficient (10.22) that we found, we get

$$\overline{z^2} = N^2 \frac{w^9}{2520} = \frac{840}{w^7} \times \frac{w^9}{2520} = \frac{w^2}{3}. \tag{10.32}$$

Regrouping the above results, the standard deviation of the position in the trial wave function (10.18) is given by

$$\Delta z = \sqrt{\overline{z^2} - (\bar{z})^2} = \sqrt{\frac{w^2}{3} - \left(\frac{w}{2} \right)^2} = \frac{w}{\sqrt{12}}. \tag{10.33}$$

Next, we calculate the average value of the momentum in state (10.18):

$$\bar{p}_z = \int_0^w \psi^*(z) \left(-\frac{\hbar}{i} \frac{d\psi(z)}{dz} \right) dz. \tag{10.34}$$

Using the fact that

$$\frac{d\psi(z)}{dz} = \frac{1}{2} N(w^2 - 6wz + 6z^2), \tag{10.35}$$

we get

$$\bar{p}_z = \frac{\hbar N^2}{2i} \int_0^W \left[z \left(z - \frac{w}{2} \right) (z - w) \right] (w^2 - 6wz + 6z^2) dz. \tag{10.36}$$

Carrying out the integrations leads to

$$\bar{p}_z = \frac{\hbar N^2}{2i} \left(\frac{w^6}{4} - \frac{3w^6}{2} + \frac{13w^6}{4} - 3w^6 + w^6 \right) = 0. \tag{10.37}$$

This is to be expected since, even in the first excited state, the particle bounces back and forth between the walls of the box (suffering multiple reflections) and hence the average momentum must be zero. The particle is a "standing" wave with no translational motion. Mathematically, this is linked to the fact that the integrand in Equation (10.36) is odd with respect to the center of the box.

To calculate the standard deviation of the momentum in the trial wave function (10.36), we first calculate

$$\overline{p_z^2} = \int_0^w \psi^* \left(-\hbar^2 \frac{d^2\psi(z)}{dz^2} \right) dz. \tag{10.38}$$

Since

$$\frac{d^2\psi(z)}{dz^2} = -3Nw + 6Nz = N(6z - 3w), \tag{10.39}$$

we get

$$\overline{p_z^2} = -\hbar^2 N^2 \int_0^w z \left(z - \frac{w}{2} \right) (z - w)(6z - 3w)dz. \tag{10.40}$$

Carrying out the integration leads to

$$\overline{p^2} = -\hbar^2 N^2 \int_0^w \left(\frac{-3w^3 z}{2} + \frac{15w^2 z^2}{2} - 12wz^3 + 6z^4 \right) dz. \tag{10.41}$$

This leads to the final result:

$$\overline{p_z^2} = \frac{42\hbar^2}{w^2}. \tag{10.42}$$

Hence,

$$\Delta p_z = \sqrt{\overline{p_z^2} - (\bar{p}_z)^2} = \sqrt{\frac{42}{w^2}\hbar^2 - (0)^2} = \sqrt{42}\frac{\hbar}{w}. \tag{10.43}$$

Using the results above, we found that $\Delta z \Delta p_z$ is equal to $3.74\frac{\hbar}{2}$, which is 65% larger than the same product calculated for the exact eigenfunction associated with the first excited state of the particle in a box (see Problems 3.5 and 4.3).

---

* Problem 10.3: Ground state of the harmonic oscillator

Suppose we start with the following trial wave function for the harmonic oscillator ground state:

$$\phi(z) = \frac{N}{z^2 + a^2}, \tag{10.44}$$

where a is a variational parameter and N is the normalization coefficient.

This choice seems reasonable since the wave function is peaked at $z = 0$ and decays for large values of z without having any node. Moreover, this form of $\phi(z)$ allows an exact calculation of the functional E_{min} appearing in the inequality (10.1).

Apply the variational principle and find E_{min}. How does it compare with the exact value of the energy of the ground state for the one-dimensional harmonic oscillator, which is $\frac{\hbar\omega}{2}$, where ω is the angular frequency appearing in the expression for the potential energy $V(z) = \frac{1}{2}m^*\omega^2 z^2$? (See Appendix B).

Solution: The normalization is determined by enforcing that $\int_{-\infty}^{\infty} \psi^* \psi dz = 1$. This leads to

$$N = \sqrt{\frac{2a^3}{\pi}}. \tag{10.45}$$

Similarly, starting with the expression of the Hamiltonian of the one-dimensional harmonic oscillator (see Appendix B), the energy $E_{\min}$ on the right-hand side of inequality (10.1) is found to be

$$E_{\min} = \frac{h^2 + 2a^4 m^{*2} \omega^2}{4m^* a^2}. \tag{10.46}$$

The latter reaches a minimum for

$$a_{\min} = \frac{1}{2^{1/4}} \sqrt{\frac{h}{m^* \omega}}. \tag{10.47}$$

The corresponding minimum energy $E_{\min}$ is

$$E_{\min}(a_{\min}) = \sqrt{2} \frac{\hbar \omega}{2}, \tag{10.48}$$

which is $\sqrt{2}$ times larger than the exact result $\hbar \omega / 2$.

** Problem 10.4: Variational method in a triangular well

This problem describes the use of the variational technique to model the two lowest eigenstates in the two-dimensional electron gas formed at the interface of a HEMT (see Appendix F). Close to the heterointerface, the conduction band in the semiconductor substrate is modeled as a triangular potential well (see Figure 3.8). The trial wave functions for the ground and first excited states are

$$\xi_0(z) = \left(\frac{b_0^3}{2} \right)^{\frac{1}{2}} z e^{\frac{-b_0 z}{2}} \tag{10.49}$$

for the ground state and

$$\xi_1(z) = \left(\frac{3b_1^5}{2[b_0^2 + b_1^2 - b_0 b_1]} \right)^{\frac{1}{2}} z \left(1 - \frac{b_0 + b_1}{6} z \right) e^{-\frac{b_1 z}{2}}, \tag{10.50}$$

for the first excited state, where b_0 and b_1 are two variational parameters, and $z = 0$ at the heterointerface.

Notice that both $\xi_0(z)$ and $\xi_1(z)$ are zero at $z = 0$, as they should be since neither wave function can penetrate the barrier at the heterointerface if it is very high. Since the potential energy increases linearly with distance into the substrate, the choice of exponentially decaying trial wave functions seems appropriate.

Show that $\xi_0(z)$, $\xi_1(z)$ are an appropriate set of trial wave functions for the ground state and first excited states, i.e., $\xi_0(z)$ and $\xi_1(z)$ are both normalized, so that $\int_0^{\infty} |\xi_i(z)|^2 dz = 1$, and ξ_0 and ξ_1 are orthogonal, meaning $\int_0^{\infty} \xi_0(z) \xi_1(z) dz = 0$.

Solution: Performing the integration over the probability in the ground state (magnitude of the square of the wave function), we get

$$\int_0^{+\infty} |\xi_0(x)|^2 dz = \int_0^{+\infty} \left(\frac{b_0^3}{2}\right) z^2 e^{-b_0 z} dz = \frac{b_0^3}{2} \frac{2}{b_0^3} = 1. \tag{10.51}$$

So ξ_0 is indeed normalized. Similarly,

$$\int_0^{+\infty} |\xi_1(z)|^2 dz = \frac{3b_1^5}{2(b_0^2 + b_1^2 - b_0 b_1)} [I_2 - I_3 + I_4], \tag{10.52}$$

where

$$I_2 = \int_0^{+\infty} z^2 e^{-b_1 z} dz, \tag{10.53}$$

$$I_3 = \frac{(b_0 + b_1)}{3} \int_0^{+\infty} z^3 e^{-b_1 z} dz, \tag{10.54}$$

$$I_4 = \left(\frac{b_0 + b_1}{6}\right)^2 \int_0^{+\infty} z^4 e^{-b_1 z} dz. \tag{10.55}$$

We find:

$$I_2 = \frac{2}{b_1^3}, \tag{10.56}$$

$$I_3 = \frac{2(b_0 + b_1)}{b_1^4}, \tag{10.57}$$

$$I_4 = \frac{2}{3} \frac{(b_0 + b_1)^2}{b_1^5}. \tag{10.58}$$

Hence,

$$\int_0^{+\infty} |\xi_1(z)|^2 dz = \frac{3}{2} \frac{b_1^5}{[b_0^2 + b_1^2 - b_0 b_1]} \left[\frac{2}{b_1^3} - \frac{2(b_0 + b_1)}{b_1^4} + \frac{2(b_0 + b_1)^2}{3b_1^5}\right]. \tag{10.59}$$

Simplifying,

$$\int_0^{+\infty} |\xi_1(z)|^2 dz = \frac{3}{2} \frac{1}{[b_0^2 + b_1^2 - b_0 b_1]} \frac{2}{3} [b_0^2 + b_1^2 - b_0 b_1] = 1. \tag{10.60}$$

So $\xi_1(z)$ is also normalized.

To prove the orthogonality of ξ_0 and ξ_1, we must show that the following integral is equal to zero:

$$\int_0^{+\infty} \xi_0^*(z)\xi_1(z) dz = \left(\frac{b_0^3}{2}\right)^{\frac{1}{2}} \left(\frac{3b_1^5}{2[b_0^2 + b_1^2 - b_0 b_1]}\right)^{\frac{1}{2}} (J_2 - J_3), \tag{10.61}$$

where

$$J_2 = \int_0^{+\infty} z^2 e^{-\left(\frac{b_0 + b_1}{2}\right) z} dz, \tag{10.62}$$

$$J_3 = \left(\frac{b_0 + b_1}{6}\right) \int_0^{+\infty} z^3 e^{-\left(\frac{b_0 + b_1}{2}\right) z} dz. \tag{10.63}$$

We find

$$J_2 = J_3 = \frac{16}{(b_0 + b_1)^3}. \tag{10.64}$$

Hence,

$$\int_0^{+\infty} \xi_0(z)\xi_1(z)\mathrm{d}z = 0, \tag{10.65}$$

and ξ_0, ξ_1 are indeed orthogonal.

** Problem 10.5: General criterion for a bound state using the variational principle

According to the variation principle, the true bound state energy E_0 in a one-dimensional potential $V(z)$ is such that the following inequality must be satisfied:

$$E_0 \leq E(\alpha) = \int \phi_\alpha^*(z)\left[-\frac{\hbar^2}{2m^*}\frac{\mathrm{d}^2}{\mathrm{d}z^2} + V(z)\right]\phi_\alpha(z)\mathrm{d}z, \tag{10.66}$$

where $\phi_\alpha(z)$ is a trial function describing the ground state and α is the variational parameter.

By using the normalized Gaussian trial function

$$\phi_\alpha(z) = \left(\frac{2\alpha}{\pi}\right)^{1/4} e^{-\alpha z^2}, \tag{10.67}$$

show that the right-hand side of the inequality (10.66) can be made negative for some positive value of α. Negative E_0 corresponds to a "bound state."

Solution: We calculate the average value of the kinetic energy operator by first calculating the first and second derivatives of the trial function (10.67). This leads to

$$\frac{\mathrm{d}\phi_\alpha(z)}{\mathrm{d}z} = -2\alpha z \left(\frac{2\alpha}{\pi}\right)^{1/4} e^{-\alpha z^2}, \tag{10.68}$$

$$\frac{\mathrm{d}^2\phi_\alpha}{\mathrm{d}z^2} = -2\alpha \left(\frac{2\alpha}{\pi}\right)^{1/4}\left(e^{-\alpha z^2} - 2\alpha z^2 e^{-\alpha z^2}\right). \tag{10.69}$$

The average value of the kinetic energy operator is given by

$$\int \phi^*\left(-\frac{\hbar^2}{2m^*}\frac{\mathrm{d}^2\phi^2}{\mathrm{d}z^2}\right)\mathrm{d}z$$

$$= \frac{\hbar^2}{m^*}\alpha\left(\frac{2\alpha}{\pi}\right)^{1/2}\left[\int_{-\infty}^{+\infty}\mathrm{d}z e^{-2\alpha z^2} - 2\alpha\int_{-\infty}^{+\infty}\mathrm{d}z z^2 e^{-\alpha z^2}\right]. \tag{10.70}$$

Using the integrals

$$\int_0^{+\infty} dz e^{-2\alpha z^2} = \frac{\sqrt{\pi}}{2\sqrt{2\alpha}}, \tag{10.71}$$

$$\int_0^{+\infty} dz z^2 e^{-2\alpha x^2} = \frac{\sqrt{\pi}}{4(2\alpha)^{3/2}}, \tag{10.72}$$

the average value $E(\alpha)$ on the right-hand side of the inequality (10.66) is given by

$$E(\alpha) = \frac{\alpha \hbar^2}{2m^*} + \sqrt{\frac{2\alpha}{\pi}} I_1, \tag{10.73}$$

where I_1 is defined by

$$I_1 = \int_{-\infty}^{+\infty} e^{-2\alpha z^2} V(z) dz. \tag{10.74}$$

The energy $E(\alpha)$ is found to be minimum when

$$\frac{dE}{d\alpha} = \frac{\hbar^2}{2m^*} + \frac{1}{\sqrt{2\alpha\pi}} I_1 - \sqrt{\frac{2\alpha}{\pi}} \int_{-\infty}^{+\infty} 2z^2 e^{-2\alpha z^2} V(z) dz = 0. \tag{10.75}$$

This last equation is equivalent to

$$0 = \frac{\hbar^2}{2m^*} + \frac{1}{\sqrt{2\alpha\pi}} I_1 - 2\sqrt{\frac{2\alpha}{\pi}} I_2, \tag{10.76}$$

where we have introduced the shorthand notation

$$I_2 = \int_{-\infty}^{+\infty} z^2 e^{-2\alpha z^2} V(z) dz. \tag{10.77}$$

Using this definition, we rewrite Equation (10.76) as

$$0 = \frac{1}{\alpha} \left[\frac{\hbar^2}{2m^*} \alpha + \sqrt{\frac{\alpha}{2\pi}} I_1 \right] - 2\sqrt{\frac{2\alpha}{\pi}} I_2. \tag{10.78}$$

This last equation can be rewritten in terms of the energy $E(\alpha)$ defined in Equation (10.73) as

$$0 = \frac{1}{\alpha} \left[E(\alpha) - \sqrt{\frac{2\alpha}{\pi}} I_1 + \sqrt{\frac{\alpha}{2\pi}} I_1 \right] - 2\sqrt{\frac{2\alpha}{\pi}} I_2. \tag{10.79}$$

Solving for $E(\alpha)$, we get

$$E(\alpha) = \sqrt{\frac{\alpha}{2\pi}} [I_1 + 4\alpha I_2]. \tag{10.80}$$

Since α must be positive for the trial function (10.67) to be normalized, $E(\alpha) < 0$ if both I_1 and I_2 are negative. Since both $e^{-\alpha z^2}$ and $z^2 e^{-\alpha z^2}$ are positive for all z, both I_1 and I_2, and hence $E(\alpha)$, are negative if the following condition is satisfied:

$$\int_{-\infty}^{+\infty} V(z) dz < 0. \tag{10.81}$$

This is a *sufficient* condition for the existence of a bound state in a one-dimensional potential with finite range. It is not a necessary condition since, for instance, the one-dimensional harmonic oscillator is characterized by a potential energy that does not satisfy the inequality (10.81). In fact, it is well known that the one-dimensional harmonic oscillator has an infinite number of bound states (see Appendix B).

**** Problem 10.6: Improvement of the criterion for existence of a bound state in one dimension

This problem considers an improvement of the criterion (10.81) for the existence of a bound state in a one-dimensional potential $V(z)$ with finite extent (i.e., assuming $V(z) = 0$ outside some interval $[-a, +a]$). Starting with a trial function of the form

$$\phi_\lambda(z) = P(z) + \lambda V(z), \tag{10.82}$$

where λ is a real parameter and $P(z)$ is the function defined as

$$P(z) = 1 \text{ for } z \in [-a, +a], \tag{10.83}$$

$$P(z) = e^{-\frac{(z-a)^2}{L^2}} \text{ for } z > a, \tag{10.84}$$

$$P(z) = e^{-\frac{(z+a)^2}{L^2}} \text{ for } z < -a, \tag{10.85}$$

show that $V(z)$ will have at least one bound state if $\int_{-a}^{+a} V(z)\mathrm{d}z \leq 0$ is satisfied. This is a weaker condition compared to the one derived in the previous problem, since even potential whose average value $\int_{-a}^{+a} V(z)\mathrm{d}z$ is equal to zero will have at least one bound state.

Solution: The following proof is an adaptation of the one given in Ref. [2], which is based on a trial function of the form (10.82) but with the function $P(z)$ approximated by the tent function

$$P(z) = 1 \text{ for } z \in [-a, a], \tag{10.86}$$
$$P(z) = 1 - (|z| - a)/L \text{ for } a < |z| < L + a, \tag{10.87}$$
$$P(z) = 0 \text{ for } L + a < |z| < \infty. \tag{10.88}$$

This choice of $P(z)$ makes the derivative of the wave function $\phi_\lambda(z)$ discontinuous at $z = \pm a$, which violates one of the principles of quantum mechanics (see Appendix A). A discontinuity at $z = \pm a$ is acceptable only if there is a delta function potential existing at these locations, as shown in Problem 3.1.

To prove that the Hamiltonian possesses a bound state, we must show that the average value of this Hamiltonian in the trial wave function (10.82) can be made negative. This average value is given by

$$\langle H \rangle = \frac{\int \phi_\lambda H \phi_\lambda \mathrm{d}z}{\int \phi_\lambda \phi_\lambda \mathrm{d}z}, \tag{10.89}$$

where the integration is from $-\infty$ to $+\infty$. The denominator in the last expression is always positive, so we must show that the numerator can be made negative. Since the potential $V(z)$ is assumed to be of finite range (or to decay rapidly enough to zero at $\pm\infty$), we can integrate the numerator by parts and use the fact that both the wave function and its derivative decay to zero at $\pm\infty$ to rewrite the numerator as follows (using the fact that $\phi_\lambda(z)$ is real):

$$\text{num} = \int_{-\infty}^{\infty} \left[\frac{\hbar^2}{2m^*} \left(\frac{d\phi_\lambda}{dz} \right)^2 + V\phi_\lambda^2 \right] dz, \tag{10.90}$$

where we carried out the integration by parts and used the fact that ϕ_λ vanishes at $\pm\infty$.

Substituting the (normalizable) trial function (10.82) in this last expression, we get

$$\text{num} = \int_{\infty}^{+\infty} \frac{\hbar^2}{2m^*} [P' + \lambda V']^2 + V(P + \lambda V)^2] dz. \tag{10.91}$$

Taking into account the explicit expression of the function $P(z)$, we can rewrite Equation (10.90) as

$$\text{num} = A\lambda^2 + B\lambda + C + \frac{\hbar^2}{2m^*} \sqrt{\frac{\pi}{2}} \frac{1}{L}, \tag{10.92}$$

where the following quantities were introduced:

$$A = \int_{-a}^{+a} dz \left[V^3(z) + \frac{\hbar^2}{2m^*} V'^2(z) \right], \tag{10.93}$$

$$B = 2 \int_{-a}^{+a} dz V^2(z), \tag{10.94}$$

$$C = \int_{-a}^{+a} dz V(z). \tag{10.95}$$

Since A, B, and C do not depend on L, for large enough L, the condition for the existence of a bound state requires the following inequality to be satisfied:

$$A\lambda^2 + B\lambda + C < 0. \tag{10.96}$$

The problem is to find a λ which satisfies this inequality. Notice that $\lambda = 0$ does if C is strictly negative. This corresponds to the criterion for the existence of a bound state found in the previous problem.

The left-hand side of the inequality (10.96) is a polynomial of second degree in λ. We first consider the case where $A < 0$. The A versus λ curve corresponds to a parabola (Equation (10.96)) oriented toward the negative axis and therefore the inequality (10.96) is satisfied for some λ of sufficiently large magnitude. This holds true even if $C = \int_{-a}^{+a} V(z) dz \leq 0$.

If $A = 0$, $B\lambda + C < 0$ can be realized if $\lambda < -\frac{C}{B}$, which is always possible even if $C = \int_{-a}^{+a} V(z)dz \leq 0$.

Finally, we consider the case $A > 0$. In this case, the polynomial of second degree on the left-hand side of (10.96) can only be made negative between the two λ roots of the polynomial. These two roots are real if the following condition is satisfied:

$$B^2 - 4AC < 0. \tag{10.97}$$

This amounts to $C < \frac{B^2}{4A}$. Using the definitions of A, B, and C above, this last inequality can be rewritten as

$$\int_{-a}^{+a} V(z)dz < \frac{\left(\int_{-a}^{+a} dz V^2(z)\right)^2}{\int_{-a}^{+a} dz \left[V^3(z) + \frac{\hbar^2}{2m^*}V'^2(z)\right]}. \tag{10.98}$$

Since the right-hand side of this last inequality is always strictly positive, this last inequality will be satisfied for any potential $V(z)$ obeying the condition

$$\int_{-a}^{+a} V(z)dz \leq 0. \tag{10.99}$$

Regrouping the results above, we see that the inequality (10.98) is satisfied for any value of the potential $V(z)$ that satisfies the condition (10.99). This is a weaker condition for the existence of a bound state than the one derived in Problem 10.5. Indeed, even a potential whose average value $\int_{-a}^{+a} V(z)dz$ is equal to zero will have at least one bound state.

Suggested problems

- Use Matlab to plot the probability densities associated with the trial wave functions for the ground and first excited states given in Problems 10.1 and 10.2, respectively. For comparison, also plot the probability densities associated with the exact eigenfunctions for the ground and first excited states of the particle in a box (see Problem 3.5). Assume a well width of 100 Å.

- Armed with the insight gained in Problems 10.1 and 10.2, derive an analytical expression for a trial wave function as a polynomial of *fourth* degree in z suitable to describe the second excited state of the particle in a box. Make sure that the trial function is normalized. Then calculate the expectation value of the kinetic energy and compare it with the true eigenenergy associated with the second excited state of the particle in a box (see Problem 3.5).

- Using the results of the previous problem, calculate the standard deviation of the position and the momentum starting with the trial wave function you derived for the second excited state of the particle in a box and compare the value of that product with its exact value (i.e., its value for the exact eigenstate associated with the second excited state), which was derived in Problem 4.3.

- Prove that the normalization coefficient of the trial wave function (10.44) for the ground state of the harmonic oscillator is given by Equation (10.45). Calculate the average value of the total energy of the one-dimensional harmonic oscillator in its ground state when its wave function is approximated by the trial function (10.44).

- Starting with the trial wave function (10.49), calculate an upper bound for the energy of the ground state in a triangular well in which the potential is assumed to be equal to infinity for $z < 0$ and for which the potential energy is given by

$$E_c(z) = \beta z \text{ for } z > 0.$$

This problem can actually be solved exactly with the use of Airy functions, as shown in Problem 3.11. The exact result for the ground state energy is given by

$$E_0 = 1.857 \left(\frac{\beta^2 \hbar^2}{m^*} \right)^{1/3}.$$

Show that the expectation value $E_{\min}$ for the ground state energy in the trial wave function (10.49) is minimum for

$$b_{0,\min} = \left(\frac{12\beta m^*}{\hbar^2} \right)^{1/3},$$

leading to an upper bound on the ground state energy of

$$E_{\min}(b_{0,\min}) = 1.966 \left(\frac{\beta^2 \hbar^2}{m^*} \right)^{1/3},$$

which is only about 6% larger than the exact result.

- For the trial wave functions of the ground and first excited states of a particle in a triangular potential, calculate the average values and standard deviations of the position and momentum in terms of the parameters b_0 and b_1. Plot the product $\Delta z \Delta p_z$ as a function of b_0 and b_1 and calculate the value of that product for the values of b_0 and b_1 which minimize the total energy in the ground and first excited states.

- By using a trial function

$$\phi_\alpha(z) = Nze^{-\alpha z^2},$$

show that a one-dimensional potential of finite range will have a first excited bound state if the following (sufficient) condition is satisfied: $\int_{-\infty}^{+\infty} \left[V(z) + \frac{zV'(z)}{2} \right] dz < 0.$

- By using a trial function

$$\phi_\alpha(z) = Nze^{-\alpha z^2},$$

show that a one-dimensional potential of finite range will have a ground state if the following (sufficient) condition is satisfied: $\int_0^{+\infty} \left[V(z) + \frac{zV'(z)}{2} \right] dz < 0.$

Using this result, show that the potential $V(z) = -V_0 z e^{-\beta z}$ has a ground state for all values of V_0 and β.

References

[1] Kroemer, H. (1994) *Quantum Mechanics for Engineering, Materials Science, and Applied Physics*, Chapter 12, Prentice Hall, Englewood Cliffs, NJ.

[2] Brownstein, K. R. (2000) Criterion for existence of a bound state in one dimension. *American Journal of Physics*, 68, p. 160.

Suggested Reading

- Srivastava, M. K. and Bhaduri, R. K. (1977) Variational method for two-electron atoms. *American Journal of Physics* 45, pp. 462–463.

- Nandi, S. (2010) The quantum Gaussian well. *American Journal of Physics* 78, pp. 1341–1344.

- Mur-Petit, J., Polls, A., and Mazzanti, F. (2002) The variational principle and simple properties of the ground state wave function. *American Journal of Physics* 70, pp. 808–810.

- Bastard, G., Mendez, E. E., Chang, L. L., and Esaki, L. (1983) Variational calculations of a quantum well in an electric field. *Physical Review B* 28, pp. 3241–3245.

- Coutinho, F. A. B. (1996) Bound states in two dimensions and the variational principle. *American Journal of Physics* 64, p. 818.

- Buell, W. F. and Shadwick, B. A. A. (1995) Potential and bound states. *American Journal of Physics* 63, pp. 256–258.

- Kocher, C. A. (1977) Criteria for bound-state solutions in quantum mechanics. *American Journal of Physics* 45, pp. 71–74.

- Yang, Y. and de Llano, M. (1989) Simple variational proof that any two-dimensional well supports at least one bound state. *American Journal of Physics* 57, pp. 85–86.

- Simon, B. (1976) The bound state of coupled Schrödinger operators in one and two dimensions. *Annals of Physics* 97, pp. 279–288.

- Hushwater, V. (1994) Application of the variational principle to perturbation theory. *American Journal of Physics* 62, pp. 379–380.

- Lee, J. (1986) The upper and lower bounds of the ground state energy using the variational principle. *American Journal of Physics* 55, pp. 1039–1040.

Chapter 11: Electron in a Magnetic Field

Many important phenomena in condensed matter physics, such as the quantum Hall effect, require an understanding of the quantum mechanical behavior of electrons in a magnetic field. In this chapter, we introduce the concept of a vector potential and gauge to incorporate magnetic fields in the Hamiltonian of an electron. We then study quantum-confined systems and derive the eigenstates of an electron in such systems subjected to a magnetic field, an example being the formation of Landau levels in a two-dimensional electron gas with a magnetic field directed perpendicular to the plane of the electron gas. The effect of a magnetic field (other than lifting spin degeneracy via the Zeeman effect) is to modify the momentum operator through the introduction of a magnetic vector potential. We study properties of the transformed momentum operator and conclude by deriving the polarizability of an electron in a three-dimensional harmonic potential in the presence of a uniform magnetic field.

Preliminary: Choice of vector potential $\vec{A}$ associated with a constant uniform magnetic field $\vec{B} = \vec{\nabla} \times \vec{A}$

As shown in Appendix F, to describe electrons in the presence of an external magnetic field, the following Hamiltonian must be used:

$$H = \frac{\left[\vec{p} - q\vec{A}(x,y,z)\right]^2}{2m} + V(x,y,z), \qquad (11.1)$$

where $\vec{A}(x,y,z)$ is the vector potential due to the magnetic field.

For a uniform magnetic field $\vec{B} = \vec{\nabla} \times \vec{A}$ along the z-axis, the gauge $\vec{A}$ is typically selected from one of the following three gauges:

$$\vec{A} = \left(-\frac{By}{2}, \frac{Bx}{2}, 0\right) \qquad \text{(symmetric gauge)}, \qquad (11.2)$$

$$\vec{A} = (0, Bx, 0), \qquad \text{(asymmetric Landau gauge)}, \qquad (11.3)$$

$$\vec{A} = (-By, 0, 0) \qquad \text{(asymmetric Landau gauge)}. \qquad (11.4)$$

Starting with the relation $\vec{B} = \vec{\nabla} \times \vec{A}$, it is easy to show that each of the three gauges given above indeed corresponds to a constant magnetic field $\vec{B}$ along the z-direction.

---

* Problem 11.1: General gauge

Find a way to write a general gauge $\vec{A}$ to generate a uniform magnetic field $\vec{B}$ along the z-axis such that any of the three gauges above can be generated by the appropriate selection of a parameter.

Problem Solving in Quantum Mechanics: From Basics to Real-World Applications for Materials Scientists, Applied Physicists, and Devices Engineers, First Edition.
Marc Cahay and Supriyo Bandyopadhyay.

Solution: With

$$\vec{A} = [-(1-\xi)By, \xi Bx, 0],$$ (11.5)

where ξ is a real parameter in the interval $[0, 1]$, it is easy to show that

$$\vec{\nabla} \times \vec{A} = B\hat{z},$$ (11.6)

where $\hat{z}$ is the unit vector along the z-axis. The vector potential $\vec{A}$ reduces to the different gauges (11.2), (11.3), and (11.4) listed above when ξ is set equal to $\frac{1}{2}$, 1, and 0, respectively.

** Problem 11.2: Landau levels in a general gauge

Find the energy levels of a free particle in a uniform magnetic field along the z-axis using the general gauge in Equation (11.5).

Solution: Our starting point is the Hamiltonian of a free particle in a magnetic field given by

$$H = \frac{1}{2m^*}(\vec{p} - q\vec{A})^2.$$ (11.7)

For a uniform magnetic field along the z-axis, we use the general gauge in Equation (11.5). The Hamiltonian becomes

$$H = \frac{1}{2m^*}[p_x - (1-\xi)qBy]^2 + \frac{1}{2m^*}(p_y + \xi qBx)^2 + \frac{1}{2m^*}p_z{}^2.$$ (11.8)

Next, we define the new variables

$$P = p_x - (1-\xi)qBy,$$ (11.9)

$$Q = \xi x + \frac{p_y}{qB},$$ (11.10)

$$P' = p_x + (1-\xi)qBy,$$ (11.11)

$$Q' = \xi x - \frac{p_y}{qB}.$$ (11.12)

Using the commutation rules governing position and momentum operators (see Appendix B), it is easy to show the following results:

$$[Q, Q'] = [P, P'] = [Q, P'] = [Q', P] = 0,$$ (11.13)

$$[Q, P] = [Q', P'] = i\hbar.$$ (11.14)

Indeed,

$$[P, Q] = \left[p_x - (1-\xi)qBy, \xi x + \frac{p_y}{qB}\right]$$

$$= \xi[p_x, x] - (1-\xi)[y, p_y] = -\xi i\hbar - (1-\xi)i\hbar = -i\hbar,$$ (11.15)

and

$$[Q', P'] = \left[\xi x - \frac{p_y}{qB}, p_x + (1 - \xi)qBy\right]$$
$$= \xi[x, p_x] - (1 - \xi)[p_y, y] = \xi i\hbar + (1 - \xi)i\hbar = i\hbar. \tag{11.16}$$

In terms of the new variables, the Hamiltonian becomes

$$H = \frac{p_z^{\,2}}{2m^*} + \frac{P^2}{2m^*} + \frac{1}{2}m^*\omega_c^{\,2}Q^2, \tag{11.17}$$

where $\omega_c = \frac{qB}{m^*}$ is the cyclotron frequency.

Since $[H, p_z] = 0$, p_z is a good quantum number. Furthermore, the second and third terms of H describe the Hamiltonian of a 1D harmonic oscillator. The energy eigenvalues of H are therefore given by

$$E_n(k_z) = \frac{\hbar^2 k_z^{\,2}}{2m^*} + \left(n + \frac{1}{2}\right)\hbar\omega_c, \tag{11.18}$$

a result independent of the parameter ξ in the general expression of the gauge $\vec{A}$. Here, ω_c is the cyclotron frequency given by $\omega_c - qB/m^*$. These energy levels are referred to as *Landau levels*.

Physical significance: The Landau levels are very important in understanding the electrical and optical properties of electron gases in the presence of a uniform magnetic field. Any observable physical quantity must be independent of the choice of gauge, and here we showed that the energies of the eigenstates in a magnetic field, which are physical observables, are independent of the gauge.

---

* Problem 11.3: Current density operator in a uniform magnetic field

Starting with the time-dependent Schrödinger equation in a uniform static magnetic field, follow the procedure outlined in Problem 3.1 to derive a current continuity equation and find the modified expression of the current density in the presence of a magnetic field. Assume a constant effective mass m^* throughout.

Solution: In the presence of a magnetic field, the Hamiltonian of the Schrödinger equation in the Coulomb gauge (i.e., $\vec{\nabla} \cdot \vec{A} = 0$) was derived in Appendix F, leading to the time-dependent Schrödinger equation

$$-\frac{\hbar^2}{2m^*}\nabla^2\psi - \frac{q\hbar}{m^*}\vec{A} \cdot \vec{\nabla}\psi + \left[\frac{q^2 A^2}{2m^*} + V\right]\psi + \frac{\hbar}{i}\frac{\partial\psi}{\partial t} = 0. \tag{11.19}$$

The complex conjugate of this equation is

$$-\frac{\hbar^2}{2m^*}\nabla^2\psi^* - \frac{q\hbar}{m^*}\vec{A}\cdot\vec{\nabla}\psi^* + \left[\frac{q^2A^2}{2m^*}+V\right]\psi^* - \frac{\hbar}{i}\frac{\partial\psi^*}{\partial t} = 0, \qquad (11.20)$$

where we (correctly) assumed that the scalar and vector potentials are real quantities. If they were imaginary or complex, they would have made the Hamiltonian non-Hermitian, which is not allowed.

Multiplying Equation (11.19) on the left by ψ^* and Equation (11.20) on the left by ψ, and then subtracting the second from the first, we get

$$\frac{\hbar^2}{2m^*}(\psi^*\nabla^2\psi - \psi\nabla^2\psi^*) - \frac{i\hbar}{m}q\vec{A}\cdot\vec{\nabla}(\psi^*\psi) + i\hbar\frac{\partial}{\partial t}(\psi^*\psi) = 0. \qquad (11.21)$$

Next, we use the identities

$$\psi^*\nabla^2\psi - \psi\nabla^2\psi^* = \vec{\nabla}\cdot(\psi^*\vec{\nabla}\psi - \psi\vec{\nabla}\psi^*), \qquad (11.22)$$

$$\vec{A}\cdot\vec{\nabla}(\psi^*\psi) = \vec{\nabla}\cdot(\psi^*\psi\vec{A}), \qquad (11.23)$$

since $\vec{\nabla}\cdot\vec{A} = 0$ in the Coulomb gauge.

Equation (11.21) therefore becomes

$$\frac{\partial\rho}{\partial t} + \vec{\nabla}\cdot\left[\frac{\hbar}{2m^*i}(\psi^*\vec{\nabla}\psi - \psi\vec{\nabla}\psi^*) - \frac{q}{m^*}\psi^*\psi\vec{A}\right] = 0, \qquad (11.24)$$

with $\rho = \psi^*\psi$, the probability density. The term in the square brackets is the generalized form of the probability current density in the presence of a uniform static magnetic field, i.e.,

$$\vec{J}(\vec{A}\neq 0) = \vec{J}(\vec{A}=0) - \frac{q}{m^*}\psi^*\psi\vec{A}. \qquad (11.25)$$

---

* Problem 11.4: An electron in a 2DEG with parabolic shape of the confining potential and a magnetic field in the plane of the electron gas

Find the eigenstates and corresponding eigenvalues for an electron confined in the z-direction by a parabolic potential well $E_c(z) = \frac{1}{2}m^*\omega^2 z^2$ in the presence of a uniform and constant magnetic field $\vec{B} = (B,0,0)$. Use the gauge $\vec{A} = (0,-Bz,0)$.

Solution: Using the given gauge in the Hamiltonian (11.8), we obtain that the one-electron Hamiltonian is

$$H_0 = \frac{1}{2m^*}(p_x^2+p_y^2+p_z^2) + \left(\frac{q^2B^2}{2m^*}+\frac{1}{2}m^*\omega^2\right)z^2 - \left(\frac{qB}{m^*}\right)p_y z. \qquad (11.26)$$

The subband eigenfunctions are the solutions of the Schrödinger equation with the above Hamiltonian and can therefore be written as

$$\psi_n(\vec{k}) = \frac{1}{\sqrt{A}} e^{i\vec{k}\cdot\vec{\rho}} \phi_n(z),$$ (11.27)

since the Hamiltonian depends only on the z-coordinate. Here, A is a normalizing area in the x–y plane, $\vec{k} = (k_x, k_y)$, and $\vec{\rho} = (x, y)$. Plugging $\psi_n(\vec{k})$ into the time-independent Schrödinger equation, we get

$$\left[-\frac{\hbar^2}{2m^*}\frac{d^2}{dz^2} + \frac{m^*}{2}\omega^2(z - z_0)^2 \right] \phi_n(z) = \epsilon_n \phi_n(z),$$ (11.28)

where

$$\omega = \left[\omega_c^2 + \omega_0^2 \right]^{\frac{1}{2}},$$
$$\omega_c = qB/m^* \quad \text{(cyclotron frequency)},$$
$$z_0 = \left[\frac{\sqrt{1-\gamma}}{\omega} \right] \left(\frac{\hbar k_y}{m^*} \right),$$
$$\gamma = \left(\frac{\omega_0}{\omega} \right)^2,$$
$$\epsilon_n = \left(n + \frac{1}{2} \right) \hbar\omega.$$ (11.29)

The eigenenergies of the Hamiltonian H_0 are then

$$E_n(\vec{k}) = \frac{\hbar^2 k^2}{2m^*} + \epsilon_n = \frac{\hbar^2 k^2}{2m^*} + \left(n + \frac{1}{2} \right) \hbar\omega,$$ (11.30)

and the $\phi_n(z)$ in Equation (11.28) are 1D harmonic oscillator wave functions given by

$$\phi_n(z) = \left[\frac{1}{2^n n!} \left(\frac{m^* w}{\pi\hbar} \right)^{\frac{1}{2}} \right]^{\frac{1}{2}} \exp\left[-\frac{m^* w}{2\hbar}(z - z_0)^2 \right] H_n \left[\left(\frac{m^* w}{\hbar} \right)^{\frac{1}{2}} (z - z_0) \right],$$ (11.31)

where H_n is the Hermite polynomial of the nth order.

** Problem 11.5: Spin-1/2 particle in a magnetic field

Taking into account the spin of the electron, the Hamiltonian of a free electron of charge q in a magnetic field is given by

$$H_0 = \frac{1}{2m_0}(\vec{p} + q\vec{A})^2 I,$$ (11.32)

where I is the 2×2 identity matrix and $\vec{A}$ is the vector potential.

Show that

$$\left[\vec{\sigma} \cdot (\vec{p} + q\vec{A})\right]^2 = (\vec{p} + q\vec{A})^2 I + 2m_0 \mu_B \vec{B} \cdot \vec{\sigma}, \qquad (11.33)$$

where $\mu_B = q\hbar/2m_0$ is the Bohr magneton and and $\vec{B} = \vec{\nabla} \times \vec{A}$.

Solution: Carrying out the expansion, we get

$$\begin{aligned}
\left[\vec{\sigma} \cdot (\vec{p} + q\vec{A})\right]^2 &= [\sigma_x(p_x + qA_x) + \sigma_y(p_y + qA_y) + \sigma_z(p_z + qA_z)] \\
&\quad \times [\sigma_x(p_x + qA_x) + \sigma_y(p_y + qA_y) + \sigma_z(p_z + qA_z)] \\
&= \sigma_x^2(p_x + qA_x)^2 + \sigma_y^2(p_y + qA_y)^2 + \sigma_z^2(p_z + qA_z)^2 \\
&\quad + \sigma_x\sigma_y(p_x + qA_x)(p_y + qA_y) \\
&\quad + \sigma_y\sigma_x(p_y + qA_y)(p_x + qA_x) \\
&\quad + \sigma_y\sigma_z(p_y + qA_y)(p_z + qA_z) \\
&\quad + \sigma_z\sigma_y(p_z + qA_z)(p_y + qA_y) \\
&\quad + \sigma_z\sigma_x(p_z + qA_z)(p_x + qA_x) \\
&\quad + \sigma_x\sigma_z(p_x + qA_x)(p_z + qA_z). \qquad (11.34)
\end{aligned}$$

Using the properties of the Pauli matrices in Appendix B, we get

$$\begin{aligned}
\left[\vec{\sigma} \cdot (\vec{p} + q\vec{A})\right]^2 &= (p_x + qA_x)^2 + (p_y + qA_y)^2 + (p_z + qA_z)^2 \\
&\quad + i\sigma_z\left(-iq\hbar\frac{\partial A_y}{\partial x} + iq\hbar\frac{\partial A_x}{\partial y}\right) \\
&\quad + i\sigma_y\left(-iq\hbar\frac{\partial A_x}{\partial z} + iq\hbar\frac{\partial A_z}{\partial x}\right) \\
&\quad + i\sigma_x\left(-iq\hbar\frac{\partial A_z}{\partial y} + iq\hbar\frac{\partial A_y}{\partial z}\right) \\
&= (p_x + qA_x)^2 + (p_y + qA_y)^2 + (p_z + qA_z)^2 \\
&\quad + q\hbar\sigma_z\left(\frac{\partial A_y}{\partial x} - \frac{\partial A_x}{\partial y}\right) \\
&\quad + q\hbar\sigma_y\left(\frac{\partial A_x}{\partial z} - \frac{\partial A_z}{\partial x}\right) \\
&\quad + q\hbar\sigma_x\left(\frac{\partial A_z}{\partial y} - \frac{\partial A_y}{\partial z}\right) \\
&= \left[\vec{p} + q\vec{A}\right]^2 + q\hbar\left(\vec{\nabla} \times \vec{A}\right) \cdot \vec{\sigma}, \qquad (11.35)
\end{aligned}$$

where we have used the fact that $\vec{p}_n = -i\hbar(\partial/\partial\vec{x}_n)$.

Using the definition of the Bohr magneton and $\vec{B} = \vec{\nabla} \times \vec{A}$, we finally get the identity (11.33).

** Problem 11.6: Wave function associated with Landau levels

Let us consider a 2DEG in the x–y plane with a uniform magnetic field perpendicular to it, as shown in Figure 11.1.

We choose the gauge

$$A_x = By, A_y = A_z = 0. \tag{11.36}$$

Find the wave functions of an electron in the 2DEG.

Solution: In this case, the Schrödinger equation describing the electron in the 2DEG is given by

$$\left[-\frac{\left(-i\hbar\frac{\partial}{\partial x} + qBy\right)^2}{2m*} - \frac{\hbar^2}{2m*}\left(\frac{\partial^2}{\partial y^2} + \frac{\partial^2}{\partial z^2}\right) + V(z) \right] \Psi(x,y,z)$$

$$= E\Psi(x,y,z), \tag{11.37}$$

where $V(z)$ is the confining potential in the z-direction.

We can expand the last equation as

$$\left\{ -\frac{\hbar^2}{2m*}\left[\frac{\partial^2}{\partial x^2} + i\hbar\frac{qB}{m*}y\frac{\partial}{\partial x} + \frac{(qBy)^2}{2m*}\right] \right.$$

$$\left. -\frac{\hbar^2}{2m*}\left[\frac{\partial^2}{\partial y^2} + \frac{\partial^2}{\partial z^2}\right] - \frac{\partial^2}{\partial z^2} + V(z) \right\} \Psi = E\Psi. \tag{11.38}$$

Since there are no *mixed terms* in the Hamiltonian in the above equation, meaning that there is not a single term that depends on more than one coordinate, the wave function can be written as the product of an x-dependent wave function, a y-dependent wave function, and a z-dependent wave function:

$$\Psi(x,y,z) = \phi(x)\eta(y)\xi(z). \tag{11.39}$$

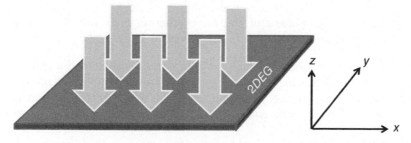

Figure 11.1: Illustration of a 2DEG with a uniform magnetic field applied perpendicular to it (pointing in the z-direction).

Furthermore, since the Hamiltonian is invariant in x, meaning that it does not depend on the x-coordinate, the x-component of the wave function is a plane wave. Hence,

$$\Psi(x, y, z) = \frac{1}{\sqrt{L_x}} e^{ik_x x} \eta(y) \xi(z), \tag{11.40}$$

where L_x is the normalizing length in the x-direction.

Substituting the last result in the Schrödinger equation, multiplying by the complex conjugate of the x-component of the wave function, and then integrating over the x-coordinate, we get

$$\left[\frac{\hbar^2 k_x^2}{2m^*} - \hbar k_x \frac{qB}{m^*} y + \frac{(qBy)^2}{2m*} - \frac{\hbar^2}{2m^*} \left(\frac{\partial^2}{\partial y^2} + \frac{\partial^2}{\partial z^2} \right) + V(z) \right] \eta(y) \xi(z)$$

$$= E \eta(y) \xi(z). \tag{11.41}$$

We can break this equation up to generate two equations:

$$\left[\frac{\hbar^2 k_x^2}{2m^*} - \hbar k_x \frac{qB}{m*} y + \frac{(qBy)^2}{2m*} \right] \eta(y) \xi(z) - \frac{\hbar^2}{2m^*} \frac{\partial^2 \eta(y)}{\partial y^2} \xi(z) = \Lambda \eta(y) \xi(z),$$

$$-\frac{\hbar^2}{2m^*} \frac{\partial^2 \xi(z)}{\partial z^2} \eta(y) + V(z) \eta(y) \xi(z) = \Xi \eta(y) \xi(z), \tag{11.42}$$

where $\Lambda + \Xi = E$. This allows us to decouple the y- and z-components of the dynamics.

Dividing Equation (11.41) throughout by $\xi(z)$ and Equation (11.42) throughout by $\eta(y)$, we find two equations which depend exclusively on y and z coordinates, respectively, meaning that the y- and z-components of motion have been decoupled. These equations are

$$\left[-\frac{\hbar^2}{2m^*} \frac{\partial^2 \eta(y)}{\partial y^2} + \frac{1}{2} m^* \omega_c^2 (y - y_0)^2 \right] \eta(y) = \Lambda \eta(y), \tag{11.43}$$

$$\left[-\frac{\hbar^2}{2m^*} \frac{\partial^2}{\partial z^2} + V(z) \right] \xi(z) = \Xi \xi(z), \tag{11.44}$$

where $\omega_c = qB/m^*$, $y_0 = \hbar k_x/(qB)$.

The solutions for the z-component of the wave function $\xi(z)$ and the corresponding eigenenergies Ξ depend on the confining potential $V(z)$.

The solution for the y-component of the wave function $\eta(y)$ is that of a simple harmonic motion oscillator oscillating along the y-coordinate with motion centered at $y = y_0$. Hence, the solution for the z-component of the wave function depends on the confining potential $V(z)$. The solution for the y-component of the wave function is

$$\eta(y) = \left(\frac{\alpha}{\sqrt{\pi} 2^m m!} \right)^{1/2} H_m[\alpha(y - y_0)] e^{-(1/2)\alpha^2 (y - y_0)^2}, \tag{11.45}$$

where m is an integer, H_m is the Hermite polynomial of the mth order (see Appendix B), and $\alpha = \sqrt{m^* \omega_c / \hbar}$. The solution for the eigenenergy is

$$\Lambda = \left(m + \frac{1}{2} \right) \hbar \omega_c. \tag{11.46}$$

---

* Problem 11.7: Average kinetic energy of an electron in a 2DEG in the presence of a perpendicular magnetic field

Starting with the results of the previous problem, find the expectation values of the kinetic energies associated with the x-component of motion in a 2DEG (in the x–y plane) subjected to a perpendicular magnetic field. Find also the kinetic energy for the y-component of the motion.

Solution: With the results from the previous problem, we have $\langle y^2 \rangle = \langle \eta(y) | y^2 | \eta(y) \rangle = \left(m + \frac{1}{2} \right) \frac{\hbar}{m^* \omega_c} + y_0{}^2$ and $\langle y \rangle = y_0$.

The velocity operator becomes $\frac{\vec{p}_{op} - q\vec{A}}{m^*}$ in a magnetic field. Hence, the expectation value is given by

$$
\begin{aligned}
\langle v_x{}^2 \rangle &= \left\langle \left(\frac{p_x - qA_x}{m^*} \right)^2 \right\rangle = \frac{\langle p_x^2 \rangle - 2\langle p_x A_x + A_x p_x \rangle + q^2 A_x{}^2}{m^{*2}} \\
&= \frac{\hbar^2 k_x^2 - 2\hbar k_x qB \langle y \rangle + q^2 B^2 \langle y^2 \rangle}{m^{*2}} \\
&= \frac{\hbar^2 k_x^2 - 2\hbar k_x qB y_0 + q^2 B^2 \langle y^2 \rangle}{m^{*2}} \\
&= \frac{q^2 B^2 \left(\langle y^2 \rangle - y_0{}^2 \right)}{m^{*2}} \\
&= \omega_c{}^2 \left(\langle y^2 \rangle - y_0{}^2 \right), \tag{11.47}
\end{aligned}
$$

where we used the facts that $A_x = By$, the operators A_x and p_x commute, $y_0 = \hbar k_x / (qB)$, $\langle p_x \rangle = \hbar k_x$, $\langle p_x^2 \rangle = \hbar^2 k_x{}^2$, and $\omega_c = qB/m^*$.

Using the given expression for $\langle y^2 \rangle$, we immediately find

$$\langle v_x{}^2 \rangle = \left(m + \frac{1}{2} \right) \frac{\hbar \omega_c}{m^*}. \tag{11.48}$$

Hence, the kinetic energy associated with the x-component of motion is

$$\frac{1}{2} m^* \langle v_x^2 \rangle = \frac{1}{2} \left(m + \frac{1}{2} \right) \hbar \omega_c. \tag{11.49}$$

Since the total energy $\Lambda = \frac{1}{2} m^* \left(\langle v_x^2 \rangle + \langle v_y^2 \rangle \right) = \left(m + \frac{1}{2} \right) \hbar \omega_c$, the kinetic energy of the y-component of motion will be the same as that of the x-component, i.e.,

$$\frac{1}{2} m^* \langle v_y^2 \rangle = \frac{1}{2} \left(m + \frac{1}{2} \right) \hbar \omega_c. \tag{11.50}$$

* Problem 11.8: Kinematic momentum operator in a magnetic field

Show that the kinematic momentum operator $\vec{\Pi}_{op} = \vec{p}_{op} - q\vec{A}$ satisfies the relation

$$\vec{\Pi}_{op} \times \vec{\Pi}_{op} = iq\hbar\vec{B}, \tag{11.51}$$

where $\vec{B}$ is the magnetic flux density.

Solution: We have

$$\left[\vec{\Pi}_{op} \times \vec{\Pi}_{op}\right]\Psi = \left(-i\hbar\vec{\nabla} - q\vec{A}\right) \times \left(-i\hbar\vec{\nabla} - q\vec{A}\right)\Psi$$

$$= -\hbar^2\vec{\nabla} \times \vec{\nabla}\Psi + iq\hbar\vec{\nabla} \times \left(\vec{A}\Psi\right)$$

$$+ iq\hbar\vec{A} \times \vec{\nabla}\Psi + q^2\vec{A} \times \vec{A}\Psi$$

$$= iq\hbar\left[\vec{\nabla} \times \left(\vec{A}\Psi\right) + \vec{A} \times \vec{\nabla}\Psi\right]. \tag{11.52}$$

Using the vector identity $\vec{\nabla} \times \left(\vec{A}\Psi\right) = \vec{\nabla}\Psi \times \vec{A} + \Psi\left(\vec{\nabla} \times \vec{A}\right)$, Equation (11.52) becomes

$$\left[\vec{\Pi}_{op} \times \vec{\Pi}_{op}\right]\Psi = iq\hbar\left[-\vec{A} \times \vec{\nabla}\Psi + \Psi\left(\vec{\nabla} \times \vec{A}\right) + \vec{A} \times \vec{\nabla}\Psi\right] = iq\hbar\vec{B}\Psi. \tag{11.53}$$

Hence,

$$\vec{\Pi}_{op} \times \vec{\Pi}_{op} = iq\hbar\vec{B}. \tag{11.54}$$

*** Problem 11.9: Wave functions in a 2DEG in the presence of constant electric and magnetic fields

Consider a 2DEG in the x–y plane with an electric field E directed in the $-y$-direction. A magnetic field of flux density B is present in the $-z$-direction. The coordinate system is the same as in Figure 11.1. Show that the wave function of an electron in this system is given by

$$\Psi(x, y) = e^{ik_x x}\eta_m(y), \tag{11.55}$$

where $\eta_m(y)$ is given by Equation (11.45), except that y_0 is replaced by $y_0{}'$, where

$$y_0{}' = \frac{1}{\omega_c}\left[\frac{\hbar k_x}{m^*} - \frac{E}{B}\right]. \tag{11.56}$$

Also, show that the eigenenergies of the electron are

$$\Lambda_m(k_x, E, B) = \left(m + \frac{1}{2}\right)\hbar\omega_c - qEy_0{}' + \frac{1}{2}m^*\left(\frac{E}{B}\right)^2, \tag{11.57}$$

where m is an integer.

Solution: The Hamiltonian for this system is

$$H = \frac{\left(\vec{p} - q\vec{A}\right)^2}{2m^*} + qEy.$$ (11.58)

Starting with the time-independent Schrödinger equation and using Equation (11.55), this yields the following equation for the y-component of the wave function:

$$\left[\frac{\hbar^2 k_x^2}{2m^*} - \left(\hbar k_x \frac{qB}{m^*} - qE\right)y + \frac{(qBy)^2}{2m*} - \frac{\hbar^2}{2m^*}\frac{\partial^2}{\partial y^2}\right]\eta(y) - \Lambda\eta(y).$$ (11.59)

The above equation can be recast as

$$\left[-\frac{\hbar^2}{2m^*}\frac{\partial^2}{\partial y^2} + \frac{1}{2}m^*\omega_c^2\left(y - y_0'\right)^2 - qEy_0 - \frac{1}{2}m^*\left(\frac{E}{B}\right)^2\right]\eta(y) = \Lambda\eta(y),$$ (11.60)

where $y_0' = \frac{1}{\omega_c}\left[\frac{\hbar k_x}{m^*} - \frac{E}{B}\right]$ and $y_0 = \frac{1}{\omega_c}\frac{\hbar k_x}{m^*}$. Equation (11.60) can then be recast as

$$\left[-\frac{\hbar^2}{2m^*}\frac{\partial^2\eta}{\partial y^2} + \frac{1}{2}m^*\omega_c^2\left(y - y_0'\right)^2\right]\eta(y)$$

$$= \left[\Lambda + qEy_0' - \frac{1}{2}m^*\left(\frac{E}{B}\right)^2\right]\eta(y).$$ (11.61)

This is the same equation as Equation (11.43), with y_0' replacing y_0 and $\Lambda' = \Lambda + qEy_0' - \frac{1}{2}m^*\left(\frac{E}{B}\right)^2$ replacing Λ.

Hence, the solutions for the wave function $\eta(y)$ in perpendicular electric and magnetic fields, sometimes referred to as magneto-electric states in a 2DEG, are given by Equation (11.45) with y_0' replacing y_0.

Furthermore, since $\Lambda' = \left(m + \frac{1}{2}\hbar\omega_c\right)$, we get that the eigenenergies of the magneto-electric states in a 2DEG are given by

$$\Lambda_m\left(k_x, E, B\right) = \left(m + \frac{1}{2}\hbar\omega_c\right) - eEy_0' + \frac{1}{2}m^*\left(\frac{E}{B}\right)^2.$$ (11.62)

***** Problem 11.10: Polarizability α of a three-dimensional harmonic oscillator in a uniform external magnetic field**

Prove that the polarizability of a three-dimensional harmonic oscillator is independent of the angle between the external magnetic and electric fields. Furthermore, the polarizability is independent of the magnetic field and is given by the well-known zero field value, $\alpha = \frac{q^2}{m^*\omega_0^2}$.

Solution: We consider the Hamiltonian of a three-dimensional harmonic oscillator in a static magnetic field ($\vec{B}$ along z),

$$H_0 = \frac{(\vec{p} - q\vec{A})^2}{2m^*} + \frac{1}{2}m^*\omega_0^2 r^2,$$ (11.63)

which can be written

$$H_0 = \frac{p^2}{2m^*} + \frac{m^*\omega_0^2 r^2}{2} - \frac{q\vec{A}.\vec{p}}{m^*} + \frac{q^2\vec{A}.\vec{A}}{2m^*}. \tag{11.64}$$

Using the symmetric gauge

$$\vec{A} = -\frac{1}{2}\vec{r}X\vec{B}, \tag{11.65}$$

the Hamiltonian (11.64) can be rewritten as

$$H_0 = \frac{p^2}{2m^*} + \frac{m^*}{2}\omega_0^2 r^2 + \frac{m^*}{4}\omega_L^2(x^2 + y^2) + \vec{\omega}_L.\vec{L}, \tag{11.66}$$

where

$$\vec{\omega}_L = \frac{-q}{2m^*}\vec{B}. \tag{11.67}$$

We then rewrite Equation (11.66) as

$$H_0 = H_\| + H_\perp, \tag{11.68}$$

where

$$H_\| = \frac{p_z^2}{2m^*} + \frac{m^*}{2}\omega_0^2 z^2, \tag{11.69}$$

$$H_\perp = \frac{p_x^2 + p_y^2}{2m^*} + \frac{m^*\omega^2}{2}(x^2 + y^2) + \omega_L L_z. \tag{11.70}$$

In Equation (11.70), $\omega_L = \frac{-qB}{2m^*}$ and $\omega = \sqrt{\omega_0^2 + \omega_L^2}$. We define the annihilation and creation operators

$$a_z = \frac{1}{\sqrt{2}}\left(\beta_0 z + \frac{i}{\hbar\beta_0}p_z\right), \tag{11.71}$$

$$a_z^\dagger = \frac{1}{\sqrt{2}}\left(\beta_0 z - \frac{i}{\hbar\beta_0}p_z\right), \tag{11.72}$$

with

$$\beta_0 = \left(\frac{m^*\omega_0}{\hbar}\right)^{1/2}. \tag{11.73}$$

The operators a_z, $a_z^\dagger$ satisfy the commutation rule

$$[a_z, a_z^\dagger] = 1. \tag{11.74}$$

We also introduce the operators

$$a_r = \frac{1}{\sqrt{2}}(a_x - ia_y), \tag{11.75}$$

$$a_\ell = \frac{1}{\sqrt{2}}(a_x + ia_y), \tag{11.76}$$

where a_x and a_y are defined using Equation (11.71), changing z to x and y, respectively.

It is easily seen that the following commutation rules are satisfied:

$$[a_\ell, a_\ell^\dagger] = 1, \tag{11.77}$$

$$[a_r, a_r^\dagger] - 1. \tag{11.78}$$

Using the definitions, the Hamiltonian (11.66) can be rewritten as

$$H_0 = H_\parallel + H_\perp, \tag{11.79}$$

where

$$H_\parallel = \hbar\omega_0 \left(N_z + \frac{1}{2} \right), \tag{11.80}$$

$$H_\perp = \hbar\omega(N_r + N_\ell + 1) + \hbar\omega_L(N_r - N_\ell), \tag{11.81}$$

and we have introduced the occupation number operators

$$N_z = a_z^\dagger a_z, \tag{11.82}$$

$$N_\ell = a_\ell^\dagger a_\ell, \tag{11.83}$$

$$N_r = a_r^\dagger a_r. \tag{11.84}$$

In the presence of an external (uniform) electric field, we must add the Stark interaction to the Hamiltonian (11.63). The total Hamiltonian can be written as

$$H = H_0 - q\vec{E}.\vec{r}. \tag{11.85}$$

Case 1: Polarizability of a harmonic oscillator in parallel configuration ($\vec{E} \parallel \vec{B}$)

In this case, Equation (11.85) becomes

$$H = H_0 - qEz. \tag{11.86}$$

Next, we perform a unitary transform on H_0 with $U = e^S$, where

$$S = -\lambda(a_z - a_z^\dagger), \tag{11.87}$$

λ being some parameter to be determined later. The transformed Hamiltonian H can be written as

$$H = UHU^\dagger = e^S He^{-S}. \tag{11.88}$$

Next, we make use of the following identity derived in Problem 2.5:

$$e^{\xi\hat{A}}\hat{B}e^{\xi\hat{A}} = \hat{B} + \xi[\hat{A},\hat{B}] + \frac{E^2}{2!}[\hat{A},[\hat{A},\hat{B}]] + \frac{\xi^3}{3!}[\hat{A},[\hat{A},[\hat{A},\hat{B}]]] + \cdots \tag{11.89}$$

For $\xi = 1$, $\hat{B} = H_0$, and $\hat{A} = S$, Equation (11.89) gives

$$e^{S}H_0 e^{-S} = H_0 + [S,H_0] + \frac{1}{2!}[S,[S,H_0]] + \cdots \tag{11.90}$$

First, we calculate $[S,H_0]$:

$$[S,H_0] = [-\lambda(a_z - a_z^\dagger), H_\perp] = -\lambda\hbar\omega_0[a_z - a_z^\dagger, a_z^\dagger a_z]. \tag{11.91}$$

Making use of the relations

$$[a_z^\dagger, a_z^\dagger a_z] = [a_z^\dagger, a_z^\dagger]a_z + a_z^\dagger[a_z^\dagger, a_z] = -a_z^\dagger, \tag{11.92}$$

we finally obtain

$$[S,H_0] = -\lambda\hbar\omega_0(a_z + a_z^\dagger). \tag{11.93}$$

Therefore,

$$\begin{aligned}
[S,[S,H_0]] &= \lambda^2\hbar\omega_0[a_z - a_z^\dagger, a_z + a_z^\dagger] \\
&= \lambda^2\hbar\omega_0\left\{[a_z, a_z^\dagger] - [a_z^\dagger, a_z]\right\} \\
&= 2\lambda^2\hbar\omega_0.
\end{aligned} \tag{11.94}$$

We therefore conclude that all the terms after the third one in the expansion (11.89) are identically zero. Grouping the previous results, the transformed Hamiltonian is therefore

$$H = e^{S}H_0 e^{-S} = H_0 - \lambda\hbar\omega_0(a_z + a_z^\dagger) + \lambda^2\hbar\omega_0. \tag{11.95}$$

Since

$$z = \frac{1}{2\beta_0}(a_z + a_z^\dagger), \tag{11.96}$$

the second term in Equation (11.95) is equal to the Stark shift $-qEz$ if we choose λ such that

$$\lambda = \frac{qE}{\sqrt{2m^*\hbar\omega_0}}\frac{1}{\omega_0}. \tag{11.97}$$

Equation (11.95) can then be rewritten as

$$H - \frac{q^2}{2m^*\omega_0^2}E^2 = H_0 - qEz. \tag{11.98}$$

In other words, the eigenvalues of $H_0 - qEz$ are the same as the eigenvalues of $H - \frac{q^2}{2m^*\omega_0^2}E^2$. Calling $|n\rangle$ the eigenstates of H_0, we therefore deduce that the eigenstates of $H_0 - qEz$ are

$$e^{S}|n\rangle, \tag{11.99}$$

with eigenvalues

$$E_n = E_n^0 - \frac{q^2}{2m^*\omega_0^2}E^2, \tag{11.100}$$

E_n^0 being the eigenvalues of H_0, given by

$$E_n^0 = \hbar\omega_0\left(n_0 + \frac{1}{2}\right) + \hbar\omega(n_r + n_\ell + 1) + \hbar\omega_L(n_r - n_\ell), \tag{11.101}$$

where n_0, n_r, n_ℓ are integers. By definition, the polarizability α is such that

$$E_n = E^0{}_n - \frac{1}{2}\alpha_n E^2. \tag{11.102}$$

By comparing Equations (11.100) and (11.102), the polarizability of each level is therefore deduced to be

$$\alpha_\| = \frac{q^2}{m^*\omega_0^2}, \tag{11.103}$$

which does not depend on the magnetic field intensity.

Case 2: Polarizability of a harmonic oscillator in the perpendicular configuration ($\vec{E} \perp \vec{B}$)

If the electric field is selected along the y-axis, the Stark shift can be written as

$$-qEy. \tag{11.104}$$

From Equations (11.75) and (11.76), we have

$$y = \frac{1}{\sqrt{2}\beta_0}(a_y + a_y^\dagger) = \frac{1}{2\beta_0 i}[(a_\ell - a_\ell^+) + (a_r^+ - a_r)]. \tag{11.105}$$

We now perform two unitary transforms on the Hamiltonian H_0, i.e.,

$$H = UH_0U^\dagger, \tag{11.106}$$

where $U = U_r U_\ell$ with

$$U_r = e^{-(\lambda a_r - \lambda^* a_r^\dagger)}, \tag{11.107}$$

$$U_l = e^{-(\mu a_\ell - \mu^* a_\ell^\dagger)}, \tag{11.108}$$

λ, μ are two complex numbers to be determined later, and the $*$ stands for complex conjugate. We have

$$H = UH_0U^\dagger = (U_\ell U_r)H_\perp(U_r^\dagger U_\ell^\dagger) + H_\|, \tag{11.109}$$

since $H_\parallel$ is not affected by either U_ℓ nor U_r. We first calculate

$$U_r H_\perp U_r^\dagger = \hbar(\omega + \omega_L) U_r N_r U_r^\dagger + \hbar\omega + \hbar(\omega - \omega_L) N_\ell. \tag{11.110}$$

Since U_r does not affect the operator N_ℓ, we have

$$U_r N_r U_r^\dagger = (U_r a_r^\dagger U_r^\dagger)(U_r a_r U_r^\dagger) \tag{11.111}$$

and

$$U_r a_r^\dagger U_r^\dagger = e^{S_r} a_r^\dagger e^{-S_r}, \tag{11.112}$$

with

$$S_r = -(\lambda a_r - \lambda^* a_r^\dagger). \tag{11.113}$$

Using once again the identity (11.89), we have

$$\begin{aligned}
U_r a_r^\dagger U_r^\dagger &= a_r^\dagger + [S_r, a_r^\dagger] + \frac{1}{2!}[S_r, [S_r, a_r^\dagger]] + \cdots \\
&= a_r^\dagger - [\lambda a_r - \lambda^* a_r^\dagger, a_r^\dagger] + 0 + \cdots \\
&= a_r^\dagger - \lambda,
\end{aligned} \tag{11.114}$$

and similarly

$$U_r a_r U_r^\dagger = a_r - \lambda^*. \tag{11.115}$$

Therefore,

$$\begin{aligned}
U_r H_\perp U_r^\dagger &= \hbar(\omega + \omega_L)(a_r^\dagger - \lambda)(a_r - \lambda^*) + \hbar\omega + \hbar(\omega - \omega_L) N_\ell \\
&\quad + \hbar(\omega + \omega_L) N_r + \hbar(\omega - \omega_L) N_\ell \\
&\quad - \hbar(\omega + \omega_L)(\lambda a_r + \lambda^* a_r^\dagger) + \hbar\omega + \hbar(\omega + \omega_L)|\lambda|^2.
\end{aligned} \tag{11.116}$$

Next, we perform the second unitary transform,

$$\begin{aligned}
U_\ell(U_r H_\perp U_r^\dagger) U_\ell &= \hbar(\omega - \omega_L) N_r - \lambda\hbar(\omega - \omega_L)(a_r + a_r^\dagger) \\
&\quad + \hbar(\omega - \omega_L)|\lambda|^2 \\
&\quad + \hbar\omega + U_\ell[\hbar(\omega - \omega_L) N_\ell] U_\ell^\dagger.
\end{aligned} \tag{11.117}$$

Following the derivation leading to Equations (11.114) and (11.115), we get

$$U_\ell a_\ell^\dagger U_\ell^\dagger = a_\ell^\dagger - \mu, \tag{11.118}$$

$$U_\ell a_\ell U_\ell^\dagger = a_\ell - \mu^*. \tag{11.119}$$

Therefore,

$$\begin{aligned}
U_\ell(U_r H_\perp U_r^\dagger) U_\ell^\dagger &= \hbar(\omega + \omega_L) N_r + \hbar(\omega - \omega_L) N_\ell \\
&\quad - \hbar(\omega + \omega_L)(\lambda a_r + \lambda^* a_r^\dagger) \\
&\quad - \hbar(\omega - \omega_L)(\mu a_\ell + \mu^* a_\ell^\dagger) + \hbar\omega \\
&\quad + \hbar(\omega - \omega_L)|\mu|^2 + \hbar(\omega + \omega_L)|\lambda|^2.
\end{aligned} \tag{11.120}$$

Choosing λ and μ such that

$$-\hbar(\omega + \omega_L)(\lambda a_r + \lambda^* a_r^\dagger) = \frac{-qE}{2\beta} i(a_r - a_r^\dagger), \tag{11.121}$$

$$-\hbar(\omega - \omega_L)(\mu a_\ell + \mu^* a_\ell^\dagger) = \frac{-qE}{2\beta i} i(a_\ell - a_\ell^\dagger), \tag{11.122}$$

we get

$$\lambda = \frac{iqE}{2(\omega + \omega_L)} \frac{1}{\sqrt{m\hbar\omega}}, \tag{11.123}$$

$$\mu = \frac{-qE}{2(\omega - \omega_L)} \frac{1}{\sqrt{m\hbar\omega}}. \tag{11.124}$$

We therefore obtain

$$U_\ell U_r H_\perp U_r^\dagger U_\ell^\dagger = \hbar\omega(N_r + N_\ell + 1) + \hbar\omega_L(N_r - N_\ell) - qEy$$
$$+ \hbar(\omega + \omega_L)|\lambda|^2 + \hbar(\omega - \omega_L)|\mu|^2, \tag{11.125}$$

and using Equation (11.109) we obtain

$$H = U_\ell U_r H_0 U_r^\dagger U_\ell^\dagger - \hbar(\omega + \omega_L)|\lambda|^2 - \hbar(\omega - \omega_L)|\mu|^2$$
$$= \hbar\omega(N_r + N_\ell + 1) + \hbar\omega_L(N_r - N_\ell) + \hbar\omega_0 \left(N_z + \frac{1}{2} \right) - qEy. \tag{11.126}$$

The eigenstates of H are

$$U_\ell U_r |n\rangle, \tag{11.127}$$

with eigenvalues

$$E_n = E_n^0 - \hbar(\omega + \omega_L)|\lambda|^2 - \hbar(\omega - \omega_L)|\mu|^2, \tag{11.128}$$

where E_n^0 is given by Equation (11.101).

Using Equations (11.123) and (11.124), we finally obtain

$$E_n = E_n^0 - \frac{1}{2} \frac{q^2 E^2}{m} \frac{1}{(\omega^2 - \omega_L^2)}. \tag{11.129}$$

Then, using $\omega^2 = \omega_0^2 + \omega_L^2$, we get

$$E_n = E_n^0 - \frac{1}{2} \left(\frac{q^2}{m^* \omega_0^2} \right) E^2, \tag{11.130}$$

which means that in the perpendicular configuration the polarizability is also given by

$$\alpha_\perp = \frac{q^2}{m^* \omega_0^2}, \tag{11.131}$$

and does not depend on the magnetic field intensity.

Case 3: Polarizability of a harmonic oscillator for an arbitrary angle between $\vec{E}$ and $\vec{B}$

Assuming the $\vec{E}$ field to be in the y–z plane at an angle θ with the z-axis, the total Hamiltonian can be written as

$$H = H_0 - qE\cos\theta - qE\sin\theta. \tag{11.132}$$

Therefore, the polarizability is given by

$$\alpha = \alpha_\perp \sin^2\theta + \alpha_\| \cos^2\theta. \tag{11.133}$$

Since $\alpha_\| = \alpha_\perp$, we get

$$\alpha = \frac{q^2}{m^*\omega_0^2}, \tag{11.134}$$

independent of the angle θ between the electric and magnetic fields and the intensity of the magnetic field.

Suggested problems

- Find a general expression for a gauge which would be associated with a homogeneous (i.e., independent of position) magnetic field in the (θ, ϕ) direction. Find the components A_x, A_y, A_z of $\vec{A}$ such that $\vec{\nabla} \times \vec{A} = \vec{B}$.

- Show that the gauge $\vec{A} = (0, -Bz, 0)$ corresponds to a uniform magnetic field along the x-axis, i.e., $\vec{B} = (B, 0, 0)$.

- Starting with the Hamiltonian for a free electron in a uniform magnetic field,

$$H = \frac{(\vec{p} - q\vec{A})^2}{2m^*},$$

show that in the Coulomb gauge, for which $\vec{\nabla} \cdot \vec{A} = 0$, H can be rewritten as

$$H = \frac{p^2}{2m^*} - \frac{q\vec{p} \cdot \vec{A}}{m^*} + \frac{q^2 \vec{A} \cdot \vec{A}}{2m^*}.$$

- Find the analytical expression for the wave functions of the Landau levels derived in the general gauge in Problem 11.1.

- An electron has its spin parallel to $\hat{n} = (\sin\theta\cos\phi, \sin\theta\sin\phi, \cos\theta)$ at time $t = 0$ and is in the presence of a magnetic field $\vec{B} = (0, 0, B)$. The Hamiltonian of this system is

$$H = \frac{1}{2}\hbar\omega_B\sigma_z,$$

with $\omega_B = g\mu_B B$.

(1) Write the initial state in terms of the eigenvectors $|\alpha\rangle$ and $|\beta\rangle$ of σ_z, i.e., $\sigma_z|\alpha\rangle = +1|\alpha\rangle$ and $\sigma_z|\beta\rangle = -1|\beta\rangle$.

(2) Find the state $\phi(t)$ for $t \geq 0$.

(3) What is the probability that a measurement of S_z gives $+\frac{\hbar}{2}$ at time t?

- Suppose an electron is trapped in a 2DEG (in the x–y plane) and that the confinement in the z-direction can be approximated by a harmonic potential of the form $V(z) = \frac{1}{2}m^*\omega^2 z^2$.

  (1) What is the Hamiltonian associated with the electron in the presence of a uniform tilted magnetic field with components $\vec{B} = (B_x, 0, B_z)$? You must first find a suitable gauge to describe this tilted magnetic field.

  (2) Using the fact that p_x is a constant of motion, show that the Hamiltonian is of the form $\frac{\hbar^2 k_x{}^2}{2m^*} + H(y, p_z)$, where $H(y, p_z)$ is the Hamiltonian associated with two coupled harmonic oscillators.

- Derive an analytical expression for the energy eigenvalues of the Hamiltonian derived in the previous step. Your expression should depend on the angle θ of the magnetic field defined such that $\sin\theta = B_x/B$, where $B = \sqrt{B_x^2 + B_y^2}$ is the external magnetic field's flux density.

- Calculate the explicit form of the unitary operator for the case of a spin-1/2 particle (electron) in a spatially uniform and time-independent magnetic field along the z-axis.

- Find the eigenvalues and eigenfunctions of an electron in a uniform magnetic field $\vec{B} = (0, 0, B)$ taking into account the spin of the electron. Use the gauge $\vec{A} = (0, Bx, 0)$.

Suggested Reading

- Kroemer, H. (1994) *Quantum Mechanics for Engineering, Materials Science, and Applied Physics*, Chapter 8, Prentice Hall, Englewood Cliffs, NJ.

- Merlin, R. (1987) Subband Landau level coupling in tilted magnetic fields: exact results for parabolic wells. *Solid State Communications* 64, pp. 99–101.

- Rensik, M. E. (1969) Electron eigenstates in uniform magnetic fields. *American Journal of Physics* 37, pp. 900–904.

- Rosas, R., Riera, R., Marin, J. L., and Leon, H. (2000) Energy spectrum of a confined two-dimensional particle in an external magnetic field. *American Journal of Physics* 68, pp. 835–840.

- Silverman, M. P. (1981) An exactly soluble quantum model of an atom in an arbitrarily strong uniform magnetic field. *American Journal of Physics* 49, pp. 546–551.

- Liu, C. T., Nakamura, K., Tsui, D. C., Ismail, K., Antoniadis, D. A., and Smith, H. I. (1989). Magneto-optics of a quasi-zero-dimensional electron gas. *Applied Physics Letters* 55, pp. 168–170.

- Gov, S., Shtrikman, S., and Thomas, H. (2000) 1D toy model for magnetic trapping. *American Journal of Physics* 68, pp. 334–343.

Chapter 12: Electron in an Electromagnetic Field and Optical Properties of Nanostructures

Many important quantum mechanics problems deal with the interaction of an electron with an electromagnetic wave such as light. They include problems dealing with absorption and stimulated emission of light, and the behavior of an electron in the presence of light [1–3]. In this chapter, we describe how to treat an electron's interaction with an electromagnetic wave, followed by some important problems and solutions.

Preliminary

An electromagnetic field has a space- and time-varying electric field $\vec{E}(\vec{r}, t)$ and a space- and time-varying magnetic field of flux density $\vec{B}(\vec{r}, t)$. They generate both vector and scalar potentials $\vec{A}(\vec{r}, t)$ and $V(\vec{r}, t)$, respectively, such that the electric and magnetic fields can be related to these potentials as [4]

$$\vec{E}(\vec{r}, t) = -\vec{\nabla} V(\vec{r}, t) - \frac{\partial \vec{A}(\vec{r}, t)}{\partial t}, \tag{12.1}$$

$$\vec{B}(\vec{r}, t) = \nabla \times \vec{A}(\vec{r}, t), \tag{12.2}$$

where $V(\vec{r}, t)$ is the (space- and time-varying) scalar potential and $\vec{A}(\vec{r}, t)$ is the (space- and time-varying) vector potential.

Obviously there are many different vector and scalar potentials that will satisfy the above equations for a given electric or magnetic field. Each such choice is called a "gauge," and all choices satisfying the above relations are legitimate, although some may be preferred over others for no other reason than mathematical convenience. No matter what choice of gauge we adopt, the final observable must always be gauge independent. Consider the following transformations to the scalar and vector potentials:

$$V(\vec{r}, t) \rightarrow V(\vec{r}, t) - \frac{\partial \Lambda(\vec{r}, t)}{\partial t}, \tag{12.3}$$

$$\vec{A}(\vec{r}, t) \rightarrow \vec{A}(\vec{r}, t) + \vec{\nabla} \Lambda(\vec{r}, t), \tag{12.4}$$

where $\Lambda(\vec{r}, t)$ is an arbitrary space- and time-dependent scalar. These potential transformations, which are called "gauge transformations," do not affect the electric field and the magnetic flux densities at all, as can be seen from Equations (12.1) and (12.2). Therefore, the electric and magnetic flux densities are gauge invariant. Any physical quantity must be gauge invariant.

Problem Solving in Quantum Mechanics: From Basics to Real-World Applications for Materials Scientists, Applied Physicists, and Devices Engineers, First Edition.
Marc Cahay and Supriyo Bandyopadhyay.
© 2017 John Wiley & Sons Ltd. Published 2017 by John Wiley & Sons Ltd.

Problem 12.1: Electron–photon interaction Hamiltonian

The Hamiltonian of an electron in an electromagnetic field is of the form

$$H = \frac{1}{2m}\left|\vec{p} - e\vec{A}(t)\right|^2 + V(\vec{r},t) + V_0(\vec{r},t), \tag{12.5}$$

where e is the electronic charge, $\vec{p} = -i\hbar\vec{\nabla}$, and $V_0(\vec{r},t)$ is the scalar potential that the electron experiences in the absence of the wave. The electromagnetic field is, of course, composed of photons. From the above, derive the electron–photon interaction operator if the electromagnetic wave is a transverse electric and magnetic (TEM) wave.

Solution: In the absence of the electromagnetic wave, the Hamiltonian is

$$H_0 = \frac{1}{2m^*}|\vec{p}|^2 + V_0(\vec{r},t) = -\frac{\hbar^2}{2m^*}\nabla^2 + V_0(\vec{r},t). \tag{12.6}$$

Therefore, the electron–photon interaction Hamiltonian is

$$H_{\text{e-p}} = H - H_0 = -\frac{e}{2m^*}\left[\vec{p}\cdot\vec{A}(\vec{r},t) + \vec{A}(\vec{r},t)\cdot\vec{p}\right]$$

$$+ \frac{e^2}{2m^*}\vec{A}(\vec{r},t)\cdot\vec{A}(\vec{r},t) + V(\vec{r},t). \tag{12.7}$$

Note that, in general, the momentum operator and the vector potential will not commute, since

$$\vec{p}\cdot\vec{A}(\vec{r},t) - \vec{A}(\vec{r},t)\cdot\vec{p} = -i\hbar\vec{\nabla}\cdot[\vec{A}(\vec{r},t)] \neq 0. \tag{12.8}$$

However, for a TEM wave, the divergence of the vector potential can be shown to be zero as follows:

First, since scalar potential is always undefined to the extent of an arbitrary potential, let us set it equal to zero and write $\vec{E}(\vec{r},t) = -\frac{\partial\vec{A}(\vec{r},t)}{\partial t}$. For a TEM wave,

$$\vec{E}(\vec{r},t) = \hat{\mu}E_0\sin(\omega t - \vec{q}\cdot\vec{r}), \tag{12.9}$$

where $\hat{\mu}$ is the unit vector in the direction of the wave's polarization, E_0 is the amplitude of the electric field in the wave, ω is the wave frequency, and $\vec{q}$ is the wavevector. Therefore, from Equation (12.1), we get

$$\vec{A}(\vec{r},t) = \hat{\mu}\frac{E_0}{\omega}\cos(\omega t - \vec{q}\cdot\vec{r}) + \lambda(\vec{r}), \tag{12.10}$$

where $\lambda(\vec{r})$ is an arbitrary time-independent quantity. We will choose the gauge to make $\lambda(\vec{r}) = 0$.

The divergence of the vector potential is

$$\vec{\nabla}\cdot[\vec{A}(\vec{r},t)] = -\hat{\mu}\cdot\vec{q}\frac{E_0}{\omega}\cos(\omega t - \vec{q}\cdot\vec{r}). \tag{12.11}$$

For a TE (transverse electric) or TEM wave, $\hat{\mu}\cdot\vec{q} = 0$. Hence $\text{div}[\vec{A}(\vec{r},t)] = 0$.

Returning to the electron–photon interaction Hamiltonian, it now becomes

$$H_{\text{e-p}} = -\frac{e}{m^*}\vec{A}(\vec{r},t)\cdot\vec{p} + \frac{e^2}{2m^*}|\vec{A}(\vec{r},t)|^2 + V(\vec{r},t). \tag{12.12}$$

Unless the wave amplitude is very large, we can neglect the square term involving the vector potential in comparison with the linear term and write (setting $V(\vec{r},t) = 0$)

$$H_{\text{e-p}} \approx -\frac{e}{m^*}\vec{A}(\vec{r},t)\cdot\vec{p} = \frac{ie\hbar}{2m^*}\frac{E_0}{\omega}\left[e^{i(\omega t - \vec{q}\cdot\vec{r})} + e^{-i(\omega t - \vec{q}\cdot\vec{r})}\right]\hat{\mu}\cdot\vec{\nabla}. \tag{12.13}$$

We can go one step further and eliminate the electric field amplitude. Equating the energy density in the wave to the photon energy density (which is somewhat equivalent to quantizing the wave or recognizing that it consists of discrete quanta of energy in the form of photons), we get

$$\frac{1}{2}\epsilon E_0{}^2 = \frac{\hbar\omega}{\Omega}, \tag{12.14}$$

where Ω is the normalizing volume. We therefore obtain

$$H_{\text{e-p}} == \frac{ie\hbar}{2m^*}\sqrt{\frac{\hbar}{2\epsilon\omega\Omega}}\left[e^{i(\omega t - \vec{q}\cdot\vec{r})} + e^{-i(\omega t - \vec{q}\cdot\vec{r})}\right]\hat{\mu}\cdot\vec{\nabla}. \tag{12.15}$$

Absorption in a quantum well

The electron–photon interaction will cause a photon to induce a transition between one electron energy state and another, resulting in stimulated emission or absorption of a photon. Consider a quantum well in the x–y plane, as shown in Figure 12.1.

An electron in the quantum well is free to move in the x–y plane, but its motion is restricted in the z-direction. The confinement causes every band (conduction

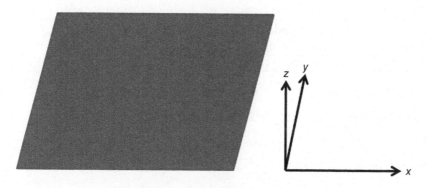

Figure 12.1: Quantum well in the x–y plane.

band, valence band, etc.) to break up into discrete subbands. The electron's wave function in the lth subband of the sth band having the wavevector $\vec{k}_t$ in the plane of the well is given by

$$\psi_{s,l}(\vec{r}, \vec{k}_t) = e^{i\vec{k}_t\cdot\vec{\rho}}\phi_{s,l}(z)u_{\vec{k}_t,s}(\vec{r}) = e^{i\vec{k}_t\cdot\vec{\rho}}\phi_l(z)u_{\vec{k}_t,s}(\vec{\rho}, z), \qquad (12.16)$$

where the vector $\vec{\rho}$ denotes the radial coordinate in the x–y plane. The quantity $\phi_{s,l}(z)$ is the z-component of the wave function in the lth subband of the sth band, and the quantity $u_{\vec{k}_t,s}\vec{r}$ is the Bloch part of the electron's wave function in the sth band for the wavevector $\vec{k}_t$.

The absorption coefficient associated with a photon exciting an electron from the mth subband of the ith band to the nth subband of the jth band is proportional to the magnitude square of the quantity

$$\left|\langle\psi_{j,l}(\vec{r}, \vec{k}_t)|H_{\text{e-p}}|\psi_{i,m}(\vec{r}, \vec{k}_t)\rangle\right|^2 = \left|\int d^3\vec{r}\,\psi_{j,l}^*(\vec{r}, \vec{k}_t)H_{\text{e-p}}\psi_{i,m}^*(\vec{r}, \vec{k}_t)\right|^2. \qquad (12.17)$$

In the case of light polarized in the plane of the quantum well, the last quantity can be shown to be proportional to [2]

$$\left|\int_{-\infty}^{+\infty} dz\phi_{j,n}^*(z)\phi_{i,m}(z)\right|^2 \times \left|ik_\mu\delta_{i,j} + \int_0^\infty d^3\vec{r}\frac{1}{\Omega}u_{k_i,j}^*(\vec{r})\left(\hat{\mu}\cdot\vec{\nabla}u_{k_t,j}(\vec{r})\right)\right|^2$$

$$= \left|\int_{-\infty}^{+\infty} dz\phi_{j,n}^*(z)\phi_{i,m}(z)\right|^2 \times \left|ik_\mu\delta_{i,j} + \frac{1}{\hbar}^2\,\hat{\mu}\cdot\vec{\Gamma}_{ij}\right|^2, \qquad (12.18)$$

where k_μ is the component of $\vec{k}_t$ along the direction of light polarization in the plane of the well, $\delta_{i,j}$ is the Kronecker delta, and $\vec{\Gamma}_{ij} = \int_0^\infty d^3\vec{r}\frac{1}{\Omega}u_{k_i,j}^*(\vec{r})(-i\hbar\vec{\nabla})u_{k_t,j}(\vec{r})$.

Table 12.1 lists the values of the vector $\vec{\Gamma}_{ij}$ for various types of transitions in a quantum well. The row headings denote the initial states and the column headings denote the final states. These states are spin dependent.

Table 12.1: The quantity $\vec{\Gamma}_{ij}$ in a quantum well. Note that the quantity is electron spin dependent.

	Conduction band (up spin)	Conduction band (down spin)
Heavy hole band (up spin)	$(\gamma/\sqrt{2})(\hat{x} + i\hat{y})$	0
Heavy hole band (down spin)	0	$i\gamma/\sqrt{2}(\hat{x} - i\hat{y})$
Light hole band (up spin)	$i\gamma\sqrt{2/3}\hat{z}$	$\frac{i\gamma}{\sqrt{6}}(\hat{x} + i\hat{y})$
Light hole band (up spin)	$\frac{\gamma}{\sqrt{6}}(\hat{x} - i\hat{y})$	$\gamma\sqrt{2/3}\hat{z}$
Split off band (up spin)	$\gamma\sqrt{1/3}\hat{z}$	$\frac{\gamma}{\sqrt{3}}(\hat{x} + i\hat{y})$
Split off band (down spin)	$\frac{-i\gamma}{\sqrt{3}}(\hat{x} - i\hat{y})$	$\frac{i\gamma}{\sqrt{1/3}}\hat{z}$

Similarly, in the case of light polarized perpendicular to the plane of the quantum well, it can be shown that the absorption is proportional to [2]

$$\left| \hat{z} \cdot \vec{\Gamma}_{ij} \int_{-\infty}^{+\infty} \phi_{j,n}^*(z)\phi_{i,m}(z) + \delta_{i,j} \int_{-\infty}^{+\infty} \phi_{j,n}^*(z) \left[-i\hbar \frac{\partial}{\partial z} \right] \phi_{i,m}(z) \right|^2. \qquad (12.19)$$

Problem 12.2: Absorption coefficient in a quantum well

Show that if light is polarized in the plane of a quantum well, then the strength of absorption involving an electron transitioning from the highest heavy hole subband to the lowest conduction band subband is three times stronger than the strength of absorption involving an electron transitioning from the highest light hole subband to the lowest conduction band subband. Also show that:

(a) The heavy hole transition preserves spin while the light hole transition flips spin.

(b) Intraband intersubband absorption due to transition between two subbands in the same band is not possible.

Solution: For light polarized in the plane of the quantum well, the unit polarization vector is given by $\hat{\nu} = a\hat{x} + b\hat{y}$, with $|a|^2 + |b|^2 = 1$. For interband transitions ($i \neq j$), $\delta_{i,j} = 0$. Therefore, the absorption coefficient

$$\alpha \sim \left| \int_{-\infty}^{+\infty} \phi_{j,1}^*(z)\phi_{i,1}(z) \right|^2 \times \left| \hat{\nu} \cdot \vec{\Gamma}_{ij} \right|^2. \qquad (12.20)$$

Using Table 12.1, we see that:

(a) For a heavy hole transition, $\left| \hat{\nu} \cdot \vec{\Gamma}_{ij} \right|^2 \sim \left| (a\hat{x} + b\hat{y}) \cdot \frac{1}{\sqrt{2}} (\hat{x} + i\hat{y}) \right|^2 = \frac{1}{2}|a \pm ib|^2 = \frac{1}{2}\left(|a|^2 + |b|^2\right) = \frac{1}{2}$ for a spin-conserving transition, and exactly zero for a spin-flip transition.

For a light hole transition, $\left| \hat{\nu} \cdot \vec{\Gamma}_{ij} \right|^2 \sim \left| (a\hat{x} + b\hat{y}) \cdot \frac{1}{\sqrt{6}}(\hat{x} + i\hat{y}) \right|^2 = \frac{1}{6}|a \pm ib|^2 = \frac{1}{6}\left(|a|^2 + |b|^2\right) = \frac{1}{6}$ for a spin-flip transition. For a spin-conserving transition, $\left| \hat{\nu} \cdot \vec{\Gamma}_{ij} \right|^2 \sim |(a\hat{x} + b\hat{y}) \cdot \hat{z}|^2 = 0$.

Therefore, the heavy hole transition is spin conserving and the corresponding absorption is three times stronger than that due to the light hole transition, which is spin flipping.

(b) For an intraband transition ($i = j$),

$$\left| \int_{-\infty}^{+\infty} \phi_{j,m}^*(z)\phi_{i,n}(z) \right|^2 = \left| \int_{-\infty}^{+\infty} \phi_{i,m}^*(z)\phi_{i,n}(z) \right|^2 = 0, \qquad (12.21)$$

since the wave functions $\phi_{i,m}(z)$ and $\phi_{i,n}(z)$ are orthogonal. Therefore, light polarized in the plane of the quantum well cannot induce transitions between subbands in the same band.

Problem 12.3:

Show that if light is polarized perpendicular to the plane of the quantum well, then no heavy hole to conduction band transition is possible, but light hole to conduction band transition is possible. Also show that intraband intersubband transition is possible, unlike in the case of light polarized in the plane of the quantum well.

Solution: In this case, the absorption coefficient is [2]

$$\alpha \sim \left| \hat{z} \cdot \vec{\Gamma}_{ij} \int_{-\infty}^{+\infty} \phi_{j,n}^*(z)\phi_{i,m}(z) + \delta_{i,j} \int_{-\infty}^{+\infty} \phi_{j,n}^*(z)\left[-i\hbar\frac{\partial}{\partial z}\right]\phi_{i,m}(z) \right|^2. \quad (12.22)$$

For interband transition, $i \neq j$ and therefore $\alpha \sim \left| \hat{z} \cdot \vec{\Gamma}_{ij} \int_{-\infty}^{+\infty} \phi_{j,n}^*(z)\phi_{i,m}(z) \right|^2$. For a heavy hole transition, $\hat{z} \cdot \vec{\Gamma}_{ij} = \hat{z} \cdot (\hat{x} \pm i\hat{y}) = 0$. Hence, no heavy hole transition is allowed.

For a light hole transition, $\hat{z} \cdot \vec{\Gamma}_{ij} \sim \hat{z} \cdot \hat{z} \neq 0$. Hence, light hole transitions are allowed.

For an intraband transition between two subbands ($i = j$, $m \neq n$),

$$\alpha \sim \left| \int_{-\infty}^{+\infty} dz \phi_{j,n}^*(z)\frac{\partial\phi_{j,m}(z)}{\partial z} \right|^2 \neq 0, \quad (12.23)$$

even though $\int_{-\infty}^{+\infty} dz \phi_{j,n}^*(z)\phi_{j,m}(z) = 0$. This means intraband and intersubband transitions are allowed.

Problem 12.4: Photoluminescence from a quantum well

A photoluminescence experiment in a CdTe/InSb quantum well was carried out with polarized light. The bulk bandgaps of InSb and CdTe are 0.17 eV and 1.56 eV, respectively. The electrons were excited with polarized light from the valence to the conduction band, and as they decayed back to the valence band they emitted photoluminescence light. When the light was polarized in the plane of the quantum well, two photoluminescence peaks were found separated by an energy of 30 meV and the height of the peak at the higher photon energy was one-third that at the lower photon energy, suggesting that perhaps they were due to light and heavy hole transitions.

(a) What other experiment should you carry out to verify this, and what would you expect to observe if these were indeed the two transitions?

(b) Can you estimate the width of the quantum well if the effective masses of heavy holes and light holes (in units of free electron mass) in the narrow gap semiconductor InSb are $m_{hh}^* = 0.244$ and $m_{lh}^* = 0.021$, respectively?

Solution:

(a) One should also measure photoluminescence with light polarized perpendicular to the plane of the quantum well. If the initial guess was correct, then the lower photon energy peak will disappear because the heavy hole transition corresponds to that peak and it should not be present when light is polarized perpendicular to the well. Furthermore, the higher energy peak should increase in strength if it is due to light hole transition. The light hole transition with in-plane light polarization has a relative strength of

$$\left|\hat{\mu}\cdot\vec{\Gamma}_{ij}\right|^2 \sim \left|(a\hat{x}+b\hat{y})\cdot\frac{1}{\sqrt{6}}(\hat{x}+i\hat{y})\right|^2 = \frac{1}{6}|a+ib|^2 = \frac{1}{6}\left(|a|^2+|b|^2\right) = \frac{1}{6}, \quad (12.24)$$

and with perpendicular-to-plane polarization, the relative strength is

$$\left|\hat{\mu}\cdot\vec{\Gamma}_{ij}\right|^2 \sim \left|\hat{x}\cdot\sqrt{\frac{2}{3}}\right|^2 = \frac{2}{3}. \quad (12.25)$$

Therefore, (i) the height of the higher photon energy peak should increase four times, and (ii) the lower energy peak should disappear when the incident light polarization is changed from in-plane to perpendicular-to-plane.

(b) Since the bandgap of InSb is so much smaller than that of CdTe, we can view the confining potential of electrons in the InSb layer as an infinite square well potential. Hence, the energy separation between the light hole and heavy hole peaks is

$$\Delta E = \frac{\hbar^2}{2}\left(\frac{1}{m_{lh}^*} - \frac{1}{m_{hh}^*}\right)\left(\frac{\pi}{W}\right)^2, \quad (12.26)$$

where W is the quantum well width. Since $\Delta E = 30\,\text{meV}$, the well width is 23.2 nm.

Problem 12.5: Quantum efficiency of a semiconductor quantum wire

A light-emitting device (e.g., a laser or a light-emitting diode) emits light when an electron decays from a higher energy state to a lower energy state, in the process emitting a photon. The quantum efficiency of the device is defined as

$$Q = \frac{\tau_{nr}}{\tau_{nr} + \tau_r}, \quad (12.27)$$

where $1/\tau_r$ is the rate with which electrons decay by emitting photons (radiative decay) and $1/\tau_{nr}$ is the rate with which the electron decays by emitting phonons (non-radiative decay).

In a semiconductor quantum wire (quasi one-dimensional system) with rectangular cross section, the conduction band is discretized into subbands (particle-in-a-box states) whose energies are given by

$$E_{i,j} = \frac{\hbar^2}{2m^*}\left[\left(\frac{i\pi}{W_y}\right)^2 + \left(\frac{j\pi}{W_z}\right)^2\right],$$
(12.28)

and the (unnormalized) wave functions of the states are given by

$$\Phi_{i,j} = e^{ik_x x}\sin\left(\frac{i\pi y}{W_y}\right)\sin\left(\frac{j\pi z}{W_z}\right),$$
(12.29)

where i and j are integers (called the "transverse subband indices in the y- and z-directions"), W_y is the thickness along the y-direction, and W_z is the width along the z-direction.

The rate with which a photon *polarized in the y-direction is emitted by an electron decaying from one subband to another with the same transverse subband index in the z-direction but different transverse subband index in the y-direction is*

$$\frac{1}{\tau_r} \sim \left|\int_0^{W_y} dy \sin\left(\frac{i\pi y}{W_y}\right)\frac{\partial \sin\left(\frac{i'\pi y}{W_y}\right)}{\partial y}\right|^2,$$
(12.30)

where i and i' are the subband indices corresponding to y-confinement. On the other hand, the rate with which a phonon *is emitted by transition between the same two states is given by [2]*

$$\frac{1}{\tau_{nr}} \sim \left|\int_0^{W_y} dy \sin\left(\frac{i\pi y}{W_y}\right)\sin\left(\frac{i'\pi y}{W_y}\right)\right|^2.$$
(12.31)

Find the quantum efficiency associated with this process.

Solution: The non-radiative rate is exactly zero since the wave functions are mutually orthogonal because they are eigenstates of the Hermitian operator describing the Hamiltonian of the quantum wire. Thus, no phonon emission can occur between these two states.

The radiative rate, however, is proportional to

$$\left|\frac{\cos(i+i')\pi - 1}{i+i'} + \frac{\cos(i-i')\pi - 1}{i-i'}\right|^2.$$
(12.32)

This quantity is non-zero unless both $i + i'$ and $i - i'$ are even. Since the radiative rate can be non-zero and the non-radiative rate is zero, this process can yield 100% quantum efficiency.

Problem 12.6: Atoms in a super intense laser field [5–7]

For an atom in the presence of an electrostatic potential $V(\vec{r})$ and a spatially uniform circular polarized laser beam characterized by the time-dependent vector potential $\vec{A}(t) = A(\hat{x}\cos\omega t + \hat{y}\sin\omega t)$, show that the time-dependent Schrödinger equation

$$i\hbar\frac{\partial}{\partial t}\psi(\vec{r},t) = H_0\psi(\vec{r},t) = \left[\frac{1}{2m}\left(\vec{p} - e\vec{A}(t)\right)^2 + V(\vec{r})\right]\psi(\vec{r},t) \qquad (12.33)$$

can be reduced to the form

$$i\hbar\frac{\partial}{\partial t}\phi(\vec{r},t) = H\phi(\vec{r},t) = \left[\frac{p^2}{2m} + V(\vec{r}+\vec{R})\right]\phi(\vec{r},t), \qquad (12.34)$$

where

$$\vec{R} = -\int_{-\infty}^{t}\frac{e}{m}\vec{A}(\tau)\mathrm{d}\tau, \qquad (12.35)$$

by performing the canonical transformation

$$\phi(\vec{r},t) = \Omega\psi(\vec{r},t) = \mathrm{e}^{iS}\psi(\vec{r},t), \qquad (12.36)$$

with

$$S = \frac{1}{\hbar}\int_{-\infty}^{t}\left[-\frac{e}{m}\vec{A}(\tau)\cdot\vec{p} + \frac{e^2}{2m}A^2(\tau)\right]\mathrm{d}\tau, \qquad (12.37)$$

where $\vec{p} = \frac{\hbar}{i}\vec{\nabla}$ is the momentum operator.

Solution: Since the operator S is Hermitian, the operator Ω is unitary. Using Ω to perform a unitary transformation on the Schrödinger Equation (12.34), we obtain

$$\left[\mathrm{e}^{+iS}H_0\mathrm{e}^{-iS}\right]\phi(\vec{r},t) = i\hbar\mathrm{e}^{iS}\frac{\partial}{\partial t}\left[\mathrm{e}^{-iS}\phi(\vec{r},t)\right]. \qquad (12.38)$$

We first rewrite e^{iS} as

$$\Omega = \mathrm{e}^{iS} = \mathrm{e}^{i(S_1+S_2)}, \qquad (12.39)$$

where

$$S_1 = -\frac{1}{\hbar}\int_{-\infty}^{t}\frac{e}{m}\vec{A}\cdot\vec{p}\mathrm{d}\tau, \qquad (12.40)$$

$$S_2 = \frac{e^2}{2m}\int_{-\infty}^{t}A^2\mathrm{d}\tau. \qquad (12.41)$$

Since S_1 and S_2 commute, use of the Glauber identity (see Problem 2.4) leads to

$$\Omega = \mathrm{e}^{iS_1}\mathrm{e}^{iS_2}. \qquad (12.42)$$

Under this unitary transformation, the transformed Hamiltonian becomes

$$H = e^{+iS} H_0 e^{-iS} = e^{-iS_2} e^{-iS_1} \frac{(\vec{p} - e\vec{A})^2}{2m} e^{iS_1} e^{iS_2} = \frac{1}{2m} \left(\vec{p} - e\vec{A} \right)^2, \qquad (12.43)$$

since both S_1 and S_2 commute with $\frac{1}{2m}(\vec{p} - e\vec{A})^2$. Furthermore, we get

$$e^{-iS_2} e^{-iS_1} V(\vec{r}) e^{iS_1} e^{iS_2} = V(\vec{r} + \vec{R}), \qquad (12.44)$$

where

$$\vec{R} = -\int_{-\infty}^{t} \frac{e}{m} \vec{A}(\tau) \mathrm{d}\tau. \qquad (12.45)$$

The right-hand side of Equation (12.38) becomes

$$\text{R.H.S.} = i\hbar e^{iS} \frac{\partial}{\partial t} \left(e^{-iS} \phi \right) = -\hbar \dot{S} \phi + i\hbar \frac{\partial \phi}{\partial t}, \qquad (12.46)$$

where the dot denotes the first derivative with respect to time.

So,

$$\text{R.H.S.} = \left[-\frac{e}{m} \vec{A} \cdot \vec{p} + \frac{e^2}{2m} A^2 \right] \phi. \qquad (12.47)$$

Note that

$$\frac{(\vec{p} - e\vec{A})^2}{2m} = \frac{p^2}{2m} - \frac{e}{m} \vec{A} \cdot \vec{p} + \frac{e^2}{2m} \vec{A} \cdot \vec{A}, \qquad (12.48)$$

because $\vec{A}$ is spatially invariant.

Regrouping the previous results, the wave function $\phi(\vec{r}, t)$ is found to satisfy the Schrödinger equation

$$-\frac{\hbar^2}{2m} \nabla^2 \phi(\vec{r}, t) + V(\vec{r} + \vec{R}) \phi(\vec{r}, t) = i\hbar \frac{\partial}{\partial t} \phi(\vec{r}, t). \qquad (12.49)$$

Suggested problems

- Consider light polarized in the plane of a quantum well in the x–y plane. Derive an expression for the absorption coefficient as a function of photon frequency associated with electron excitation from the first heavy hole subband to the first electron subband. Assume that the heavy hole subband is completely filled with electrons and the conduction band subband is completely empty.

  Hint: The absorption coefficient is

  $$\alpha \sim \left| \int_{-\infty}^{+\infty} \mathrm{d}z \phi_{\text{electron}}^*(z) \phi_{\text{heavy hole}}(z) \mathrm{d}z \right|^2$$

  $$\times \left| \vec{\mu} \cdot \vec{\Gamma}_{\text{electron:heavy hole}} \right|^2 \sum_{\vec{k}_t} \delta \left(E_{\text{electron}} - E_{\text{heavy hole}} - \hbar\omega \right),$$

  where $\vec{\mu}$ is the photon polarization unit vector, E_{electron} is the energy of an electron in the lowest conduction subband, $E_{\text{heavy hole}}$ is the energy of an electron in the highest heavy hole subband, and ω is the photon frequency.

- Repeat the above problem for a quantum wire whose axis is along the x-direction.

  Hint: The absorption coefficient is

  $$
  \alpha \sim \left| \int_{-\infty}^{+\infty} dz \phi^*_{\text{electron}}(z) \phi_{\text{heavy hole}}(z) dz \right|^2
  $$

  $$
  \times \left| \int_{-\infty}^{+\infty} dy \phi^*_{\text{electron}}(z) \phi_{\text{heavy hole}}(y) dy \right|^2
  $$

  $$
  \times \left| \vec{\mu} \cdot \vec{\Gamma}_{\text{electron:heavy hole}} \right|^2 \sum_{\vec{k}_x} \delta \left(E_{\text{electron}} - E_{\text{heavy hole}} - \hbar \omega \right).
  $$

- For light polarized in the plane of a quantum well, show that the strengths of absorption from the light hole band to the conduction band, and from the split-off band to the conduction band, bear the ratio 1:2. Show also that for light polarized in the z-direction, that ratio is reversed and becomes 2:1.

References

[1] Kroemer, H. (1994) *Quantum Mechanics for Engineering, Materials Science, and Applied Physics*, Chapter 2, Prentice Hall, Englewood Cliffs, NJ.

[2] Bandyopadhyay, S. (2012) *Physics of Nanostructured Solid State Devices*, Chapter 6, Springer, New York.

[3] Ridley, B. K. (1988) *Quantum Processes in Semiconductors*, 2nd edition, Oxford University Press, New York.

[4] Jackson, J. D. (1975) *Classical Electrodynamics*, 2nd edition, Wiley, New York.

[5] Lobo, R. and Hipólito, O. (1987) Electrons bound on a liquid-helium surface in the presence of a laser field. *Physical Review B* 35, pp. 8755–8758.

[6] Galvaão, R. M. O and Miranda, L. C. M. (1983) Quantum theory of an electron in external fields using unitary transformations. *American Journal of Physics* 51, pp. 729–733.

[7] Faisal, F. H. M. (1973) Multiple absorption of laser photons by atoms. *Journal of Physics B: Atomic, Molecular and Optical Physics* 6, pp. L89–L92.

Chapter 13: Time-Dependent Schrödinger Equation

This chapter shows how to solve the time-dependent Schrödinger equation and calculate the probability of transitions between eigenstates of a given Hamiltonian due to the presence of a time-dependent potential that *cannot* be considered as a weak perturbation. Several properties of one-dimensional Gaussian wave packets [1–4] are studied, including a calculation of the spatio-temporal dependence of their probability current and energy flux densities and a proof that their average kinetic energy is a constant of motion. An algorithm to study the time evolution of wave packets, based on the Crank–Nicholson scheme [5], is discussed for the cases of totally reflecting and absorbing boundaries at the ends of the simulation domain.

** Problem 13.1: Time-dependent states of a harmonic oscillator

The eigenenergies of a particle in a one-dimensional parabolic potential $m^*\omega^2 z^2/2$ are expressed as $E_n = (n + \frac{1}{2})\hbar\omega$ (see Appendix B). Suppose that at time $t = 0$, the particle has equal probability of being in the ith and jth eigenstate and zero probability of being in any other state. Derive an expression for the time dependence of the velocity $v(t)$.

Hint: *Do not try to use the Ehrenfest theorem (see Problem 2.20). You will not need to know the wave functions of the electrons in the ith or jth eigenstate to solve this problem, but will need to know that they obey the following relation:*

$$\int_{-\infty}^{\infty} \phi_i^*(z) z \phi_j(z) \mathrm{d}z = \begin{cases} \sqrt{\frac{\hbar}{m^*\omega}\left(\frac{i+1}{2}\right)} & \text{for } j = i+1, \\ \sqrt{\frac{\hbar}{m^*\omega}\left(\frac{i}{2}\right)} & \text{for } j = i-1, \\ 0 & \text{otherwise.} \end{cases} \tag{13.1}$$

Solution: At any time, the wave function of the electron is a linear superposition of all eigenstates and hence can be written as

$$\psi(z, t) = \sum_n C_n(t)\phi_n(z). \tag{13.2}$$

The probability of finding the electron in the nth eigenstate at time t is $|C_n(t)|^2$. Obviously, $\sum_n |C_n(t)|^2 = 1$, at all t, for the sake of normalization (the electron has unit probability of being in a mixture of one or more allowed states at all times).

The time-dependent Schrödinger equation is

$$i\hbar\frac{\partial\psi(z, t)}{\partial t} = H\psi(z, t), \tag{13.3}$$

Problem Solving in Quantum Mechanics: From Basics to Real-World Applications for Materials Scientists, Applied Physicists, and Devices Engineers, First Edition.
Marc Cahay and Supriyo Bandyopadhyay.
© 2017 John Wiley & Sons Ltd. Published 2017 by John Wiley & Sons Ltd.

where

$$H = -\frac{\hbar^2}{2m^*}\frac{d^2}{dz^2} + \frac{1}{2}m^*\omega^2 z^2. \tag{13.4}$$

Substituting Equation (13.2) in Equation (13.3), we get

$$i\hbar\sum_n \frac{\partial C_n(t)}{\partial}\phi_n(z) = \sum_n C_n(t)H\phi_n(z) = \sum_n C_n(t)E_n\phi_n(z), \tag{13.5}$$

where we have used the fact that H is time independent, and we have also made use of the time-independent Schrödinger equation $H\phi_n(z) = E_n\phi_n(z)$.

Multiplying the above equation throughout by $\phi_n^*(z)$ and integrating over all z, we get

$$i\hbar\sum_n \frac{\partial C_n(t)}{\partial t}\int_{-\infty}^{+\infty} \phi_m^*(z)\phi_n(z)\mathrm{d}z = \sum_n C_n(t)E_n\int_{-\infty}^{+\infty} \phi_m^*(z)\phi_n(z). \tag{13.6}$$

Since the ϕ are eigenfunctions of the Hamiltonian, which is always a Hermitian operator, they are orthonormal, meaning $\int_{-\infty}^{+\infty}\phi_m^*(z)\phi_n(z)\mathrm{d}z = \delta_{m,n}$, where the delta is the Kronecker delta, i.e., it is 1 when $m = n$ and 0 otherwise. Therefore, the last equation reduces to

$$i\hbar\sum_n \frac{\partial C_n(t)}{\partial t}\delta_{m,n} = \sum_n C_n(t)E_n\delta_{m,n}. \tag{13.7}$$

The Kronecker delta removes the summation since only one term (corresponding to $m = n$) will survive in the sum and all the other terms will vanish. This means,

$$i\hbar\frac{\partial C_m(t)}{\partial t} = C_m(t)E_m. \tag{13.8}$$

The solution of the last equation is

$$C_m(t) = C_m(0)\mathrm{e}^{-\frac{iE_m t}{\hbar}}. \tag{13.9}$$

Substituting this result in Equation (13.2), we obtain

$$\psi(z,t) = \sum_n \mathrm{e}^{-iE_n t/\hbar}C_n(0)\phi_n(z). \tag{13.10}$$

The problem states that at time $t = 0$, there was equal probability of being in the ith and jth state and zero probability of being in any other state. Hence, $C_i(0) = C_j(0) = 1/\sqrt{2}$, and all the other $C(0)$s are zero. Substituting these results in Equation (13.10), we get

$$\begin{aligned}\psi(z,t) &= \mathrm{e}^{-\frac{iE_i t}{\hbar}}C_i(0)\phi_i(z) + \mathrm{e}^{-\frac{iE_j t}{\hbar}}C_j(0)\phi_j(z) \\ &= \frac{1}{\sqrt{2}}\left[\mathrm{e}^{-\frac{iE_i t}{\hbar}}\phi_i(z) + \mathrm{e}^{-\frac{iE_j t}{\hbar}}\phi_j(z)\right].\end{aligned} \tag{13.11}$$

Hence,

$$
\begin{aligned}
\langle z \rangle(t) &= \int_{-\infty}^{+\infty} \psi^*(z,t) z \psi(z,t) \mathrm{d}z \\
&= \frac{1}{2} \int_{-\infty}^{+\infty} \left[e^{\frac{iE_i t}{\hbar}} \phi_i{}^*(z) + e^{\frac{iE_j t}{\hbar}} \phi_j{}^*(z) \right] z \left[e^{-\frac{iE_i t}{\hbar}} \phi_i(z) + e^{-\frac{iE_j t}{\hbar}} \phi_j(z) \right] \mathrm{d}z, \\
&= \frac{1}{2} \left[\int_{-\infty}^{\infty} \phi_i^*(z) z \phi_i(z) \mathrm{d}z + \int_{-\infty}^{\infty} \phi_j^*(z) z \phi_j(z) \mathrm{d}z \right. \\
&\quad \left. + e^{i(E_i - E_j)t/\hbar} \int_{-\infty}^{\infty} \phi_i^*(z) z \phi_j(z) \mathrm{d}z + e^{i(E_j - E_i)t/\hbar} \int_{-\infty}^{\infty} \phi_j^*(z) z \phi_i(z) \mathrm{d}z \right] \\
&= \int_{-\infty}^{+\infty} \psi^*(z,t) z \psi(z,t) \mathrm{d}z \left[\frac{e^{i(E_i - E_j)t/\hbar} + e^{-i(E_i - E_j)t/\hbar}}{2} \right],
\end{aligned} \tag{13.12}
$$

where we have used the fact that in any eigenstate, $\langle z \rangle = \int_{-\infty}^{\infty} \phi_i^*(z) z \phi_i(z) \mathrm{d}z = 0$ and $\int_{-\infty}^{\infty} \phi_i^*(z) z \phi_j(z) \mathrm{d}z = \int_{-\infty}^{\infty} \phi_j^*(z) z \phi_i(z) \mathrm{d}z$, since the operator z is Hermitian.

Therefore,

$$
\langle z \rangle(t) = \int_{-\infty}^{+\infty} \mathrm{d}z \, \phi_i^*(z) z \phi_j(z) \cos\left[(i-j)\omega t\right], \tag{13.13}
$$

since $E_n = (n + \frac{1}{2})\hbar\omega$.

If $i = j + 1$, then

$$
\int_{-\infty}^{+\infty} \mathrm{d}z \, \phi_i^*(z) z \phi_j(z,t) = \sqrt{\frac{i}{2} \frac{\hbar}{m^*\omega}}, \tag{13.14}
$$

and

$$
\langle z \rangle(t) = \sqrt{\frac{i}{2} \frac{\hbar}{m^*\omega}} \cos(\omega t). \tag{13.15}
$$

If $i = j - 1$, then

$$
\int_{-\infty}^{+\infty} \mathrm{d}z \, \phi_i^*(z) z \phi_j(z) = \sqrt{\frac{j}{2} \frac{\hbar}{m^*\omega}}, \tag{13.16}
$$

and

$$
\langle z \rangle(t) = \sqrt{\frac{j}{2} \frac{\hbar}{m^*\omega}} \cos(\omega t). \tag{13.17}
$$

In all the other cases, $\langle z \rangle(t) = 0$.

Therefore,

$$
v(t) = \frac{\mathrm{d}\langle z \rangle(t)}{\mathrm{d}t} = \begin{cases} -\sqrt{\frac{i\hbar\omega}{2m^*}} \sin(\omega t) & \text{for } i = j \pm 1, \\ 0, & \text{otherwise.} \end{cases} \tag{13.18}
$$

The physical interpretation of this result is that classical simple harmonic motion, where the velocity of the oscillator varies sinusoidally with time, can be initiated by placing the quantum mechanical oscillator in a mixture of two adjacent eigenstates.

*** Problem 13.2: Evolution of a one-dimensional harmonic oscillator in a time-varying electric field

Suppose a one-dimensional harmonic oscillator is in its ground state at time $t = 0$ when an external spatially invariant time-dependent force $F(t)$ is turned on. Find the probability that the oscillator will be in its nth state at time t for the following two cases:

(a) $F(t) = F_0 e^{-t^2/\tau^2}$,

(b) $F(t) = F_0 \left[1 + t^2/\tau^2\right]$.

Solution: The 1D harmonic oscillator is described by the Hamiltonian

$$H = -\frac{\hbar^2}{2m^*}\frac{d^2}{dz^2} + \frac{1}{2}m^*\omega^2 z^2. \tag{13.19}$$

In the presence of a spatially uniform but time-varying force $F(t)$, the Hamiltonian becomes

$$H = H_0 - zF(t). \tag{13.20}$$

This last Hamiltonian can be written in terms of the creation and annihilation operators defined in Problem 2.12. We get

$$H = \hbar\omega\left(a^\dagger a + \frac{1}{2}\right) - (a^\dagger + a)\sqrt{\hbar/2m^*\omega}F(t). \tag{13.21}$$

We seek a solution $\phi(t)$ to the time-dependent Schrödinger equation

$$i\hbar\frac{\partial\phi(t)}{\partial t} = H\phi(t) = (H_0 + H')\phi(t), \tag{13.22}$$

where

$$H_0 = \hbar\omega\left(a^\dagger a + \frac{1}{2}\right), \tag{13.23}$$

$$H' = -(a^\dagger + a)\sqrt{\frac{\hbar}{2m^*\omega}}F(t). \tag{13.24}$$

Using the commutation relation

$$[a, a^\dagger] = 1, \tag{13.25}$$

we get

$$[a, H] = \hbar\omega a - \sqrt{\frac{\hbar}{2m^*\omega}}F(t), \tag{13.26}$$

$$[a^\dagger, H] = -\hbar\omega a^\dagger - \sqrt{\frac{\hbar}{2m^*\omega}}F(t). \tag{13.27}$$

Next, we calculate the time evolution of the average value of the annihilation operator $A(t) = \langle \phi(t)|a|\phi(t) \rangle$. Using Ehrenfest's theorem (see Problem 2.20), we get

$$\frac{\mathrm{d}A}{\mathrm{d}t} = \frac{\mathrm{d}\langle a \rangle}{\mathrm{d}t} = -\frac{i}{\hbar}\langle \phi[a, H]\phi \rangle = -i\omega A + i(2m^*\hbar\omega)^{-\frac{1}{2}}F(t). \tag{13.28}$$

The general solution of this first-order differential equation is

$$A(t) = A(-\infty) + i(2m^*\hbar\omega)^{-\frac{1}{2}}\int_{-\infty}^{t} F(t')e^{-i\omega(t-t')}\mathrm{d}t'$$

$$= A(-\infty) + ie^{-i\omega t}\lambda(t), \tag{13.29}$$

where the following function was introduced:

$$\lambda(t) = (2m^*\hbar\omega)^{-\frac{1}{2}}\int_{-\infty}^{t} F(t')e^{i\omega t'}\,\mathrm{d}t'. \tag{13.30}$$

The probability for the oscillator to be in its nth eigenstate at time t is

$$P_n = |\langle n|\phi(t) \rangle|^2. \tag{13.31}$$

Since for the 1D harmonic oscillator we have

$$|n\rangle = (n!)^{-1/2}(a^\dagger)^n|0\rangle, \tag{13.32}$$

Equation (13.31) can be rewritten as

$$P_n = \frac{1}{n!}|\langle 0|a^n|\phi(t) \rangle|^2. \tag{13.33}$$

Next, we consider the state

$$\chi(t) = (a - A(t))\phi(t). \tag{13.34}$$

Using Equations (13.28) and (13.34), we get

$$i\hbar\frac{\mathrm{d}\chi(t)}{\mathrm{d}t} = (a - A(t))H\phi(t) - i\hbar\frac{\mathrm{d}A(t)}{\mathrm{d}t}\phi = (H + \hbar\omega)\chi. \tag{13.35}$$

Since $H + \hbar\omega$ is Hermitian, $\langle \chi|\chi \rangle$ is a constant of motion. At time $t = -\infty$, $\phi(-\infty)$ is the ground state of the harmonic oscillator. Therefore, $A(-\infty) = 0$. Hence, $\chi(-\infty) = 0$. From Equation (13.35), we then get $\frac{\mathrm{d}\chi}{\mathrm{d}t}(-\infty) = 0$. As a result, $\chi(t) = 0$ for all time and therefore

$$a\phi(t) = A(t)\phi(t). \tag{13.36}$$

Using Equations (13.33) and (13.36), we get

$$P_n = \frac{|A(t)|^2}{n!}|\langle 0|\phi(t) \rangle|^2 = \frac{|\lambda(t)|^2}{n!}|\langle 0|\phi(t) \rangle|^2. \tag{13.37}$$

Finally, we calculate $\langle 0|\phi(t)\rangle$.

$$i\hbar\frac{\mathrm{d}}{\mathrm{d}t}\langle 0|\phi(t)\rangle = \langle 0|H|\phi(t)\rangle$$

$$= \frac{1}{2}\hbar\omega\langle 0|\phi(t)\rangle - \sqrt{\frac{\hbar}{2m^*}}F(t)\langle 0|(a^\dagger + a)\phi(t)\rangle. \tag{13.38}$$

This yields

$$\frac{\mathrm{d}}{\mathrm{d}t}\langle 0|\phi(t)\rangle = \left[-\frac{i}{2}\omega + i(2m^*\hbar\omega)^{-1/2}F(t)A(t)\right]\langle 0|\phi(t)\rangle. \tag{13.39}$$

Integrating from $-\infty$ to t, and using the fact that $\phi(-\infty) = |0\rangle$, we get

$$\langle 0|\phi(t)\rangle = \exp\left(-\frac{i}{2}\omega t + i(2m^*\hbar\omega)^{-1/2}\int_{-\infty}^{t}F(t')A(t')\mathrm{d}t'\right). \tag{13.40}$$

Furthermore, since $A(-\infty) - 0$, we get

$$A(t) = ie^{-i\omega t}\lambda(t). \tag{13.41}$$

Therefore,

$$|\langle 0|\phi(t)\rangle|^2 = \exp\left[-(2m^*\hbar\omega)^{-1/2}\int_{-\infty}^{t}\mathrm{d}t'F(t')e^{-i\omega t'}\right.$$

$$\left.\times\int_{-\infty}^{t'}\mathrm{d}t''F(t'')e^{i\omega t''} + \text{c.c.}\right], \tag{13.42}$$

where c.c. stands for complex conjugate.

By first swapping the dummy indices t' and t'', and then swapping the order of integration, we get

$$\int_{-\infty}^{t}\mathrm{d}t'\int_{-\infty}^{t'}\mathrm{d}t''F(t')F(t'')e^{-i\omega(t'-t'')}$$

$$= \int_{-\infty}^{t}\mathrm{d}t'\int_{t'}^{t}\mathrm{d}t''F(t')F(t'')e^{i\omega(t'-t'')}. \tag{13.43}$$

Taking this last result into account in Equation (13.42) leads to

$$|\langle 0|\phi(t)\rangle|^2 = \exp\left[-|\lambda(t)|^2\right]. \tag{13.44}$$

Hence,

$$P_n = \frac{|\lambda(t)|^2}{n!}e^{-|\lambda(t)|^2}. \tag{13.45}$$

At $t = +\infty$, for $F(t) = F_0 e^{-t^2/\tau^2}$ we get

$$|\lambda(\infty)|^2 = \frac{\pi F_0^2\tau^2}{2m^*\hbar\omega}e^{-\frac{\omega^2\tau^2}{2}}, \tag{13.46}$$

and for $F(t) = F_0[1 + (t^2/\tau^2)]$ we find

$$|\lambda(\infty)|^2 = \frac{\pi^2 F_0^2\tau^2}{2m^*\hbar\omega}e^{-2\omega\tau}. \tag{13.47}$$

*** Problem 13.3: Time dependence of the probability current density associated with a one-dimensional Gaussian wave packet

Preliminary: *A plane wave solution to the time-dependent Schrödinger equation associated with a free particle moving in one dimension is given by*

$$\Phi_k(x,t) = A e^{i(k_x x - \omega(k_x)t)}, \tag{13.48}$$

where $\omega(k_x) = \hbar k_x^2 / 2m^*$.

A wave packet is a weighted superposition of such plane waves with components that can move in either direction. The most general one-dimensional wave packet associated with a free particle is of the form [1]

$$\Phi(x,t) = \frac{1}{\sqrt{2\pi}} \int_{-\infty}^{+\infty} g(k_x) e^{i(k_x x - \omega(k_x)t)} dk_x. \tag{13.49}$$

At $t = 0$, we have

$$\Phi(x,0) = \frac{1}{\sqrt{2\pi}} \int_{-\infty}^{+\infty} g(k_x) e^{ik_x x} dk_x, \tag{13.50}$$

and $g(k_x)$ can be obtained by simply taking the Fourier transform of the original wave packet, i.e.,

$$g(k_x) = \frac{1}{\sqrt{2\pi}} \int_{-\infty}^{+\infty} \Phi(x,0) e^{-ik_x x} dx. \tag{13.51}$$

For an initial Gaussian wave packet [6],

$$\Psi(x,0) = \left(\frac{2}{\pi a^2}\right)^{1/4} e^{i\phi} e^{ik_{x,0}x} e^{-\frac{x^2}{a^2}}, \tag{13.52}$$

where ϕ is an arbitrary phase and a is a parameter characterizing the initial spread of the wave packet; $k_{x,0}$ is the momentum characterizing the initial average momentum of the wave packet (it is left as an exercise to show that the average value of the momentum operator along the x-axis of the wave packet given by Equation (13.52) is $\langle p_x \rangle = \hbar k_0$).

The Fourier transform of this Gaussian wave packet is

$$g(k_x) = \frac{\sqrt{a}}{2\pi^{1/4}} e^{i\phi} e^{-\frac{a^2}{4}(k_x - k_{x,0})^2}. \tag{13.53}$$

The wave packet at time t can then be calculated using Equation (13.49):

$$\Psi(x,t) = \left(\frac{2a^2}{\pi}\right)^{\frac{1}{4}} \frac{e^{i\phi} e^{ik_0 x}}{\left(a^4 + \frac{4\hbar^2 t^2}{m^{*2}}\right)^{\frac{1}{4}}} \times \exp\left[\frac{-\left(x - \frac{\hbar k_{x,0} t}{m^*}\right)^2}{a^2 + \frac{2i\hbar t}{m^*}}\right]. \tag{13.54}$$

Compute the spatio-temporal dependence of the probability current density $J_x(x,t)$ associated with the one-dimensional Gaussian wave packet (13.54).

Solution: In Problem 5.1, we showed that the probability current density is given by

$$J_x(x,t) = \frac{\hbar}{2m^*i} \left[\psi^*(x,t)\frac{d\psi(x,t)}{dx} - \psi(x,t)\frac{d\psi^*(x,t)}{dx} \right]. \tag{13.55}$$

This last expression can be rewritten as

$$J_x(x,t) = \frac{\hbar}{m}\text{Im}\left[\psi^*(x,t)\frac{d\psi(x,t)}{dx} \right], \tag{13.56}$$

where Im stands for imaginary part.

The Gaussian wave packet (13.54) is first written as $\psi(x,t) = Ne^{ik_0x}\phi(x,t)$, where

$$N = \left(\frac{2a^2}{\pi}\right)^{\frac{1}{4}} \frac{e^{i\phi}}{\left[a^4 + \frac{4\hbar^2t^2}{m^{*2}},\right]^{\frac{1}{4}}}, \tag{13.57}$$

$$\phi(x,t) = \exp\left[\frac{\left(x - \frac{\hbar k_0 t}{m}\right)^2}{a^2 + \frac{2i\hbar t}{m^*}} \right]. \tag{13.58}$$

Hence,

$$\frac{d\psi(x,t)}{dx} = N\left[ik_0\phi(x,t) + \frac{d\phi(x,t)}{dx} \right] e^{ik_0x}, \tag{13.59}$$

and

$$\psi^*(x,t) = N^*e^{-ik_0x}\phi^*(x,t). \tag{13.60}$$

Therefore,

$$\psi^*(x,t)\frac{d\psi(x,t)}{dx} = N^2\left[ik - \frac{2}{f(t)}\left(x - \frac{\hbar kt}{m^*}\right) \right]\phi(x,t)\phi^*(x,t), \tag{13.61}$$

where we have evaluated the derivative $\frac{d\phi(x,t)}{dx}$ from Equation (13.58), and introduced the function

$$f(t) = a^2 + \frac{2i\hbar t}{m^*}. \tag{13.62}$$

Therefore,

$$\psi^*(x,t)\frac{d\psi(x,t)}{dx} = |N|^2\left[ik_0 - 2\frac{\left(a^2 - \frac{2i\hbar t}{m^*}\right)}{\left(a^4 + \frac{4\hbar^2t^2}{m^{*2}}\right)}\left(x - \frac{\hbar k_0 t}{m^*}\right) \right]\phi(x,t)\phi^*(x,t), \tag{13.63}$$

and the probability current density becomes

$$J_x(x,t) = \frac{\hbar k_0}{m^*}|N|^2\left[\frac{a^4 + \frac{4\hbar^2t^2}{m^{*2}}\left(x/(\hbar k_0 t/m^*)\right)}{a^4 + \frac{4\hbar^2t^2}{m^{*2}}} \right]\phi^*(x,t)\phi(x,t). \tag{13.64}$$

Using Equations (13.57) and (13.58), we obtain the final expression:

$$J_x(x,t) = \frac{\hbar k_0}{m^*} \sqrt{\frac{2}{\pi a^2}} \frac{\left[1 + \frac{4\hbar^2 t^2}{m^*}(x/(\hbar k_0 t))\right]}{\left[1 + \frac{4\hbar^2 t^2}{m^* a^4}\right]^{3/2}} \exp\left[\frac{-2\left(x - \frac{\hbar k_0 t}{m^*}\right)^2}{a^2\left(1 + \frac{4\hbar^2 t^2}{m^* a^4}\right)}\right]. \tag{13.65}$$

---

* Problem 13.4: Time dependence of the energy flux density associated with a one-dimensional Gaussian wave packet

For the one-dimensional Gaussian wave packet (13.59) propagating in free space, compute the time dependence of the energy flux density

$$S_x = -\frac{\hbar^2}{2m^*}\left(\frac{\partial \psi^*}{\partial t}\frac{d\psi}{dx} + \frac{\partial \psi}{\partial t}\frac{d\psi^*}{dx}\right) \tag{13.66}$$

introduced in Problem 5.8.

Solution: As shown in Problem 5.9, the energy flux density of a free particle solution to the one-dimensional Schrödinger equation can be written as

$$S_x(x,t) = \frac{\hbar^3}{2m^{*2}}\text{Im}\left[\left(\frac{d^2\psi(x,t)}{dx^2}\right)\left(\frac{d\psi^*(x,t)}{dx}\right)\right], \tag{13.67}$$

where Im stands for imaginary part.

Proceeding as in the previous problem, we first rewrite the one-dimensional Gaussian wave packet (13.54) as $\psi(x,t) = Ne^{ik_0 x}\phi(x,t)$, where N and $\phi(x,t)$ are given by Equations (13.57) and (13.58), respectively.

Next, we derive the following expressions for the derivatives of the wave function associated with the wave packet:

$$\frac{d^2\psi(x,t)}{dx^2} = N\left[\frac{d^2\phi(x,t)}{dx^2} + 2ik_0\frac{d\phi(x,t)}{dx} - k^2\phi(x,t)\right]e^{ik_0 x}, \tag{13.68}$$

$$\frac{d\phi(x,t)}{dx} = -\frac{2}{f(t)}u(x,t)\phi(x,t), \tag{13.69}$$

where $f(t)$ is given by Equation (13.62) and the function $u(x,t)$ is defined as

$$u(x,t) = x - \frac{\hbar k_0 t}{m^*}. \tag{13.70}$$

From Equation (13.69), we also get

$$\frac{d^2\phi(x,t)}{dx^2} = -\frac{2}{f(t)}\left[1 - \frac{2}{f(t)}u^2\right]\phi(x,t). \tag{13.71}$$

Starting with Equations (13.59) and (13.68), we calculate the product:

$$\left(\frac{d^2\psi(x,t)}{dx^2}\right)\left(\frac{d\psi^*(x,t)}{dx}\right) = |N|^2\left[-\frac{2}{f} + \frac{4u^2}{f^2} - \frac{4iu(x,t)k_0}{f} - k_0^2\right]$$

$$\times\left[-ik_0 - \frac{2}{f^*}\right]\phi^*(x,t)\phi(x,t). \tag{13.72}$$

Expanding this last expression and regrouping terms, we finally find

$$S(x,t) = S(k_0) \left[1 + \left\{ \frac{12u(x,t)}{k_0^2 X} + \frac{16u(x,t)^3}{k_0^5 X^4} g(u(x,t),t) \right\} \frac{\hbar k_0 t}{m^*} \right]$$
$$\times N^2 \phi^*(x,t)\phi(x,t), \tag{13.73}$$

where

$$S(k_0) = \frac{\hbar k_0}{m^*} \left(\frac{\hbar^2 k_0^2}{2m^*} \right) \tag{13.74}$$

is the energy flux density associated with the plane wave $e^{ik_0 x}$, while g is the function given by

$$g(u(x,t),t) = 2a^4 \left[a^2 + 2u(x,t)^2 \right] + 8 \frac{\hbar^2 t^2}{m^{*2}} \left[a^2 + 6u(x,t)^2 \right]. \tag{13.75}$$

---

* Problem 13.5: Average kinetic energy of a one-dimensional Gaussian wave packet in free space

As shown in Problem 5.8, the kinetic energy flux density associated with a free particle in one dimension is given by

$$W(x,t) = \frac{\hbar^2}{2m^*} \left(\frac{d\Psi^*(x,t)}{dx} \right) \left(\frac{d\Psi(x,t)}{dx} \right). \tag{13.76}$$

Calculate the explicit expression for W(x,t) starting with the time-dependent one-dimensional Gaussian wave packet given in Equation (13.54), and show that the total kinetic energy of the wave packet, i.e., $E_{\text{tot}} = \int_{-\infty}^{+\infty} W(x,t)$, is time independent and given by

$$E_{\text{tot}} = \frac{\hbar^2}{2m^*} \left(k_0^2 + \frac{1}{2\sigma^2} \right). \tag{13.77}$$

Solution: Starting with Equation (13.59), we have

$$\frac{d\psi^*(x,t)}{dx} = N \left[-ik_0 \phi^*(x,t) + \frac{d\phi^*(x,t)}{dx} \right] e^{-ik_0 x}. \tag{13.78}$$

Hence,

$$W(x,t) = \frac{\hbar^2}{2m^*} |N|^2 \phi(x,t)\phi^*(x,t) \left[k_0^2 + \frac{4u^2(x,t)}{f(t)f^*(t)} + 2ik_0 \left(\frac{1}{f(t)} - \frac{1}{f^*(t)} \right) \right], \tag{13.79}$$

where $f(t)$ is given by Equation (13.62). Using this equation, we get that

$$W(x,t) = \frac{\hbar^2}{2m^*} |N|^2 \phi(x,t)\phi^*(x,t) \left[k_0^2 + \frac{4u(x,t)\left(x + \frac{\hbar k_0 t}{m^*} \right)}{X} \right], \tag{13.80}$$

where $X = a^4 + \frac{4\hbar^2 t^2}{m^{*2}}$.

Integrating both sides of Equation (13.80) over x from $x = -\infty$ to $x = +\infty$, and using the integrals

$$\int_0^\infty dx x^2 e^{-r^2 x^2} = \frac{\sqrt{\pi}}{4r^3}, \tag{13.81}$$

$$\int_{-\infty}^{+\infty} dx x^3 e^{-r^2 x^2} = 0, \tag{13.82}$$

the total kinetic energy of the wave packet is found to be

$$E_{\text{tot}} = \frac{\hbar^2}{2m^*}\left(k_0^2 + \frac{1}{2\sigma^2}\right), \tag{13.83}$$

which is time independent. This generalizes to the well-known classical result that the kinetic energy of a free particle is independent of time because, by definition, a free particle has no force acting on it and hence its velocity does not change with time. This result can also be arrived at by a simple application of Ehrenfest's theorem, derived in Problem 2.20.

***** Problem 13.6: Numerical solution of the time-dependent one-dimensional Schrödinger equation

Starting with the Cayley approximation for the unitary operator to calculate a solution to the time-dependent Schrödinger equation (see Problem 2.24), derive a finite difference scheme to compute a numerical solution to the one-dimensional Schrödinger equation for the case of a spatially varying effective mass and either (a) perfectly reflecting or (b) absorbing boundary conditions at the ends of the solution domain.

Solution: Finite difference approximation for the one-dimensional Schrödinger equation in the case of a constant effective mass

Our starting point is the one-dimensional time-dependent Schrödinger equation for a particle with a constant effective mass m^* in an arbitrary time-independent potential $V(x)$ (see Problem 1.1):

$$H\psi(x,t) = -\frac{\hbar^2}{2m^*}\frac{\partial^2 \psi(x,t)}{\partial x^2} + V(x)\psi(x,t) = i\hbar\frac{\partial \psi(x,t)}{\partial t}. \tag{13.84}$$

For the time being, let us ignore the time dependence and consider the total derivative in space as opposed to the partial derivative. Using equally spaced intervals along the x-axis at integer multiples of a small increment Δx, we first write a finite difference version of the kinetic operator using the approximation [7]

$$\frac{d^2 \psi_i}{dx^2} = \frac{\psi_{i+1} - 2\psi_i + \psi_{i-1}}{(\Delta x)^2}. \tag{13.85}$$

At location i (i.e. in the ith mesh), the expression for the left-hand side of Equation (13.84) can be written as

$$H\psi_i = -\left(\frac{\hbar^2}{2m}\right)\left(\frac{\psi_{i+1} - 2\psi_i + \psi_{i-1}}{\Delta x^2}\right) + V_i\psi_i. \tag{13.86}$$

For a conservative system whose Hamiltonian is time independent (because the potential is time independent), a formal solution to the Schrödinger equation can be written as

$$\psi(x,t) = U(t,t_0)\psi(x,t_0) = e^{-j(t-t_0)H/\hbar}\psi(x,t_0), \tag{13.87}$$

where j is the imaginary square-root of -1.

Calling $\delta t = t - t_0$, and using the Cayley approximation for the unitary operator $e^{-j\delta t H/\hbar}$ (see Problem 2.24), we get

$$U(t - t_0) = e^{-j\delta t H/\hbar} = \frac{1 - j\delta t H/2\hbar}{1 + j\delta t H/2\hbar} = \frac{2\hbar - j\delta t H}{2\hbar + j\delta t H}. \tag{13.88}$$

To characterize the wave function at different time intervals separated by a small increment δt, we use a superscript n which takes integer multiples of δt. With the use of Equation (13.84), the wave functions at two different times separated by a time interval δt are related as follows:

$$\psi_i^{n+1} = \frac{2\hbar - j\delta t H}{2\hbar + j\delta t H}\psi_i^n. \tag{13.89}$$

This last equation can be rearranged as

$$(2\hbar + j\delta t H)\psi_i^{n+1} = (2\hbar - j\delta t H)\psi_i^n. \tag{13.90}$$

Using the expression for $H\psi_i$ in Equation (13.86), we get

$$2\hbar\psi_i^{n+1} - \frac{j\delta t\hbar^2}{2m^*}\left(\frac{\psi_{i+1}^{n+1} - 2\psi_i^{n+1} + \psi_{i-1}^{n+1}}{(\Delta x)^2} - \frac{2m^*}{\hbar^2}V_i^{n+1}\psi_i^{n+1}\right)$$
$$= 2\hbar\psi_i^n + \frac{j\delta t\hbar^2}{2m^*}\left(\frac{\psi_{i+1}^n - 2\psi_i^n + \psi_{i-1}^n}{(\Delta x)^2} - \frac{2m^*}{\hbar^2}V_i^n\psi_i^n\right). \tag{13.91}$$

The last equation leads to an implicit algorithm for obtaining the wave function at the next time step, i.e., at time $t = (n+1)\delta t$, at three discrete points in space (points $(i-1)\Delta x$, $i\Delta x$, and $(i+1)\Delta x$):

$$\psi_{i+1}^{n+1} + \psi_i^{n+1}\left(\frac{2j(\Delta x)^2\hbar}{\delta t} - 2 - (\Delta x)^2 V_i^n\right) + \psi_{i-1}^{n+1}$$
$$= -\psi_{i+1}^n + \psi_i^n\left(\frac{2j(\Delta x)^2\hbar}{\delta t} + 2 + (\Delta x)^2 V_i^n\right) - \psi_{i-1}^n. \tag{13.92}$$

This implicit algorithm is called the Crank–Nicholson scheme. The latter ensures the conservation of the probability density over time, i.e., the integral over the simulation domain of the probability density of the wave packet is conserved. This probability is equal to unity if the original wave packet is normalized.

For a given initial wave packet, repeated application of the implicit algorithm (13.92) allows one to calculate the wave packet at any later time across the entire solution domain. The implementation of the boundary conditions at both ends of the solution domain is described later.

Finite difference approximation for the one-dimensional time-dependent Schrödinger equation in the case of a spatially varying effective mass $m^*(x)$

In this case, the time-dependent Schrödinger equation is given by (see Problem 1.1)

$$i\hbar\frac{\partial\psi(x,t)}{\partial t} = -\frac{\hbar^2}{2[m^*(x)]^2}\frac{\partial m^*(x)}{\partial x}\frac{\partial\psi(x,t)}{\partial x} - \frac{\hbar^2}{2m^*(x)}\frac{\partial^2\psi(x,t)}{\partial x^2}$$
$$+ V(x)\psi(x,t). \tag{13.93}$$

We again rewrite this using the Crank–Nicholson form as

$$i\hbar\frac{\psi_i^{n+1} - \psi_i^n}{\Delta t} = \frac{\hbar^2}{2m_i^{*2}}\left(\frac{m_{i+1}^* - m_{i-1}^*}{2\Delta x}\right)\left(\frac{\psi_{i+1}^{n+1} - \psi_{i-1}^{n+1}}{4\Delta x} + \frac{\psi_{i+1}^n - \psi_{i-1}^n}{4\Delta x}\right)$$
$$- \frac{\hbar^2}{2m_i^*}\left[\frac{\psi_{i+1}^{n+1} - 2\psi_i^{n+1} + \psi_{i-1}^{n+1}}{2(\Delta x)^2} + \frac{\psi_{i+1}^n - 2\psi_i^n + \psi_{i-1}^n}{2(\Delta x)^2}\right]$$
$$+ \frac{1}{2}\left(V_i^{n+1}\psi_i^{n+1} + V_i^n\psi_i^n\right). \tag{13.94}$$

At every grid point, we must find the wave function ψ_i^{n+1} at time $t = (n+1)\delta t$ given the wave function ψ_i^n at time $t = n\delta t$. This is accomplished by deriving a difference equation for the variables representing the difference of the wave function at a given location at two successive time steps, i.e.,

$$\delta\psi_i = \psi_i^{n+1} - \psi_i^n. \tag{13.95}$$

With these new variables, Equation (13.94) can be rewritten as

$$\left(1 + \frac{m_{i+1}^* - m_{i-1}^*}{4m_i^*}\right)\delta\psi_{i-1} + \left(\frac{j4m_i^*(\Delta z)^2}{\hbar\Delta t} - 2 - \frac{2m_i^*(\Delta x)^2 V_i^{n+1}}{\hbar^2}\right)\delta\psi_i$$
$$+ \left(1 - \frac{m_{i+1}^* - m_{i-1}^*}{4m_i^*}\right)\delta\psi_{i+1} = \left(-\frac{m_{i+1}^* - m_{i-1}^*}{2m_i^*} - 2\right)\psi_{i-1}^n$$
$$+ \left(4 + \frac{2m_i^*(\Delta x)^2(V_i^{n+1} + V_i^n)}{\hbar^2}\right)\psi_i^n + \left(\frac{m_{i+1}^* - m_{i-1}^*}{2m_i^*} - 2\right)\psi_{i+1}^n. \tag{13.96}$$

If we use a simulation domain with N_x internal grid points, Equation (13.96) must be written $N_x + 2$ times, including for $i = 0$ and $i = N_x + 1$, to implement the left and right boundary conditions, respectively. The resulting set of $N_x + 2$ equations can then be written in matrix form:

$$\mathbf{A} \cdot \boldsymbol{\delta\psi} = \mathbf{b}, \tag{13.97}$$

where the matrix $\mathbf{A}$ is an $(N_x + 2) \times (N_x + 2)$ tridiagonal matrix of the form

$$\mathbf{A} = \begin{pmatrix} a_{1,1} & a_{1,2} & 0 & \cdots & 0 & 0 \\ a_{2,1} & a_{2,2} & a_{2,3} & 0 & \cdots & 0 \\ 0 & a_{3,2} & a_{3,3} & a_{3,4} & \cdots & 0 \\ \vdots & \vdots & \ddots & \ddots & \ddots & \vdots \\ 0 & 0 & \cdots & a_{N_x+1,N_r} & a_{N_x+1,N_x+1} & a_{N_x+1,N_x+2} \\ 0 & 0 & \cdots & 0 & a_{N_x+2,N_x+1} & a_{N_x+2,N_x+2} \end{pmatrix}, \tag{13.98}$$

and $\boldsymbol{\delta\psi}$ and $\mathbf{b}$ are column vectors of dimension $(N_x + 2) \times 1$.

The wave packet update after each time increment over the solution domain, including at the boundaries, is then given by

$$\psi_i^{n+1} = \psi_i^n + \delta\psi_i. \tag{13.99}$$

The elements of the matrix $\mathbf{A}$ are obtained from Equation (13.96) and are given by, for $i = 1, \ldots, N_x$:

$$a_{i+1,i+1} = -2 - \frac{2m_i^*(\Delta x)^2 V_i^{n+1}}{\hbar^2} + i \frac{4m_i^*(\Delta x)^2}{\hbar \Delta t}, \tag{13.100}$$

$$a_{i+1,i+2} = 1 - \frac{m_{i+1}^* - m_{i-1}^*}{4m_i^*}, \tag{13.101}$$

$$a_{i+1,i} = 1 + \frac{m_{i+1}^* - m_{i-1}^*}{4m_i^*}. \tag{13.102}$$

Furthermore, the inner components of the column vector $\mathbf{b}$ are given by, for $i = 1, \ldots, N_x$:

$$b_{i+1} = \left(-\frac{m_{i+1}^* - m_{i-1}^*}{2m_i^*} - 2 \right) \psi_{i-1}^n + \left(4 + \frac{2m_i^*(\Delta x)^2(V_i^{n+1} + V_i^n)}{\hbar^2} \right) \psi_i^n$$
$$+ \left(\frac{m_{i+1}^* - m_{i-1}^*}{2m_i^*} - 2 \right) \psi_{i+1}^n. \tag{13.103}$$

The matrix elements $a_{1,1}$, $a_{1,2}$, a_{N_x+2,N_x+2}, and a_{N_x+2,N_x+1} and vector components b_1 and b_{N_x+2} depend on the boundary conditions at the end of the simulation domain, as described next.

Boundary conditions

(a) **Reflecting boundaries:** The implementation of a perfectly reflecting boundary is straightforward. In this case, there is zero probability of finding the particle outside the solution domain. Hence, the wave function must therefore vanish at both boundaries, i.e.,

$$\psi_0^n = \psi_{N+x+1}^n = 0, \forall n. \tag{13.104}$$

Table 13.1: First and last row values of **A** and **b** for the case of 100% reflecting boundaries at both ends of the solution domain.

$a_{1,1}$	$a_{1,2}$	b_1	a_{N_x+2,N_x+1}	a_{N_x+2,N_x+2}	b_{N_x+2}
1	0	0	0	1	0

Numerically, this is imposed by setting the elements of the first and last rows of the tridiagonal matrix **A** and column vector **b** defined above as shown in Table 13.1.

(b) Absorbing boundaries: The implementation of absorbing boundaries is trickier, but the extra work is well worth it because it eliminates the effects of spurious reflections found with perfectly reflecting boundaries. They affect the overall shape of the wave packet as time elapses.

First, we consider the special case of plane wave solutions to the time-dependent Schrödinger equation, i.e.,

$$\psi(z,t) = e^{\pm i\left(kz - \frac{E}{\hbar}t\right)}, \tag{13.105}$$

where the $+$ and $-$ signs correspond to a plane wave propagating to the right and left, respectively.

The kinetic energy E and the momentum k of the particle are related by the energy dispersion relation

$$k = \frac{1}{\hbar}\sqrt{2m^*E}. \tag{13.106}$$

This energy dispersion relation must be satisfied everywhere and at all time. The presence of the square root makes it difficult to implement a partial differential equation that ensures the presence of absorbing boundary conditions at the ends of the solution domain. This problem can be circumvented by replacing the energy dispersion relation in Equation (13.106) by a linear approximation, as illustrated in Figure 13.1.

If the energy content of the initial wave packet is mostly contained within the energy range between α_1 and α_2, as shown in Figure 13.1, we first approximate the energy dispersion relation in Equation (13.106) by a straight line characterized by the following energy–momentum relationship:

$$E = \frac{\pm f_1}{\sqrt{2m^*}}(\hbar k) + f_2, \tag{13.107}$$

where

$$f_1 = \frac{\alpha_2 - \alpha_1}{\sqrt{\alpha_2} - \sqrt{\alpha_1}}, \tag{13.108}$$

$$f_2 = \frac{\alpha_1\sqrt{\alpha_2} - \alpha_2\sqrt{\alpha_1}}{\sqrt{\alpha_2} - \sqrt{\alpha_1}}. \tag{13.109}$$

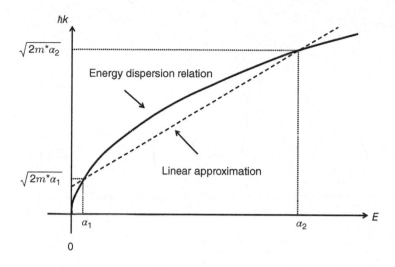

Figure 13.1: Linear approximation to the parabolic energy dispersion relation for a free particle over the energy range $[\alpha_1, \alpha_2]$ describing the spectral content of the initial wave packet.

Note that the $\pm$ sign of f_1 in Equation (13.107) is related to waves propagating in either the positive or the negative direction. The appropriate sign must be selected when implementing the reflecting boundary at the right and left ends of the solution domain.

Next, we multiply Equation (13.107) on the right by $\psi(x,t)$ and get

$$E\psi(x,t) = \left(\frac{\pm f_1}{\sqrt{2m^*}}(\hbar k) + f_2 \right) \psi(x,t). \tag{13.110}$$

We then replace E and k by their quantum mechanical operators $i\hbar\partial/\partial t$ and $-i\partial/\partial x$, respectively. This leads to the differential equation

$$i\hbar\frac{\partial}{\partial t}\psi(x,t) = \left(-i\hbar\frac{\pm f_1}{\sqrt{2m^*}}\frac{\partial}{\partial z} + f_2 \right) \psi(x,t). \tag{13.111}$$

As will be shown next, this last equation can be approximated by a difference equation, which allows a straightforward numerical implementation of absorbing boundary conditions at both ends of the solution domain.

Left absorbing boundary: The left boundary will be reached by the left-moving plane wave components associated with the Fourier transform of the moving wave packet. To implement the left absorbing boundary condition starting with Equation (13.111), we use the negative sign with f_1. The left boundary affects the first row of the matrices and vectors in Equation (13.98), i.e., the elements $a_{1,1}$, $a_{1,2}$, and b_1. Since only a first-order spatial derivative is involved in Equation (13.111), we implement it using a forward Euler difference scheme. The difference form of

Equation (13.111) becomes

$$i\hbar\left(\frac{\psi_{i+1}^{n+1} - \psi_{i+1}^n + \psi_i^{n+1} - \psi_i^n}{2\Delta t}\right) = \frac{i\hbar f_1}{\sqrt{2m_i^*}}\left(\frac{\psi_{i+1}^{n+1} - \psi_i^{n+1} + \psi_{i+1}^n - \psi_i^n}{2\Delta x}\right)$$

$$+ f_2\left(\frac{\psi_{i+1}^{n+1} + \psi_{i+1}^n + \psi_i^{n+1} + \psi_i^n}{4}\right)$$

$$+ \left(\frac{V_{i+1}^{n+1}\psi_{i+1}^{n+1} + V_i^{n+1}\psi_i^{n+1} + V_{i+1}^n\psi_{i+1}^n + V_i^n\psi_i^n}{4}\right). \qquad (13.112)$$

Using Equation (13.95), setting $i = 0$ in the previous equation, and rearranging terms, we get

$$\delta\phi_0\left(1 + \frac{f_1\delta t}{\Delta x\sqrt{2m_0^*}} + \frac{if_2\delta t}{2\hbar} + \frac{i\delta t V_0^{n+1}}{2\hbar}\right)$$

$$+ \delta\phi_1\left(1 + \frac{f_1\delta t}{\Delta x\sqrt{2m_0^*}} + \frac{if_2\delta t}{2\hbar} + \frac{i\delta t V_0^{n+1}}{2\hbar}\right)$$

$$= \phi_0^n\left(-\frac{2f_1\delta t}{\Delta x\sqrt{2m_0^*}} - \frac{if_2\delta t}{\hbar} - \frac{i\delta t(V_0^{n+1} + V_0^n)}{2\hbar}\right)$$

$$+ \phi_1^n\left(\frac{2f_1\delta t}{\Delta x\sqrt{2m_0^*}} - \frac{if_2\delta t}{\hbar} - \frac{i\delta t(V_1^{n+1} + V_1^n)}{2\hbar}\right). \qquad (13.113)$$

Using this last equation, we get the following expression for the elements in the top rows of the matrix $\mathbf{A}$ and column vector $\mathbf{b}$ needed to implement the left absorbing boundary conditions:

$$a_{1,1} = 1 + \frac{f_1\delta t}{\Delta x\sqrt{2m_0^*}} + \frac{if_2\delta t}{2\hbar} + \frac{i\delta t V_0^{n+1}}{2\hbar}, \qquad (13.114)$$

$$a_{1,2} = 1 + \frac{f_1\delta t}{\Delta x\sqrt{2m_0^*}} + \frac{if_2\delta t}{2\hbar} + \frac{i\delta t V_0^{n+1}}{2\hbar}, \qquad (13.115)$$

$$b_1 = \phi_0^n\left(-\frac{2f_1\delta t}{\Delta x\sqrt{2m_0^*}} - \frac{if_2\delta t}{2\hbar} - \frac{i\delta t(V_0^{n+1} + V_0^n)}{2\hbar}\right)$$

$$+ \phi_1^n\left(\frac{2f_1\delta t}{\Delta x\sqrt{2m_0^*}} - \frac{if_2\delta t}{2\hbar} - \frac{i\delta t(V_1^{n+1} + V_1^n)}{2\hbar}\right). \qquad (13.116)$$

Setting $i = 1$ in the last three equations yields the elements of the first rows of $\mathbf{A}$ and $\mathbf{b}$ in Equation (13.97) needed to implement a left absorbing boundary.

Right absorbing boundary: The derivation of the right absorbing boundary follows the same approach as in the previous section, except with two important differences. First, we must use a positive value of f_1 associated with the description of the right-moving components in the Fourier transform of the moving wave packet.

Second, the first-order spatial derivative appearing in Equation (13.111) must be implemented using a backward Euler difference scheme.

The difference equation to implement the right boundary condition can then be obtained from Equation (13.112) by simply flipping the sign of f_1, and changing the spatial subscripts of ψ from i and $i+1$ to $i-1$ and i, respectively. This leads to the analytical expressions for the elements in the bottom row of the matrix $\mathbf{A}$, i.e. a_{N_x+2,N_x+2} and a_{N_x+2,N_x+1}, and the bottom element in the column vector $\mathbf{b}$, i.e., b_{N_x+2}, needed to implement the right absorbing boundary condition.

Suggested problems

- For the initial one-dimensional wave packet given in Equation (13.52), show that the average value of its momentum $\langle p \rangle = \hbar k_0$. Calculate the standard deviation of both the position and momentum as a function of the parameter σ.

- Calculate the time dependence of:

  (1) the average position $\langle x(t) \rangle$,

  (2) the momentum $\langle p(t) \rangle$, and

  (3) the standard deviations $\Delta x(t)$ and $\Delta p(t)$

  of the one-dimensional wave packets given by Equation (13.54).

- Perform the integral given in Equation (13.49) by completing the square using the dispersion relation for the free particle, and show that the spatio-temporal dependence of the wave function associated with the one-dimensional Gaussian wave packet is given by Equation (13.59).

- Consider a one-dimensional harmonic oscillator initially in the state

$$\phi(x) = N(1 + \alpha^2 x^2)e^{-\frac{1}{2}x^2}, \tag{13.117}$$

  where $\alpha = (m\omega/\hbar)^{1/2}$ and N is a normalization constant.

  (1) Find N.

  (2) What possible values are obtained in a measurement of the total energy of the oscillator and with what probability?

  (3) Give an analytical expression of the wave function at time t.

- If a particle of mass m^* is in an infinite potential well described by $V(x) = 0$ for $|x| \le a$ and $+\infty$ otherwise, and its wave function at time $t = 0$ is an even function of x, then what possible values result from a measurement of the particle kinetic energy at $t = 0$?

  How soon after time $t = 0$ will the particle return to its initial state if left undisturbed?

- In going from Equation (13.38) to Equation (13.39), we used the fact that

$$\langle 0 | a^\dagger | \phi(t) \rangle = 0. \qquad (13.118)$$

  Explain why this is true.

- Consider a particle described by a one-dimensional wave packet $\phi(z)$ at $t = 0$ for which $\langle z \rangle = 0$ and $\langle p_z \rangle = 0$.

  Define the operators

$$a = \frac{z}{2\sigma} + i\frac{\sigma p_z}{\hbar}, \qquad (13.119)$$

$$a^\dagger = \frac{z}{2\sigma} - i\frac{\sigma p_z}{\hbar}. \qquad (13.120)$$

  Starting with the fact that for any $\phi(z)$, $\langle a^\dagger a \rangle \geq 0$, prove the Heisenberg uncertainty relation for position and momentum,

$$\Delta z \Delta p_z \geq \frac{\hbar}{2}. \qquad (13.121)$$

  Show that the equality sign holds for a one-dimensional Gaussian wave packet.

- Starting with the results of Problem 13.3, write Matlab code to plot the spatial variation of the probability density and probability current density $J_x(x,t)$ as a function of time. Select the appropriate time scale to display the wave packet spreading for the k_0 and a parameters you select to represent your wave packet.

- Calculate the probability density $\rho(x,t)$ associated with the Gaussian wave packet in Equation (13.54). Then, using the result of the previous exercise, show that the current continuity equation $\frac{d\rho(x,t)}{dt} + \frac{dJ_x(x,t)}{dx} = 0$ is satisfied for the one-dimensional wave packet.

- In Problem 13.4, derive the missing steps going from Equation (13.72) to Equation (13.73).

- Write Matlab code to plot the spatial variation of the probability density and energy flux density $S_x(x,t)$ as a function of time (see Problem 13.5). Select the appropriate time scale to display the wave packet spreading for your selected values of the k_0 and a describing your wave packet.

- Consider the one-dimensional problem consisting of a particle subjected to a constant force F.

  (1) Write down the Schrödinger equation in both the x and p representation (use a Fourier transform to obtain the latter—see Problem 3.2).

  (2) Starting with the Schrödinger equation in momentum space, show that a solution at time t, designated as $\phi(p,t)$, can be found from the solution at time $t = 0$, designated as $\phi_0(p) = \phi(p, t = 0)$, as follows:

$$\phi(p,t) = \phi_0(p - Ft)e^{i[(p-Ft)^3 - p^3]/6m^*\hbar F}. \qquad (13.122)$$

(3) Using this last result, show that

$$|\phi(p,t)|^2 = |\phi_0(p - Ft)|^2. \tag{13.123}$$

What is the physical significance of this result?

- Following the derivation of the difference equation for implementing a left absorbing boundary condition given in Equation (13.112), derive the explicit form of the difference equation needed to implement a right absorbing boundary condition. Using this difference equation, find the explicit values of the elements of the last rows of **A** and **b** in Equation (13.97) necessary to implement a right absorbing boundary.

References

[1] Cohen-Tannoudji, C., Diu, B., and Laloe, F. (2000) *Quantum Mechanics*, Complement G_I, Hermann, Paris.

[2] Kroemer, H. (1994) *Quantum Mechanics for Engineering, Materials Science, and Applied Physics*, Chapter 4, Prentice Hall, Englewood Cliffs, NJ.

[3] Ohanian, H. C. (1990) *Principles of Quantum Mechanics*, Sections 2.4 and 2.5, Prentice Hall, Upper Saddle River, NJ.

[4] Merzbacher, E. (1998) *Quantum Mechanics*, 3rd edition, Wiley, New York.

[5] Goldberg, A., Schey, H., and Schwartz, J. L. (1967) Computer-generated motion pictures of one-dimensional quantum mechanical transmission and reflection phenomena. *American Journal of Physics* 35, pp. 177–186.

[6] Blinder, S. M. (1968) Evolution of a Gaussian wavepacket. *American Journal of Physics* 36, pp. 525–526.

[7] Chapra, S. C. and Canale, R. P. (1988) *Numerical Methods for Engineers*, 2nd edition, McGraw-Hill, New York.

Suggested Reading

- Henneberger, W. C. (1968) Perturbation method for atoms in intense light beams. *Physical Review Letters* 21, pp. 838–841.

- Harris, E. G. (1971) Quantum theory of an electron in a uniform time-dependent field. *American Journal of Physics* 39, pp. 683–684.

- Chow, P. C. (1972) Computer solutions to the Schrödinger equation. *American Journal of Physics* 40, pp. 730–734.

- Faisal, F. H. M. (1973) Multiple absorption of laser photons by atoms. *Journal of Physics B: Atomic, Molecular, and Optical Physics* 6, pp. L89–L92.

- Feely, J. H. (1974) Quantum theory of an electron in external fields. *American Journal of Physics* 42, pp. 326–328.

- Miranda, L. C. M. (1976) Phonon damping in the simultaneous presence of intense radiation and magnetic fields. *Journal of Physics C: Solid State Physics* 9, pp. 2971–2976.

- Brandi, H. S., Koiller, B., Lins de Barros, H. G. P., Miranda, L. C. M., and Castro, J. J. (1978) Theory of electron–hydrogen atom collisions in the presence of a laser field. *Physical Review A* 17, pp. 1900–1906.

- Lima, C. A. S. and Miranda, L. C. M. (1981) Atoms in superintense laser fields. *Physical Review A* 23, pp. 3335–3337.

- Galvao, R. M. O. and Miranda, L. C. M. (1983) Quantum theory of an electron in external field using unitary transforms. *American Journal of Physics* 51, pp. 729–733.

- Lobo, R. and Hipolito, O. (1987) Electrons bound on a liquid-helium surface in the presence of a laser field. *Physical Review B* 35, pp. 8755–8758.

- Bhatt, R., Piraux, B., and Burnett, K. (1988) Potential scattering of electrons in the presence of intense laser fields using the Kramers–Henneberger transformation. *Physical Review A* 37, pp. 98–105.

- Styer, D. F. (1989) The motion of wave packets through their expectation values and uncertainties. *American Journal of Physics* 58, pp. 742–744.

- Ninio, F. (1990) The forced harmonic oscillator and the zero-phonon transition of the Mossbauer effect. *American Journal of Physics* 58, pp. 742–755.

- Parker, G. W. (1990) Evolution of the quantum states of a harmonic oscillator in a uniform time-varying electric field. *American Journal of Physics* 40, pp. 120–125.

- Kuska, J.-P. (1992) Absorbing boundary conditions for the Schroedinger equation on finite intervals. *Physical Review B* 46, pp. 5000–5003.

- Robinett, R. W. (1996) Quantum mechanical time development operator for the uniformly accelerated particle. *American Journal of Physics* 64, pp. 803–808.

- Wagner, M. (1996) Strongly driven quantum wells: an analytical solution to the time-dependent Schrödinger equation. *Physical Review Letters* 76, pp. 4010–4013.

- Unnikrishnan, K. and Rau, A. R. P. (1996) Uniqueness of the Airy wavepacket in quantum mechanics. *American Journal of Physics* 64, pp. 1034–1035.

- Chen, R. L. W. (1996) Computer graphics for solutions of time-dependent Schrödinger equations. *American Journal of Physics* 50, pp. 902–906.

- Qu, F., Fonseca, A. L. A., and Nunes, O. A. C. (1997) Intense field effects on hydrogen impurities in quantum dots. *Journal of Applied Physics* 82, pp. 1236–1241.

- Ford, G. W. and O'Connell, R. F. (2001) Wave packet spreading: temperature and squeezing effects with applications to quantum measurement and decoherence. *American Journal of Physics* 70, pp. 319–324.

- Balasubramanian, S. (2001) Time-dependent operator method in quantum mechanics. *American Journal of Physics* 69, pp. 508–511.

- Cox, T. and Lekner, J. (2008) Reflection and non-reflection of particle wave packets. *European Journal of Physics* 29, pp. 671–679.

- Lekner, J. (2009) Airy wave packet solutions to the Schrödinger equation. *European Journal of Physics* 30, pp. L43–L46.

- Blanes, S., Casas, F., Oeto, J. A., and Ros, J. (2010) A pedagogical approach to the Magnus expansion. *European Journal of Physics* 31, pp. 907–918.

- Garrix, A. E., Sztrajman, A., and Mitnik, D. (2010) Running into trouble with the time-dependent propagation of a wave packet. *European Journal of Physics* 31, pp. 785–799.

Appendix A: Postulates of Quantum Mechanics

This appendix closely follows the treatment of Nielsen and Chuang [1]. A review of fundamental concepts such as vector space, inner product, outer product, Hilbert space, linear operators, matrices, eigenvectors, adjoint, unitary, and Hermitian operators, tensor products, operator functions, commutators, and anti-commutators is given in Chapter 2, Section 2.1 of Ref. [1].

Postulate 1

Associated with any isolated physical system is a complex vector space with inner product (Hilbert space) known as the *state space* of the system. In Dirac notation, the system is completely described by its *state vector* $|\psi\rangle$, which is a unit vector in the state space, i.e., we must have $\langle\psi|\psi\rangle = 1$, which is referred to as the normalization condition.

Postulate 2

The evolution of a closed quantum system is described by a unitary transformation, i.e., the state $|\psi(t_1)\rangle$ of the system at time t_1 is related to the state $|\psi(t_2)\rangle$ at time t_2 by a unitary operator U which depends only on the initial (t_1) and final (t_2) times in the evolution history:

$$|\psi(t_2)\rangle = U(t_2, t_1)|\psi(t_1)\rangle. \tag{A.1}$$

Note that $\langle\psi(t_2)|\psi(t_2)\rangle = \langle\psi(t_1)|U^+U|\psi(t_1)\rangle = \langle\psi(t_1)|\psi(t_1)\rangle = 1$, i.e., the state vector stays normalized at all times.

In non-relativistic quantum mechanics, Postulate 2 is sometimes formulated as a prelude to the Schrödinger equation which describes the time evolution of the state of a closed system characterized by a Hamiltonian H:

$$i\hbar\frac{\mathrm{d}}{\mathrm{d}t}|\psi\rangle = H|\psi\rangle. \tag{A.2}$$

If the Hamiltonian does not depend explicitly on time, a formal solution of the Schrödinger equation can be written as

$$|\psi(t_2)\rangle = e^{\frac{-i}{\hbar}H(t_2-t_1)}|\psi(t_1)\rangle, \tag{A.3}$$

and since H is Hermitian, the operator U defined as

$$U(t_2, t_1) = e^{\frac{-i}{\hbar}H(t_2-t_1)} \tag{A.4}$$

is unitary (see Chapter 2).

Problem Solving in Quantum Mechanics: From Basics to Real-World Applications for Materials Scientists, Applied Physicists, and Devices Engineers, First Edition.
Marc Cahay and Supriyo Bandyopadhyay.
© 2017 John Wiley & Sons Ltd. Published 2017 by John Wiley & Sons Ltd.

Postulate 3: Quantum measurements

The quantum projective approach of von Neumann describes the wave function collapse associated with measurements performed on quantum mechanical systems. According to this approach, a projective measurement is described by an observable M, which is a Hermitian operator on the state space of the system. Measurements obey the following set of rules:

- Since the operator M is Hermitian, it has a spectral decomposition

$$M = \sum_m m P_m, \tag{A.5}$$

where P_m is the projector onto the eigenspace of M with eigenvalue m and corresponding eigenvector $|m\rangle$. The quantity P_m is explicitly given by the outer product

$$P_m = |m\rangle\langle m|. \tag{A.6}$$

- The possible outcomes of the measurement of the observable M correspond to the eigenvalues m of the observable.

- Upon measuring M on a quantum mechanical system characterized by the ket $|\psi\rangle$, the probability of getting the result m is given by

$$p(m) = \langle\psi|P_m|\psi\rangle = \langle\psi|m\rangle\langle m|\psi\rangle = |\langle\psi|m\rangle|^2. \tag{A.7}$$

- After the measurement is made, if outcome m has occurred, the state of the quantum system immediately after the measurement is given by (or has collapsed to)

$$P_m|\psi\rangle / \sqrt{p(m)} = \frac{\langle m|\psi\rangle|m\rangle}{\sqrt{|\langle\psi|m\rangle|^2}}, \tag{A.8}$$

where $\sqrt{p(m)}$ is introduced to renormalize the state after collapse occurs.

A useful property of projection measurements: In order to calculate the average value of repeated measurements of the observable M on a quantum mechanical system in the state $|\psi\rangle$, we must perform the summation

$$E(M) = \sum_m m p(m), \tag{A.9}$$

which is what we expect from the definition of average value associated with a random variable using probability theory. Plugging in the definition of $p(m) = \langle\psi|P_m|\psi\rangle$,

$$E(M) = \sum_m m\langle\psi|P_m|\psi\rangle = \langle\psi|\sum_m m P_m|\psi\rangle = \langle\psi|M|\psi\rangle. \tag{A.10}$$

The latter expression can usually be calculated easily for a given operator M and state $|\psi\rangle$.

Similarly, the standard deviation of a large number of measurements of the observable M is defined by

$$\Delta(M) = \left[\langle\psi|(M - \langle M\rangle)^2|\psi\rangle\right]^{\frac{1}{2}},\tag{A.11}$$

where

$$\langle M\rangle \doteq \langle\psi|M|\psi\rangle.\tag{A.12}$$

Therefore,

$$\Delta(M) = \sqrt{\langle\psi|(M^2 - 2\langle M\rangle M + \langle M\rangle^2)|\psi\rangle} = \sqrt{\langle M^2\rangle - \langle M\rangle^2},\tag{A.13}$$

where $\langle M^2\rangle = \langle\psi|(M^2|\psi\rangle$.

The standard deviation $\Delta(M)$ can therefore also be calculated easily once the operator M and the ket $|\psi\rangle$ are known.

The extension of Postulate 1 to a multiparticle system is referred to as Postulate 4 of quantum mechanics. It is based on the concept of tensor product. This book does not consider many particle problems and therefore we will not make use of Postulate 4.

References

[1] Nielsen, M. A. and Chuang, I. L., (2000) *Quantum Computation and Quantum Information*, Section 2.2, Cambridge University Press, New York.

Appendix B: Useful Relations for the One-Dimensional Harmonic Oscillator

The eigenstates of the 1D harmonic oscillator described by the Hamiltonian [1–5]

$$H = \frac{p_z^2}{2m} + \frac{1}{2}m\omega^2 z^2 \tag{B.1}$$

are given by

$$\phi_n(z) = \sqrt{\frac{\alpha}{\sqrt{\pi}2^n n!}}\, e^{-\frac{\xi^2}{2}} H_n(\xi), \tag{B.2}$$

where n is an integer, $\xi = \alpha z$, and

$$\alpha = \sqrt{\frac{m\omega}{\hbar}}. \tag{B.3}$$

The corresponding eigenvalues are

$$E_n = \left(n + \frac{1}{2}\right)\hbar\omega. \tag{B.4}$$

In Equation (B.2), the quantities $H_n(\xi)$ are the Hermite polynomials given by

$$H_n(\xi) = (-1)^n e^{\xi^2} \frac{d^n}{d\xi^n} e^{-\xi^2}. \tag{B.5}$$

Using this last relation, the first few Hermite polynomials can be written explicitly as

$$H_0(\xi) = 1, \tag{B.6}$$
$$H_1(\xi) = 2\xi, \tag{B.7}$$
$$H_2(\xi) = 4\xi^2 - 2, \tag{B.8}$$
$$H_3(\xi) = 8\xi^3 - 12\xi. \tag{B.9}$$

The Hermite polynomials can also be obtained from the generating function [5]

$$e^{-t^2+2t\xi} = \sum_{n=0}^{\infty} \frac{H_n(\xi)}{n!} t^n, \tag{B.10}$$

or

$$H_n(\xi) = \left.\left(\frac{d^n}{d\xi^n} e^{-t^2+2t\xi}\right)\right|_{t=0}, \tag{B.11}$$

i.e.,

$$H_n(\xi) = (-1)^n e^{\xi^2} \left[\frac{d^n}{d\xi^n} e^{-\xi^2}\right]. \tag{B.12}$$

Problem Solving in Quantum Mechanics: From Basics to Real-World Applications for Materials Scientists, Applied Physicists, and Devices Engineers, First Edition.
Marc Cahay and Supriyo Bandyopadhyay.
© 2017 John Wiley & Sons Ltd. Published 2017 by John Wiley & Sons Ltd.

References

[1] Cohen-Tannoudji, C., Diu, B., and Laloe, F. (2000) *Quantum Mechanics*, Chapter 5, Hermann, Paris.

[2] Kroemer, H. (1994) *Quantum Mechanics for Engineering, Materials Science, and Applied Physics*, Chapter 10, Prentice Hall, Englewood Cliffs, NJ.

[3] Ohanian, H. C. (1990) *Principles of Quantum Mechanics*, Chapter 6, Prentice Hall, Upper Saddle River, NJ.

[4] Levi, A. F. J. (2006) *Applied Quantum Mechanics*, 2nd edition, Chapter 6, Cambridge University Press, Cambridge.

[5] Arfken, G. (1970) *Mathematical Methods For Physicists*, Chapter 13, Academic Press, New York.

Appendix C: Properties of Operators [1–5]

In quantum mechanics, any physical variable has an associated operator such that if an experiment is made to "measure" the value of the physical variable, the expected result of the measurement will be the "expectation value" of the operator, defined as

$$\langle \phi | O | \phi \rangle = \int d^3 \vec{r}\, \Phi^*(\vec{r},t) O(\vec{r},t) \Phi(\vec{r},t), \tag{C.1}$$

where the asterisk denotes complex conjugate, $O(\vec{r},t)$ is the spatially and temporally varying operator describing the physical variable, and $\phi(\vec{r},t)$ is the spatially and temporally varying wave function which is an eigenfunction of the operator, meaning that the following relation is satisfied: $O(\vec{r},t)\Psi(\vec{r},t) = \lambda\Psi(\vec{r},t)$, where λ is a constant (called the "eigenvalue" of the operator). Clearly, λ is also the expectation value, or the value that is expected to be measured if an experiment were carried out to measure the physical variable represented by the operator.

All legitimate quantum mechanical operators must be Hermitian

In other words, they must satisfy the relation

$$\int d^3 \vec{r}\, \Phi^*(\vec{r},t) O(\vec{r},t) \Psi(\vec{r},t) = \int d^3 \vec{r}\, [O(\vec{r},t)\Phi(\vec{r},t)]^* \Psi(\vec{r},t), \tag{C.2}$$

because Hermitian operators have real eigenvalues. This ensures that the expectation value of the operator, or the eigenvalue, will be always real. Any imaginary or complex expectation value would have been unphysical since the measured value in any experiment must always be a real quantity.

It is easy to show that the eigenvalues of Hermitian operators are real. Indeed, if $O(\vec{r},t)$ is a Hermitian operator with eigenvalue λ and eigenfunction $\Psi(\vec{r},t)$, then, by definition,

$$O(\vec{r},t)\Psi(\vec{r},t) = \lambda\Psi(\vec{r},t). \tag{C.3}$$

According to Equation (C.2), Hermiticity implies

$$\int d^3 \vec{r}\, \Psi_2^*(\vec{r},t) O(\vec{r},t) \Psi_1(\vec{r},t) = \int d^3 \vec{r}\, [O(\vec{r},t)\Psi_2(\vec{r},t)]^* \Psi_1(\vec{r},t). \tag{C.4}$$

Using $\Psi_1(\vec{r},t) = \Psi_2(\vec{r},t) = \Psi(\vec{r},t)$, and then inserting Equation (C.3) in Equation (C.4), we obtain

$$\lambda^* \int d^3 \vec{r}\, \Psi^*(\vec{r},t) \Psi(\vec{r},t) = \lambda \int d^3 \vec{r}\, \Psi^*(\vec{r},t) \Psi(\vec{r},t), \tag{C.5}$$

or

$$\lambda = \lambda^*. \tag{C.6}$$

Hence, λ is real.

Problem Solving in Quantum Mechanics: From Basics to Real-World Applications for Materials Scientists, Applied Physicists, and Devices Engineers, First Edition.
Marc Cahay and Supriyo Bandyopadhyay.
© 2017 John Wiley & Sons Ltd. Published 2017 by John Wiley & Sons Ltd.

Physical significance of this result: The eigenvalue of a quantum mechanical operator is its expectation value, which is the value expected to be measured on average after repeated measurements. The latter must be a real quantity since one cannot measure an imaginary physical variable (e.g., imaginary position or imaginary momentum or imaginary energy). Hence, quantum mechanical operators must always possess real eigenvalues. Since Hermitian operators are guaranteed to yield real eigenvalues, all legitimate quantum mechanical operators must be Hermitian.

The eigenfunctions of a Hermitian operator corresponding to two distinct eigenvalues are mutually orthogonal

Let ϕ and ψ be two eigenfunctions of a Hermitian operator corresponding to two different eigenvalues α and β, i.e., $\alpha \neq \beta$.

By definition, $O\psi = \alpha\psi$ and $O\phi = \beta\phi$. Hermiticity implies

$$\int d^3\vec{r}(O\psi)^*\phi = \int d^3\vec{r}\psi^*O\phi, \tag{C.7}$$

or

$$\alpha^* \int d^3\vec{r}\psi^*\phi = \beta \int d^3\vec{r}\psi^*\phi. \tag{C.8}$$

But since α is real,

$$\alpha \int d^3\vec{r}\psi^*\phi = \beta \int d^3\vec{r}\psi^*\phi, \tag{C.9}$$

or

$$(\alpha - \beta) \int d^3\vec{r}\psi^*\phi = 0. \tag{C.10}$$

Since $\alpha \neq \beta$, we must have $\int d^3\vec{r}\psi^*\phi = 0$. This implies that eigenfunctions corresponding to distinct eigenvalues are orthornormal.

Physical significance of this result: The wave function of no eigenstate of a quantum mechanical system can be written as a linear combination of the wave functions of other eigenstates. If that were not true, then we could have written

$$|\phi_m\rangle = \sum_{p=1,p\neq m} C_p|\phi_p\rangle, \tag{C.11}$$

where we used the Dirac bra–ket notation to write the wave function. Evaluating the norm, we get

$$\langle\phi_m|\phi_m\rangle = \sum_{p=1,p\neq m} C_p\langle\phi_m|\phi_p\rangle = \sum_{p=1,p\neq m} C_p\delta_{p,m} = 0, \tag{C.12}$$

since the Kronecker $\delta_{p,m} = 0$ when $p \neq m$, which is an absurdity since it implies that the norm of the wave function is equal to zero everywhere and at all times. That would mean that the probability of finding the electron in that eigenstate is forever zero, i.e. the eigenstate is no eigenstate at all!

Useful operator identities

One of the fundamental tenets of quantum mechanics is the Heisenberg uncertainty principle discussed in Problem 4.1 in Chapter 4. This pertains to operators that do not commute and is based on some basic properties of the commutator of the operators associated with some physical variables. The commutator of two operators A and B is defined as

$$[A, B] = AB - BA. \tag{C.13}$$

The following properties are easily proven:

$$[A, B] = -[B, A] \tag{C.14}$$
$$[A, BC] = [A, B]C + B[A, C] \tag{C.15}$$
$$(AB)^{\dagger} = B^{\dagger} A^{\dagger} \tag{C.16}$$
$$[A, B]^{\dagger} = -[A, B] \quad \text{if } A \text{ and } B \text{ are Hermitian.} \tag{C.17}$$

References

[1] Nielsen, M. A. and Chuang, I. L. (2000) *Quantum Computation and Quantum Information*, Chapter 2, Cambridge University Press, New York.

[2] Cohen-Tannoudji, C., Diu, B., and Laloe, F. (2000) *Quantum Mechanics*, Complement B_{II}, Hermann, Paris.

[3] Kroemer, H. (1994) *Quantum Mechanics for Engineering, Materials Science, and Applied Physics*, Chapter 7, Prentice Hall, Englewood Cliffs, NJ.

[4] Ohanian, H. C. (1990) *Principles of Quantum Mechanics*, Section 4.3, Prentice Hall, Upper Saddle River, NJ.

[5] Levi, A. F. J. (2006) *Applied Quantum Mechanics*, 2nd edition, Chapter 5, Cambridge University Press, Cambridge.

Appendix D: The Pauli Matrices and their Properties [1–5]

The 2×2 Pauli matrices are defined as follows:

$$\sigma_z = \begin{pmatrix} 1 & 0 \\ 0 & -1 \end{pmatrix}, \tag{D.1}$$

$$\sigma_x = \begin{pmatrix} 0 & 1 \\ 1 & 0 \end{pmatrix}, \tag{D.2}$$

$$\sigma_y = \begin{pmatrix} 0 & -i \\ i & 0 \end{pmatrix}. \tag{D.3}$$

The following properties of the Pauli matrices can be proven easily:

$\det (\sigma_j) = -1$ for $j = x, y,$ or z

$\text{Tr} (\sigma_j) = 0$

$\sigma_x^2 = \sigma_y^2 = \sigma_z^2 = I$

$\sigma_p \sigma_q + \sigma_q \sigma_p = 0 \quad (p \neq q; \ p, q = x, y, z)$

$\sigma_x \sigma_y \sigma_z = iI$

$\sigma_x \sigma_y = -\sigma_y \sigma_x = i\sigma_z$

$\sigma_y \sigma_z = -\sigma_z \sigma_y = i\sigma_x$

$\sigma_z \sigma_x = -\sigma_x \sigma_z = i\sigma_y.$

These last three identities can be written in a more compact form as

$$\sigma_j \sigma_k = i\epsilon_{jkl} + \delta_{kj} I, \tag{D.4}$$

where $\epsilon_{jkl} = 0$ if the indices j, k, l are not distinct, $\epsilon_{jkl} = 1$ if the indices (j, k, l) are distinct and a cyclic permutation of the (x, y, z) indices, and $\epsilon_{jkl} = -1$ if the indices (j, k, l) are distinct and not a cyclic permutation of (x, y, z).

Eigenvectors of the Pauli matrices

The eigenvalues of the three Pauli spin matrices are ± 1. The corresponding eigenvectors are listed below.

Matrix σ_z: The eigenvectors of σ_z must satisfy

$$\sigma_z |\pm\rangle_z = \pm 1 |\pm\rangle_z. \tag{D.5}$$

Problem Solving in Quantum Mechanics: From Basics to Real-World Applications for Materials Scientists, Applied Physicists, and Devices Engineers, First Edition.
Marc Cahay and Supriyo Bandyopadhyay.

In a (2×1) column vector form, the normalized eigenvectors of σ_z are given by

$$|+\rangle_z = \begin{pmatrix} 1 \\ 0 \end{pmatrix}, \tag{D.6}$$

$$|-\rangle_z = \begin{pmatrix} 0 \\ 1 \end{pmatrix}. \tag{D.7}$$

It is easy to verify that these two eigenvectors are orthonormal, as they must be since they are eigenvectors of a Hermitian matrix corresponding to distinct (non-degenerate) eigenvalues.

Matrix σ_x: The eigenvectors of σ_x must satisfy

$$\sigma_x|\pm\rangle_x = \pm 1|\pm\rangle_x. \tag{D.8}$$

Starting with Equation (D.2), these eigenvectors are found to be

$$|+\rangle_x = \frac{1}{\sqrt{2}} \begin{pmatrix} 1 \\ 1 \end{pmatrix}, \tag{D.9}$$

$$|-\rangle_x = \frac{1}{\sqrt{2}} \begin{pmatrix} 1 \\ -1 \end{pmatrix}. \tag{D.10}$$

Once again, the two eigenvectors are orthonormal. As can be easily checked, these eigenvectors can also be expressed as

$$|\pm\rangle_x = \frac{1}{\sqrt{2}}[|+\rangle_z \pm |-\rangle_z]. \tag{D.11}$$

Matrix σ_y: The eigenvectors of σ_y must satisfy

$$\sigma_y|\pm\rangle_y = \pm 1|\pm\rangle_y. \tag{D.12}$$

Using Equation (D.3), these eigenvectors are found to be

$$|+\rangle_y = \frac{1}{\sqrt{2}} \begin{pmatrix} 1 \\ i \end{pmatrix}, \tag{D.13}$$

$$|-\rangle_y = \frac{1}{\sqrt{2}} \begin{pmatrix} 1 \\ -i \end{pmatrix}. \tag{D.14}$$

These eigenvectors are also orthonormal and can be expressed as

$$|\pm\rangle_y = \frac{1}{\sqrt{2}}[|+\rangle_z \pm i|-\rangle_z]. \tag{D.15}$$

The eigenvectors of the Pauli spin matrices are examples of *spinors*, which are 2×1 column vectors that represent the spin state of an electron. If we know the spinor associated with an electron in a given state, we can deduce the electron's spin orientation, i.e., find the quantities $\langle S_x \rangle$, $\langle S_y \rangle$, and $\langle S_z \rangle$, where the angular brackets $\langle \cdots \rangle$ denote expectation values.

References

[1] Cohen-Tannoudji, C., Diu, B., and Laloe, F. (1977) *Quantum Mechanics*, Chapter IV, Hermann, Paris.

[2] Nielsen, M. A. and Chuang, I. L., (2000) *Quantum Computation and Quantum Information*, Chapter 2, Cambridge University Press, New York.

[3] Bandyopadhyay, S. and Cahay, M. (2015) *Introduction to Spintronics*, 2nd edition, Chapter 2, CRC Press, Boca Raton, FL.

[4] Kroemer, H. (1994) *Quantum Mechanics for Engineering, Materials Science, and Applied Physics*, Chapter 21, Prentice Hall, Englewood Cliffs, NJ.

[5] Ohanian, H. C. (1990) *Principles of Quantum Mechanics*, Chapter 9, Prentice Hall, Upper Saddle River, NJ.

Appendix E: Threshold Voltage in a High Electron Mobility Transistor Device

Here, we derive an expression for the threshold voltage V_T of a HEMT starting with the energy band diagram shown in Figure E.1.

By analogy with MOSFETs [1–4], we expect the sheet carrier concentration n_s in the 2DEG near the AlGaAs/GaAs interface to be given by

$$en_s = \frac{\epsilon_s}{d}(V_G - V_T), \tag{E.1}$$

where ϵ_s is the electric constant of the wide gap layer and and V_G is the gate voltage.

We will show that

$$V_T = \phi_m + \frac{(E_F{}^{sub} - \Delta E_c)}{e} - V_d, \tag{E.2}$$

where $V_d = \frac{eN_D(d-w)^2}{2\epsilon_s}$, N_D is the uniform doping concentration in the region $[-d, -w]$ of the wide gap layer, $E_F{}^{sub}$ is the Fermi energy in the bulk narrow gap material measured from the bottom of the triangular well at the interface, ΔE_c is the conduction band offset between the wide and narrow gap materials, and ϕ_m is the work function of the metallic gate.

Solution: We will make the so-called depletion approximation for the wide gap layer, assuming a uniform doping density N_D for $-d < z < -w$ and $N_D = 0$ in $-w < z < 0$ (spacer layer). We first see from the energy band diagram that

$$e\phi_m - eV_G = |eV_2| + \Delta E_c - E_F{}^{sub}. \tag{E.3}$$

We solve Poisson's equation in the interval $[-w, -d]$:

$$\frac{dE}{dz} = \frac{\rho(z)}{\epsilon_s} = \frac{eN_D}{\epsilon_s}, \tag{E.4}$$

where E is the electric field related to the conduction band slope $\left(E = \frac{1}{q}\frac{dE_c}{dz}\right)$.

Integrating from $-w$ to z, we get

$$E(z) = E(-w) + \frac{eN_D}{\epsilon_s}(z + w). \tag{E.5}$$

The electrostatic potential is then obtained from

$$\frac{dV}{dz} = -E(z). \tag{E.6}$$

Problem Solving in Quantum Mechanics: From Basics to Real-World Applications for Materials Scientists, Applied Physicists, and Devices Engineers, First Edition.
Marc Cahay and Supriyo Bandyopadhyay.
© 2017 John Wiley & Sons Ltd. Published 2017 by John Wiley & Sons Ltd.

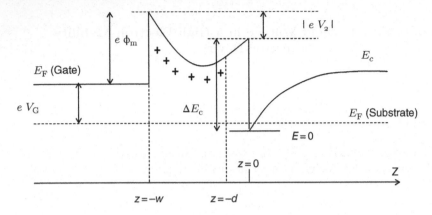

Figure E.1: A HEMT is implemented with a heterostructure comprising a narrow bandgap semiconductor and a wide bandgap semiconductor. The wide gap semiconductor is doped with donor atoms and the resulting free electrons transfer to the narrow bandgap semiconductor to minimize their potential energies. This results in spatial separation of the electrons from their parent donors, which causes the electron mobility to be high because of suppression of scattering due to ionized donor atoms. A quasi two-dimensional layer (2DEG) of electrons forms at the heterointerface. The energy band diagram along the direction perpendicular to the heterointerface is shown. There is a gate bias applied to the top metallic gate leading to a difference between the Fermi level in the bulk narrow gap semiconductor $E_F{}^{\text{sub}}$ and the Fermi level in the gate metal $E_F{}^{\text{G}}$. The quantity ϕ_{m} is the work function of the metal. The wide bandgap layer is doped uniformly with donors for $-d < z < -w$ and undoped in the region $-w < z < 0$.

Integrating this last equation from $-w$ to z using Equation (E.5), we get

$$V(z) - (-w) = -E(-w)(z + w) - \frac{eN_{\text{D}}}{\epsilon_{\text{s}}}\left(\frac{z^2}{2} + wz + \frac{w^2}{2}\right), \qquad (\text{E.7})$$

or

$$V(z) - V(-w) = -E(-w)(-d + w) - \frac{eN_{\text{D}}}{\epsilon_{\text{s}}}\left(\frac{d^2}{2} + wd + \frac{w^2}{2}\right) \qquad (\text{E.8})$$

$$= -E(-w)(w - d) - \frac{eN_{\text{D}}}{2\epsilon_{\text{s}}}(d - w)^2. \qquad (\text{E.9})$$

Similarly, in the interval $[-w, 0]$, neglecting the contribution from any residual free carriers, Poisson's equation is simply

$$\frac{dE}{dz} = 0. \qquad (\text{E.10})$$

Hence, the electric field is constant in the spacer layer, i.e.

$$E(z) = E(-w). \qquad (\text{E.11})$$

Therefore, in that interval, the electrostatic potential satisfies

$$\frac{dV}{dz} = -E(z) = -E(-w),$$ (E.12)

which leads to

$$V(z) - V(0) = -E(-w)z.$$ (E.13)

Setting the electrostatic potential at $z = 0$ to zero, we get

$$V(z) = -E(-w)z.$$ (E.14)

Hence,

$$V(-w) = E(-w)w.$$ (E.15)

From the set of equations above, we obtain

$$V(-d) = E(-w)d - \frac{eN_D}{2\epsilon_s}(d - w)^2.$$ (E.16)

Therefore, referring back to Figure E.1,

$$V_2 = -V(-d) = \frac{eN_D}{2\epsilon_s}(d - w)^2 - E(-w)d.$$ (E.17)

So,

$$V_2 = \frac{eN_D}{2\epsilon_s}(d - w)^2 - E_s d = V_d - \epsilon_s d.$$ (E.18)

From Equations (E.3)–(E.18), the electric field E_s in the spacer layer is therefore

$$E_s = \frac{(V_d - \phi_m - \frac{E_F - \Delta E_c}{e} + V_G)}{d}.$$ (E.19)

Using the continuity of the displacement vector at the interface of the wide and narrow gap semiconductors, we get $\epsilon_s E_s = \epsilon_{narrow} E_{narrow}$, where ϵ_s is the permittivity of the wide gap semiconductor in the spacer layer and ϵ_{narrow} is the permittivity in the narrow gap semiconductor. Applying Gauss's law to a box extending from the heterointerface far into the GaAs substrate leads to

$$E_s = \left(\frac{\epsilon_{narrow}}{\epsilon_s}\right) E_{narrow} = \frac{e}{\epsilon_s}(n_s + N_{depl}),$$ (E.20)

where E_{narrow} is the electric field in the narrow gap semiconductor and N_{depl} is the depletion charge due to the N_A dopants in the substrate, which we assume is negligible. Hence, $E_s = \frac{e}{E_s} n_s$. Plugging this back into Equation (E.19), we get

$$e n_s = \epsilon_s E_s = \epsilon_s \frac{(V_d - \phi_m - \frac{1}{e}(E_F - \Delta E_c) + V_G)}{d},$$ (E.21)

which can be written in the more compact form

$$e n_s = \epsilon_s \frac{(V_G - V_T)}{d}$$ (E.22)

by defining the threshold voltage V_T as

$$V_T = \phi_m + \frac{(E_F - \Delta E_c)}{e} - V_d.$$ (E.23)

References

[1] Sze, S. M. (1981) *Physics of Semiconductor Devices*, 2nd edition, Wiley, New York.

[2] Pierret, R. F. (1983) *Modular Series on Solid State Devices, Field Effect Devices* Vol. IV, Addison Wesley, Reading, MA.

[3] Taur, Y. and Ning, T. H. (1998) *Fundamentals of Modern VLSI Devices*, Cambridge University Press, Cambridge.

[4] Streetman B. G. and Banerjee S. (2000) *Solid State Electronic Devices*, 5th edition, Prentice Hall, Upper Saddle River, NJ.

Suggested Reading

- Krantz, R. J. and Bloss, W. L. (1989) The role of unintentional acceptor concentration on the threshold voltage of modulation-doped field-effect transistors. *IEEE Transactions on Electron Devices* 36, pp. 451–453.

- Krantz, R. J. and Bloss, W. L. (1989) Subthreshold I–V characteristics of AlGaAs/GaAs MODFETs: the role of unintentional doping. *IEEE Transactions on Electron Devices* 36, pp. 2593–2595.

- Krantz, R. J. and Bloss, W. L. (1990) The role of acceptor density on the high channel carrier density I–V characteristics of AlGaAs/GaAs MODFETs. *Solid-State Electronics* 33, pp. 941–945.

- Delagebeaudeuf, D. and Linh, N. T. (1982) Metal-(n) AlGaAs-GaAs two-dimensional electron GaAs FET. *IEEE Transactions on Electron Devices* 29, pp. 955–960.

- Lee, K., Schur, M. S., Drummond, T. J., and Morkoç, M. (1983) Current–voltage and capacitance–voltage characteristics of modulation doped field-effect transistors. *IEEE Transactions on Electron Devices* 30, pp. 207–212.

Appendix F: Peierls's Transformation [1, 2]

To account for an external magnetic field of flux density $\vec{B}$ and an electrostatic potential ϕ in the Schrödinger equation of a free particle given by

$$H\psi = \frac{\vec{p} \cdot \vec{p}}{2m}\psi = E\psi, \tag{F.1}$$

we must make the following two substitutions:

$$\vec{p} \rightarrow \vec{p} - q\vec{A}(\vec{r}, t), \tag{F.2}$$

referred to as Peierls's substitution or transformation, and

$$E \rightarrow E - q\phi(\vec{r}, t), \tag{F.3}$$

where $\vec{A}$ is a vector potential whose curl is the magnetic flux density, i.e., $\vec{B} = \vec{\nabla} \times \vec{A}$, and ϕ is the electrostatic potential. The latter is defined as the work done per unit charge to bring the charge to the location $\vec{r}$ from some reference point, typically selected at infinity.

The new Hamiltonian of the Schrödinger equation then becomes

$$H = \frac{\left(\vec{p} - q\vec{A}\right)^2}{2m} + q\phi. \tag{F.4}$$

The easiest way to derive the two substitutions described above is to start with a Lagrangian description of charge dynamics. In the presence of an electrostatic potential alone, the Lagrangian of a charged particle is given by the difference between the kinetic (T) and potential (V) energies:

$$L = T - V = (1/2)m\vec{v} \cdot \vec{v} - q\phi. \tag{F.5}$$

The charge dynamics are then described by Euler's equation,

$$\frac{\partial L}{\partial x_i} - \frac{\mathrm{d}}{\mathrm{d}t}\left(\frac{\partial L}{\partial \dot{x}_i}\right) = 0, \tag{F.6}$$

where the x_i are spatial coordinates of the particle, and the single dot superscript denotes the first derivative with respect to time.

With the Lagrangian above, Euler's equation becomes

$$-q\frac{\partial \phi}{\partial x_i} - \frac{\mathrm{d}}{\mathrm{d}t}\left(m\dot{x}_i\right) = 0, \tag{F.7}$$

which is equivalent to the Newton equation of motion

$$m\ddot{x}_i = -q\frac{\partial \phi}{\partial x_i}. \tag{F.8}$$

Problem Solving in Quantum Mechanics: From Basics to Real-World Applications for Materials Scientists, Applied Physicists, and Devices Engineers, First Edition.
Marc Cahay and Supriyo Bandyopadhyay.

In the presence of an external magnetic field, the force acting on the charge q should contain the Lorentz contribution, and the total force acting on the particle is given by

$$\vec{F} = q\vec{E} + q\vec{v} \times \vec{B}, \tag{F.9}$$

where the electric field $\vec{E} = -\vec{\nabla}\phi - \partial \vec{A}/\partial t$.

Next, we show that a modification of the Lagrangian in Equation (F.5) to the new form

$$L = (1/2)m\vec{v} \cdot \vec{v} - q\phi + q\vec{v} \cdot \vec{A} \tag{F.10}$$

leads to Newton's equation where the force term includes both the Lorentz force due to the magnetic field and the force due to the electric field.

Starting with this new expression for the Lagrangian, we get the canonical momentum component

$$p_i = \frac{\partial L}{\partial \dot{x}_i} = m\dot{x}_i + qA_i, \tag{F.11}$$

or, in vector form,

$$\vec{p} = m\vec{v} + q\vec{A}. \tag{F.12}$$

Now,

$$\frac{\mathrm{d}}{\mathrm{d}t}\left(\frac{\partial L}{\partial \dot{x}_i}\right) = m\ddot{x}_i + q\frac{\mathrm{d}A_i}{\mathrm{d}t}. \tag{F.13}$$

Since, in the most general case, $\vec{A}$ is a function of position and time, we have

$$\frac{\mathrm{d}A_i}{\mathrm{d}t} = \frac{\partial A_i}{\partial t} + \vec{v} \cdot \vec{\nabla}A_i. \tag{F.14}$$

Moreover, because

$$\frac{\partial L}{\partial x_i} = -q\frac{\partial \phi}{\partial x_i} + q\vec{v} \cdot \frac{\partial \vec{A}}{\partial x_i}, \tag{F.15}$$

Euler's equation becomes

$$-q\frac{\partial \phi}{\partial x_i} + q\vec{v} \cdot \frac{\partial \vec{A}}{\partial x_i} - m\ddot{x}_i - q\left(\frac{\mathrm{d}A_i}{\mathrm{d}t} + \vec{v} \cdot \vec{\nabla}A_i\right) = 0. \tag{F.16}$$

In vector notation, the right-hand side of the last equation can be expressed as a general force $\vec{F}$,

$$\vec{F} = q\left(-\vec{\nabla}\phi - \frac{\partial \vec{A}}{\partial t}\right) + q\left[\vec{\nabla}\left(\vec{v} \cdot \vec{A}\right) - \left(\vec{v} \cdot \vec{\nabla}\right)\vec{A}\right], \tag{F.17}$$

which reduces to

$$\vec{F} = q\vec{E} + q\vec{v} \times (\vec{\nabla} \times \vec{A}) = q\vec{E} + q\vec{v} \times \vec{B}, \tag{F.18}$$

where we have used the relation between the magnetic field and the vector potential

$$\vec{B} = \vec{\nabla} \times \vec{A}. \tag{F.19}$$

Starting with the modified Lagrangian, the Hamiltonian of the charge particle can then be derived as

$$H = \sum_{i=1}^{3} p_i \dot{x}_i - L = \sum_{i=1}^{3} (m\dot{x}_i + qA_i)\dot{x}_i - \frac{1}{2}m(\dot{x}_i)^2 + q\phi - qA_i\dot{x}_i, \qquad \text{(F.20)}$$

or

$$H = \sum_{i=1}^{3} \frac{1}{2}m(\dot{x}_i)^2 + q\phi, \qquad \text{(F.21)}$$

which is equivalent to

$$H = \frac{\left(\vec{p} - q\vec{A}\right)^2}{2m} + q\phi, \qquad \text{(F.22)}$$

where we have used Equation (F.11).

References

[1] Kroemer, H., (1994) *Quantum Mechanics for Engineering, Materials Science, and Applied Physics*, Chapter 10, Prentice Hall, Englewood Cliffs, NJ.

[2] Greiner, W. (2001) *Quantum Mechanics: An Introduction*, 4th edition, Section 9.2, Springer-Verlag, Berlin.

Appendix G: Matlab Code

Listed below is the Matlab code used to generate some of the figures in the book. Each program is identified with the problem number and the chapter where it was used. A copy of the source code can be requested from Marc Cahay at marc.cahay@ uc.edu.

Matlab code for Problem 3.13

```
% Finding the first three roots of the Airy function
% Written by
% Jordan Bishop
% Project for Quantum Systems class taught at UC by M. Cahay
    (Fall 2014)

clc;
clear all;
close all;

% first root of the Airy function Ai

ZetaLeft1 = 1;%Established by looking at the graph of the
    function
ZetaRight1 = 3;%Established by looking at the graph of the
    function

 % Break the function into a left and right sections
 AiryFuncLeft1 = airy(−ZetaLeft1);
 AiryFuncRight1 = airy(−ZetaRight1);

%Loop for the left side
while (AiryFuncLeft1 > 0)
    ZetaLeft1 = ZetaLeft1 + .00001;
    AiryFuncLeft1 = airy(−ZetaLeft1);
end
%Loop for the right side
while (AiryFuncRight1 < 0)
    ZetaRight1 = ZetaRight1 − .00001;
    AiryFuncRight1 = airy(−ZetaRight1);
end

%average the left and right to find the zero
root1 = (ZetaLeft1 + ZetaRight1)/2;

disp('Root 1 is:'); % displaying first root of Airy function Ai
disp(root1);
```

Problem Solving in Quantum Mechanics: From Basics to Real-World Applications for Materials Scientists, Applied Physicists, and Devices Engineers, First Edition.
Marc Cahay and Supriyo Bandyopadhyay.
© 2017 John Wiley & Sons Ltd. Published 2017 by John Wiley & Sons Ltd.

```
% second root of the Airy function

  ZetaLeft2 = 3.5;%Established by looking at the graph of the
     function
  ZetaRight2 = 5;%Established by looking at the graph of the
     function
  AiryFuncLeft2 = airy(-ZetaLeft2);
  AiryFuncRight2 = airy(-ZetaRight2);

while (AiryFuncLeft2 < 0)
    ZetaLeft2 = ZetaLeft2 + .00001;
    AiryFuncLeft2 = airy(-ZetaLeft2);
end

while (AiryFuncRight2 > 0)
    ZetaRight2 = ZetaRight2 - .00001;
    AiryFuncRight2 = airy(-ZetaRight2);
end

root2 = (ZetaLeft2 + ZetaRight2)/2;

disp('Root 2 is:'); % displaying the second root of the Airy
     function Ai
disp(root2);

% third Root of the Airy Function
ZetaLeft3 = 5;
  ZetaRight3 = 6;
  AiryFuncLeft3 = airy(-ZetaLeft3);
  AiryFuncRight3 = airy(-ZetaRight3);
while (AiryFuncLeft3 > 0)
    ZetaLeft3 = ZetaLeft3 + .00001;
    AiryFuncLeft3 = airy(-ZetaLeft3);
end

while (AiryFuncRight3 < 0)
    ZetaRight3 = ZetaRight3 - .00001;
    AiryFuncRight3 = airy(-ZetaRight3);
end

root3 = (ZetaLeft3 + ZetaRight3)/2;

disp('Root 3 is:'); % displaying the third root of the Airy
     function Ai
disp(root3);
```

Matlab code for Problem 4.3

```
% Written by
% Erik Henderson and Adam Fornalcyzk
% Project for Quantum Systems class taught at UC by M. Cahay
%    (Fall 2015)

clear all;
close all;
clc;

n = 1:5;
f = ((n .* pi) ./ sqrt(3)) .* sqrt(1 - 6 ./ (n.^2 .* pi * pi))
ratio = f ./ 2;

plot(n,ratio,'*')
xlabel('Eigen States')
ylabel('Ratio')
title('Levels of uncertainty')
```

Matlab code for Problem 5.6

```
% Matlab code to generate Figure 5.2
% written by Henry Jentz, Charlie Skipper, and M. Cahay (Fall
%    2015)
% Project for Quantum Systems class taught at UC by M. Cahay
%    (Fall 2015)
%
% This program generates plots of the transmission probability T
%    (Equation 5.77),
% reflection probability R (Equation 5.78) and absorption
%    probability A (Equation 5.79)
% as a function of the incident energy of an electron with
%    effective mass m* = 0.067 m0
% (where m0 is the free electron mass) impining on an absorbing
%    repulsive one-dimensional
% delta scatterer.
% The strength of the repulsive delta scatterer V0 is in eV-
%    angstrom
% The strength of the imaginary part of the absorbing delta W0 is
%    in eV-angstrom.
%

ev=1.602;% this is electron volt and has a 10^-19 factored into
%    the equations
W_o = [0*ev 1*ev 2*ev];% this has a angstrom along with the
%    electron volt factored into the equation
```

```
V_o=-0.1*ev;% this has a angstrom along with the electron volt
    factored into the equation
E=0:0.00001*ev:0.001*ev;% this is energy in electron volts
m=9.1;% mass of an electron has a factored 10^-31 pulled out
m_star=0.067*m;% mass of electron * effective mass of
    gallium_arsenide(GaAs)
h_bar=1.054;% this is to the power 10^-34
% k= sqrt(2Em_star)/h_bar
% |r|^2= (((m_star^2 * V_o^2)/h_bar^4)+ ((m_star^2 *
% W_o^2)/h_bar^4))/ (((m_star^2 * V_o^2)/h_bar^4) + (k + ((m_star
    *
% V_o)/h_bar^2))^2)

%|t|^2= k^2/(((m_star^2 * V_o^2)/h_bar^4)+(k+((m_star * W_o)/
    h_bar^2))^2)
%we want to simplify these equations a bit
% x= m_star*V_o^2/(2E*h_bar^2)  y= m_star*W_o^2/(2E*h_bar^2)
% |r|^2= (x + y)/(x + (1 + sqrt(y))^2
% |t|^2= 1/(x + (1 + sqrt(y))^2
% A= sqrt(y)* 1/(x + (1 + sqrt(y))^2

for iteration = 1:3
    x= ((m_star*V_o^2)*10^-2)./(2.*E.*h_bar^2);
    y= ((m_star*W_o(iteration)^2)*10^-2)./(2.*E.*h_bar^2);
    r= (x+y)./(x+(1+sqrt(y)).^2);
    t=(1)./(x+(1+sqrt(y)).^2);
    %A_1=1-r-t;
    A_2= 2.*sqrt(y)./(x+(1+sqrt(y)).^2);
    figure;
    %grid on
    hold on
    %plot(E,A_1,'r');
    %plot(E,r,'g');
    %plot(E,t);
    %plot(E,A_2,'r');
    %plot(E,r+t+A_2);
    plot(E*1e3,r,E*1e3,t,E*1e3,A_2), xlabel('Energy (meV)')
    legend('R','T','A')
end
hold off
```

Matlab code for Problem 5.7

```
% Matlab code to generate Figure 5.5
% written by Henry Jentz, Charlie Skipper, and M. Cahay (Fall
    2015)
% Project for Quantum Systems class taught at UC by M. Cahay
    (Fall 2015)
```

```
%

clc;
clear all;
close all;

m0 = 9.1;
mstar = 0.067 * m0;
hbar = 1.054;
ev = 1.602;

figure
for E = 0:0.0005:0.3
    E1 = E*ev;
    k = (0.1/hbar)*sqrt(2*mstar)*sqrt(E1);

    V0 = -0.3*ev;
    W0 = 0.1*ev;
    a = 50;

    E2 = E1 - V0 + i*W0;

    alphap =  sqrt(((k^2)*E2)/E1);
    alphar = abs(real(alphap));
    alphai = abs(imag(alphap));
    alpha = alphar + i*alphai;

    P = k/alpha;

    r = ((1-P^2)*(1-exp(-2*i*alpha*a)))/(((1+P)^2)*exp(-2*i*alpha
        *a) - ((1-P)^2));
    R = (real(r))^2+(imag(r))^2    ;

    t = (4*P)/(((1+P)^2)*exp(i*a*(k-alpha)) - ((1-P)^2)*exp(i*a*(
        k+alpha)));
    T = (real(t))^2+(imag(t))^2   ;

    M = exp(-2*alphai*a);
    N = exp(2*alphai*a);
    M1 = exp(-2*1i*alphar*a);
    N1 = exp(2*1i*alphar*a);

    C = 0.5*(1+P)*t*exp((i*k*a)-(i*alpha*a));
    D = 0.5*(1-P)*t*exp((i*k*a)+(i*alpha*a));
    Cstar = conj(C);
    Dstar = conj(D);

    A = ((k*W0)/E1)*((((C*Cstar*(M-1))/(-2*alphai))+((D*Dstar*(N
        -1))/(2*alphai))+((C*Dstar*(N1-1))/(2*1i*alphar))+((D*
        Cstar*(M1-1))/(-2*1i*alphar)));
```

```
    sum = R + T + A;

    hold on
    plot(E,R,'r')
    ylim([0,1])
    hold on
    plot(E,T,'b')
    xlim([0,0.3])
    ylim([0,1])
    hold on
    plot(E,A,'g')
    xlabel('Energy (eV)','Fontsize',16)
end
%legend('|r|^2', '|t|^2', 'A');
```

Matlab code for Problem 6.6

```
% Matlab code to generate Figures 6.4, 6.5, and 6.6
% written by Chelsea Duran and Ashley Mason (Fall 2015)
% Project for Quantum Systems class class taught at UC by
%    M. Cahay (Fall 2015)
%

clc;
clear all;
close all;

W=100e-10; %m
e=1.6e-19; %Coulombs
Nd=10^23; %m^-3
eo=8.85e-12; %F/m
er=12.9; %For GeAs

z=0:.1e-10:100e-10; %m

%Charge Concentration (Equation 6.65)
p = e*Nd*cos(2*pi.*z./W); %C/m^3

%Electric Field (Equation 6.71)
E=((e*Nd*W)/(2*pi*eo*er))*sin(2*pi.*z./W); %V/m

%Electrostatic Potential Energy (Equation 6.73)
V=((e*Nd*(W^2))/((2*pi)^2*eo*er))*(cos(2*pi.*z./W)-1); %Volts

%Changing position from meter to angstroms and electrostatic
%    potential in
```

```
%nV

z1 = z*10^10;
V = V*10^3;
%Plot the three equations
figure
%subplot(3,1,1)
plot(z1,p,'b')
%title('Total Charge Concentration in the Quantum Wire');
ylabel('Total Charge Density (C/m^3)');
xlabel('z (Angstroms)');

figure
%subplot(3,1,2)
plot(z1,E,'bl')
%title('Electric Field inside the Quantum Wire');
ylabel('Electric Field (V/m)');
xlabel('z (Angstroms)');

figure
%subplot(3,1,3)
plot(z1,V,'bl')
%title('Electrostatic Potential Energy inside the Quantum Wire');
ylabel('Electrostatic Potential (mV)');
xlabel('z (Angstroms)');
```

Matlab code for Problem 6.11

```
% Matlab code to generate Figure 6.10
% written by Chelsea Duran and Ashley Mason (Fall 2015)
% Project for Quantum Systems class taught at UC by M. Cahay
    (Fall 2015)
%

clc;
clear all;
close all;

%Normalize electron distribution
%associated to 1D electron gas
%impinging on an infinite potential (Equation 6.122)

%constants needed

hbar=1.054e-34;
m_star=0.067*9.10e-31;
Kb=1.38e-23;
```

```
T=[300  77  4.2];  % three different temperatures at which plot is
    made
i=1;
while(i~=4)
    Ef=-3*Kb*T(i);

    %finding N0 and lambda

    N0=(1/hbar)*sqrt(2*m_star*Kb*T(i)/pi);
    lambda=hbar/(sqrt(2*m_star*Kb*T(i)))*1e10;  %lambda in
        angstroms

    x=-1000:.1:0;
    n{i}=(1-exp(-x.^2/lambda.^2));  %normalized electron
        concentration

    i=i+1;
end

%Generating plot of normalized electron concentration
%at three different temperatures, 4.2, 77, and 300K.

figure(1)
plot(x, n{1}, 'r')
%str = 'T= 300 K \rightarrow ';
%text(x(9200),n{1}(9200),str)
hold on;
plot(x, n{2}, 'b')
%str = 'T=77K \rightarrow ';
%text(x(7500),n{2}(5500),str)
plot(x, n{3}, 'g')
%str = '\leftarrow T=4.2K';
%text(x(5000),n{3}(5000),str)
xlabel('z ($$\AA$$)','interpreter','latex')
ylabel('Normalized Electron Density');
ylim([0  1.1]);
```

Matlab code for Problem 6.15

```
% Matlab code to generate Figure 6.13
%Plot of normalized spectral energy distribution vs wavelength
%for blackbody emitted from a 3D box.
%Problem 6.15 Equation (6.161)
%Code writen by Chelsea Duran and Ashley Mattson - Fall 2015
%Quantum Systems class taught in Fall 2015 at UC by M. Cahay
%Plot of the distributions at T= 300K, 1000K, and 5000K.
```

```matlab
clc;
clear all;
close all;



%Constants
k = 1.38;    %E−23;                    %Boltzmann's Constant
c = 2.997;   %E8;                       %Speed of light
h = 6.626;   %E−34;                    %Planck Constant
T1 = 300;                              %Temperature: 300K
T2 = 1000;                             %Temperature: 1000K
T3 = 5000;                             %Temperature: 5000K

x = logspace(−9,−1, 300);       %limits for lambda

d = 5;
f1 = 1./(x.^d);                           %1/lambda^5
g1 = 1./(exp(((h∗c)./(x∗k∗T1)).∗1e−3)−1);      %1/(e^lambda−1)
    constants are in with lambda
u = f1.∗g1;
y1 = log10(f1.∗g1);                              %Normalized
    spectral energy distribution

f2 = 1./(x.^d);                       %function 2 (just changing T)
g2 = 1./(exp(((h∗c)./(x∗k∗T2)).∗1e−3)−1);
y2 = log10(f2.∗g2);


f3 = 1./(x.^d);                       %function 3 (just changing T)
g3 = 1./(exp(((h∗c)./(x∗k∗T3)).∗1e−3)−1);
y3 = log10(f3.∗g3);

figure('units','normalized','outerposition',[0 0 1 1])
x2=x∗10^10;
%semilogx(x2,y1,'g',x2,y2,'b',x2,y3,'r')
semilogx(x2,y1,'k',x2,y2,'k',x2,y3,'k')


hold on
plot ([3500 3500],[−300 100], 'k—')
plot ([7500 7500],[−300 100],'k—')

xlabel('$$\lambda$$ ($$\AA$$)','interpreter','latex','FontSize'
    ,20)
ylabel('G(\lambda)/(8\pi V c h)','Fontsize',20)
xlim([100 10e5])
ylim([−60 40])
```

Matlab code for solving the transcendental equations (6.167) and (6.177)

```
% written by Kelsey Baum (Fall 2014)
% Project for Quantum Systems class taught at UC by M. Cahay
    (Fall 2014)
%


clc;
clear all;
close all;

a = 3; % 3D case, solution of transcendental equation (6.167)
xo = 2.8;
xm = a*(1-exp(-xo));
xn = a*(1-exp(-xm));

while xn-xm > 0.0001
xm = xn;
xn = a*(1-exp(-xm));
ond
disp(xn)


a = 5; %1D case, solution of transcendental equation (6.177)
xo = 4.9;
xm = a*(1-exp(-xo));
xn = a*(1-exp(-xm));

while xn-xm > 0.0001
xm = xn;
xn = a*(1-exp(-xm));
end
disp(xn)
```

Matlab code for solution to Problem 7.7

```
% Matlab code to generate Figure 7.3
% Plot of the transmission probability (Equation 7.70)
% and reflection probability (Equation 7.71)
% as a function of the reduced wavevector $k/k_\delta$
% Date: 4/2/2016, M. Cahay
% x is the reduced wavevector $k/k_\delta$
% it is varied over the range [0,10]

clc;
clear all;
close all;
```

```
x = 0:0.1:10;

% The transmission probability
T = x.^2./(1.+x.^2);

% The reflection probability
R = 1./(1.+x.^2);

% R+T =1, R=T=0.5 when $k/k_\delta$ =1.

plot(x,T,'--',x,R), xlabel('$k/k_{\delta}$','interpreter',
    'latex')
legend('T','R')
```

Index

Problem Solving in Quantum Mechanics: From Basics to Real-World Applications for Materials Scientists, Applied Physicists, and Devices Engineers, First Edition.
Marc Cahay and Supriyo Bandyopadhyay.
© 2017 John Wiley & Sons Ltd. Published 2017 by John Wiley & Sons Ltd.